WEATHER, CLIMATE, AND THE
GEOGRAPHICAL IMAGINATION

INTERSECTIONS: ENVIRONMENT, SCIENCE, TECHNOLOGY

Sarah Elkind and Finn Arne Jørgensen, Editors

WEATHER, CLIMATE, AND THE GEOGRAPHICAL IMAGINATION

PLACING ATMOSPHERIC KNOWLEDGES

Edited by Martin Mahony and Samuel Randalls

UNIVERSITY OF PITTSBURGH PRESS

Published by the University of Pittsburgh Press, Pittsburgh, Pa., 15260
Copyright © 2020, University of Pittsburgh Press
All rights reserved
Manufactured in the United States of America
Printed on acid-free paper
10 9 8 7 6 5 4 3 2 1

Cataloging-in-Publication data is available from the Library of Congress

ISBN 13: 978-0-8229-4616-8
ISBN 10: 0-8229-4616-5

Cover art: "Mapping Changes in Global Temperature, 1850–2017," from Ed Hawkins's *Climate Lab Book*. Data from HadCRUT4.6. An annual average for a particular grid cell and year is only shown if 6 or more months have data, otherwise it is coloured white. The color scale runs from around -2.5°C to +2.5°C.

Cover design: Alex Wolfe

CONTENTS

ACKNOWLEDGMENTS

A VOLUME LIKE THIS is the product of much distributed labor. We'd like to thank in the first instance all those who attended our sessions at the 2015 International Conference of Historical Geographers, whose enthusiastic engagement with our topic assured us that this was something worth pursuing. The University of Nottingham, through a Fellowship awarded to Martin Mahony, generously supported a follow-up workshop at which the draft contributions to this volume were worked into shape. Isla Forsyth, Mike Heffernan, and Stephen Legg helped us along by generously giving their time to read and comment on draft papers. We'd also like to thank Vladimir Janković, Jonathan Oldfield, and Richard Powell for their insights and contributions along the way, as well as all the scholars whom we've chatted to about the project at various conferences and workshops. The International Commission on the History of Meteorology community and their in-house journal *History of Meteorology* provides a regular source of inspiration. Sarah Elkind and two anonymous reviewers provided invaluable comments on the manuscript, and Sandy Crooms at University of Pittsburgh Press has been unfailingly patient and positive with us.

WEATHER, CLIMATE, AND THE
GEOGRAPHICAL IMAGINATION

INTRODUCTION

Weather, Climate, and the Geographical Imagination

Martin Mahony and Samuel Randalls

THE INTELLECTUAL HISTORY OF CLIMATE, once perhaps a rather arcane corner of historical inquiry, is now a burgeoning, vibrant field of study. This is, in part, directed by a concern to historically situate contemporary concerns about climate change and by a renewed sense of the importance of historical scholarship in exploring the multifaceted relationships between climate and society. As global temperatures rise under the forcing hand of humanity's greenhouse gas emissions, new questions are being asked of how societies make sense of their weather, of the cultural values that are afforded to climate, and of how environmental futures are imagined, feared, predicted, and remade. The urgency of contemporary debates about global climate change—about efforts to mitigate, to adapt, perhaps to manage and control—do not often leave room for considered reflection about the values that infuse our knowledge and understanding of an object, the global climate, which seems to elude direct sensory experience and to hover somewhere above the scales of conventional humanistic engagement with the world. Yet a growing number of scholars in the humanities and social sciences are embracing weather and climate as sites of inquiry. Some are seeking to supplement, or perhaps to translate and humanize, more dominant scientific renderings of such objects; others to probe more critically, to challenge particular scientific framings of environmental change, or to situate the practices and politics of the atmospheric sciences in wider historical and cultural contexts. While the ugly residues of past environmental determinisms may have seen climate banished from much humanistic scholarship around the middle of the twentieth century, a renaissance is now underway, and the homogenizing eye of earth system science is being supplemented by a multitude of views of what weather and climate look like, feel like, and mean to people in a mounting diversity of social and cultural settings.[1]

In this book we seek to contribute to this new conversation by bringing together a range of voices from history of science, historical geography, and environmental history, each speaking to a set of questions about the role of space and place in the production, circulation, reception, and application of knowledges about weather and climate. In recent years historians of science, buoyed by new historiographical interests in the nature of scientific observation, the politics of expertise, and the cultural import of prediction, have offered important new readings of the historical development of the sciences of climate.[2] Historical and cultural geographers, throwing off an earlier reticence to engage with climate and all its environmental-determinist baggage, have shone new light on the competing narratives of climate and climate change that animate political and cultural worlds.[3] At the core of this scholarship on climate is a renewed attention to the geographies of knowledge about phenomena—weather and climate—that are themselves inherently spatial. Weather, although a product of globe-spanning dynamics, is always experienced in place, while the idea of climate has historically operated at a range of spatial scales, from the microclimate of the body, through claims about the climatic character of nations, to the notion of a global, and perhaps fragile, climate system. As horizons of expectation about climate and its changes have stretched into the far future, conceptions of regional climatic difference have given way to temporal concerns for the steady—or perhaps erratic—evolution of the global climate system under human forcing. Yet despite new understandings of the power of human agency in shaping the weather, climate still plays a powerful, sovereign role in its imagined capacity to fundamentally shape human geographies of violence, economic prosperity, and environmental vulnerability.[4] Understanding this lineage not only of climatic determinism but of climatic expectation more broadly is a critical historical task with urgent contemporary resonances.

The contributors to this volume collectively develop the concept of "geographical imagination" to address the intersecting forces of scientific knowledge, cultural politics, bodily experience, and spatial imaginaries that shape the history of knowledges about climate. In recent years the concept of geographical imagination has come to be read both as a way of describing particular suites of knowledge-making practices and as a way of describing much broader modes of comprehension and experience, where the conditions "of both the known world and the horizons of possible worlds" mingle in imaginaries of space and place.[5] The geographical imagination may be said to occupy a space between objective reality and subjective experience, where perceptions of the real and the imagined shape each other to produce influential, widely circulated, and enduring sets of knowledge and expectation—or (rendered in different theoretical terms that we develop below), geographical imaginaries shape, and are shaped by, the materials, practices, discourses, and places through which they are produced. Our aim is to critically reconsider the role of knowledges, experiences, and expectations of weather and climate in the shaping of particular geographical imaginations and to interrogate the material, cultural, and environmental geographies through which such knowledges have been produced, circulated, and put to work in human dealings with climate, at a range of spatial scales. We first review the importance of geographically and sociologically interrogating "knowledge" about weather and climate and how such knowledges have generated particular kinds of imaginations. We then loop back to examine how these imaginations have also, in turn, shaped specific knowledge-making practices themselves.

GEOGRAPHIES OF KNOWLEDGE

Scientific knowledge-making, like any other realm of human activity, has its geographies. Science proceeds in and through space and participates in the construction of cultural and political geographies by which human interactions with the nonhuman come to be known, understood, and governed. The claim that scientific knowledge is "a geographical phenomenon" is the organizing principle of a growing body of scholarship which stresses that to understand the cultural and epistemic authority of science means understanding the spatial practices by which scientific knowledge is produced, by which it circulates, and that shape its reception and interpretation.[6] Over the last three decades, historians and geographers of science have cemented a "spatial turn" in the study of scientific culture, with concepts such as space, place, network, and circulation now canonical members of the science studies lexicon.

Numerous origin stories might be told about this spatial turn. In many respects it is the logical outcome of early, post-Kuhn work on the social practices of scientific knowledge-making, which made the decisive argument that the production of universal truths could not be explained simply by appeal to those truth claims' own correspondence to an external reality. Figures like those pushing the Strong Programme in the Sociology of Scientific Knowledge (SSK) argued that the success or failure of different knowledge claims was not reducible to their facticity but, rather, was a function of social relationships and modes of persuasion.

David Bloor's "symmetry principle" called for sociologists to use the same analytical tools to explain both right and wrong knowledge; and fine-grained studies began of the social constitution of scientific worlds—their structuring by social hierarchies, group identities, and power relations.[7] In the 1980s inheritors of this tradition offered new historical sociological analyses of the spaces of scientific knowledge production, examining how the social relationships that structured scientific knowledge-making were expressed in spatial arrangements—for example, in the exclusion of certain groups from the "gentlemanly" settings of early modern laboratories, in the designation of authoritative witnesses to laboratory practices, and in the spatial extension, through the enrolment of certain "literary technologies," of networks of virtual witnesses to scientific discovery.[8]

These spatially inflected historical sociologies of science, which Richard Powell situates in a "socio-spatial" school of history of science, proceeded alongside the development of new ethnographic approaches to studying the making of scientific knowledge.[9] Laboratory ethnographies produced influential new theories of the webs of human and nonhuman relationships and agencies that shaped scientific practice and of nonhuman agency in the networks through which entities were tied together in the lab, and subsequently mobilized in the wider world.[10] If the lab was a peculiarly local place, carefully set apart from the world yet in regulated dialogue with selected parts of it, new theories of "science on the move" sought to complement analysis of this localization with analysis of science's spatialization, of the making-mobile of scientific ideas, artefacts, technologies, and tools.[11] The Latourian model of circulating "immutable mobiles" has been particularly influential in emphasizing the agency of science's inscriptions—images, graphs, texts—as solid, unchanging artefacts that belie the social contingencies of their production and act to effectively transfer scientific ideas and epistemic authority across space through ever-expanding actor-networks.[12] Yet this model has been critiqued for its "imperialistic language" of enrolment, expansion, and solidity, and for its "oddly realist bow towards bigness" in its evaluation of what makes a successful network.[13] It stands in need of "decolonisation and demasculinisation" as a model of how science transforms the material and social world.[14] Work in the postcolonial tradition of science studies has perhaps done the most to unsettle the material and semiotic formalism of actor-network theory (ANT), as well as to challenge the historiographical shortcomings of "diffusion" models of scientific expansion more broadly.[15] Postcolonial science studies position colonial spaces as (of course) sites of domination, appropriation, and control but also as sites of contestation, hybridization, and exchange, where scientific knowledge developed out of the sort of intercultural encounters and disputes that ANT and diffusionist historiographies fail to capture.

Recent scholarship on the relationships between climate, empire, and colonialism has started to shed more light on these questions of how science traveled and on the role of the sciences of climate in the imagination and production of colonial space. British colonialism has been a particular focus. Australia and New Zealand were among the first places where meteorology and climatology were quickly institutionalized within colonial government structures, answering "the calls of colonialism and modern science to know and categorize, and in so doing, control environments."[16] Institutionalized observations began to challenge imported understandings of a set of Australasian climates that might resemble those of home or that could be anticipated to vary with clockwork regularity.[17] Organized meteorology began to displace understandings of local climates that passed between indigenous inhabitants and incoming settlers in the contact zones of colonization, yet there is evidence that Maori meteorologies may have persisted as intercultural modes of weather anticipation well beyond the arrival of European instruments, charts, and predictions.[18] In South Asia, surgeons associated with the East India Company were by the late eighteenth century developing new understandings of atmospheric dynamics on land and at sea, and their exchanges with colleagues in Australasia were broadening the geographic scope of atmospheric vision and conjecture.[19] Yet such actors were not simple agents and champions of imperial power. Many were openly critical of British rule in India and elsewhere, and by the mid-nineteenth century their successors were vocal advocates of forest conservation to protect local climates from the apparently desiccating effects of both indigenous and colonial timber extraction, even if subsequent conservation laws performed their own hardening of imperial control over people, space, and natural environments.[20] Katharine Anderson's description of Indian meteorology and climatology as it became more concertedly institutionalized in the late nineteenth century offers more direct echoes of the enrolment of the sciences into the structure and functioning of colonial states in Australia and New Zealand. Meteorology offered a model of the centralized operation of a vast network of disciplined human subjects, of a new symbiosis between science and state, while the vastness of the Indian empire offered British meteorologists the kind of synoptic field of vision with which their American counterparts were blessed.[21] While continuing to piece together a picture of global climate oscillations, Indian colonial meteorology also held out the prospect of naturalizing the famines that shook the empire in the late nineteenth century, in describing natural causes that might displace nascent arguments about the state's ultimate responsibility for the death by starvation of millions of Indian citizens.[22]

Meteorology and climatology undoubtedly functioned as tools of empire, but we now have enough historical evidence to refute notions that they were sciences that simply diffused from the metropole and that their relationships to imperial power were straightforward. Deborah Coen, for instance, has emphasized how the patchwork structure of the Habsburg Empire and the pluralism of Habsburg science encouraged climatic thinking across scales and an appreciation for difference and complexity rather than centralized uniformity.[23] Imperial "peripheries" also produced meteorological knowledge claims that were not pale imitations of metropolitan science but that were original and distinctive and, in many cases, enduring.[24] They were produced by a diverse cast of actors with complicated relationships to colonial states and subjects, and in a range of spaces—observatory, the field, medical posts, army offices, and ships—that demand a more detailed appreciation of the historical geographies of weather observation and climatic thought. Although in this book we remain largely Anglophone in context, chapters in this volume by Ruth Morgan, Georgina Endfield, Meredith McKittrick, and James Kneale and Samuel Randalls diversify the cast of characters that populate our meteorological histories, rendering colonial and postcolonial spaces as sites of knowledge hybridization and situating the production and circulation of climate knowledges within multifaceted circuits of cultural exchange, economic transaction, and imperial ambition. Alongside chapters by Katharine Anderson, Martin Mahony, and Simon Naylor and Matthew Goodman, they also point to the diverse material culture of meteorological knowledge production, recalling Gregory Cushman's work on the shaping of meteorological knowledge (in his case hurricane prediction), not just by rival social groups but by assemblages of matter and media—meteorological guidebooks onboard sailing vessels, newspaper cuttings pieced together in storm scrapbooks, the rubber and copper of undersea cabling.[25] Turning to ANT insights on the agency of matter and mobile inscriptions in the functioning of chains of translation can help refashion our understanding of the history of meteorology, where hagiography and stories of linear progress are surprisingly dominant. Beyond pioneering individuals, histories attuned to the spatial and material cultures of meteorological practice can offer a more nuanced picture of human societies' efforts to come to terms with weather and climate.

Self-styled literature on the geographies of science has not often sought connections with the postcolonial, and neither has a lot of related work across STS on science and globalization.[26] The coupling of cultural and political-economic critiques of scientific imperialism that postcolonialism offers nonetheless has important implications for understandings of globality in science. While in both ANT and postcolonial approaches, globality is viewed as the achievement of multiple local transactions and accomplishments, the postcolonial sensibility arguably offers better resources for thinking through the cultural and economic forces that have produced different kinds of globality in different times and plac-es.[27] The historiography of meteorology and climatology often shapes up as the history of how those sciences became global—in the sense of their achievement of reliable, robust knowledge of global processes and of the capacity to predict the behavior of those processes into the future.[28] Such work has offered important historical depth and nuance to our understanding of the rise of climate and weather modeling, and of the institutional and cultural politics which shaped that rise. But this work also arguably internalizes a broader historiographical fetishization of a recent break to globalization, which positions the pursuit of globality in science, politics, economics, and culture as a uniquely postwar, twentieth-century phenomenon.[29] Of course, the transformations of the postwar world were intense and far-reaching, to the point where the marker of the dawn of the Anthropocene—the global "age of humans"—is likely to be set at 1950. Yet our histories of global ambitions and practices must do more than narrate teleologies of twentieth-century globalism and must work to identify sites of hybridization and resistance, and to uncover how the global, as an object and a condition, exists in different forms in a variety of historical locations.[30] For the history of the sciences of weather and climate, this means supplementing stories of the triumph of mid- to late twentieth-century scientific infrastruc-tures and institutions with stories about how global space has been imagined, worked with, traversed, and brought into being in different times and places. Jon Oldfield has offered a much needed corrective to histories of European and North American constructions of global climate by focusing on how a distinctive global climatology emerged and evolved in tsarist and Soviet Russia, shaped by an "ecological" approach to understanding the links between climates, societies, and natures that situated climate as a multi-scaler object nested within broader intellectual models of global ecological order.[31] Likewise Coen has shown how Habsburg climate scientists' appreciation of the importance of scaling provides an important corrective in rethinking contemporary assumptions that scale is something newly important to the global climate problems of the later twenti-eth century.[32] Unquestionably, more non-Anglophone examples would extend this range of stories even further. In this volume, Anderson offers a picture of historically situated globalities in the interwar period, pointing to distinctive cultural anxieties about the place of the human in the unruly spaciousness of the globe and in emerging, globe-spanning technological networks of observation, navigation, and control.[33] Adamson too adds complexity to our understanding of past imaginings of climate as global system, describing the imperial infra-structures—technological and bodily—through which new claims about the interconnectedness of a variable climate were made.

This is a project that demands closer attention to the spaces of meteorological knowledge production, from the globe down to the microgeographies of observation and back again. The observatory has recently emerged as a key site for historians of science who are concerned with making connections between the regulated spaces of scientific observation and the networks that established new claims to globality in the nineteenth century.[34] Astronomy could be described as the "pattern science" of the nineteenth century, a model of discipline, coordination, and exactitude to which lesser fields such as meteorology were compared and to which their practitioners aspired.[35] The expansion of astronomical and magnetic observatories in Europe and in European colonies was motivated both by the search for new understandings of terrestrial and heavenly forces and by the more mundane tasks of accurate timekeeping and reliable navigation.[36] In the British Empire, the observatory emerged as both an icon of Western reason and civilization, a space set apart from the world yet also a model for that world's ideal functioning (Naylor and Goodman, this volume). By keeping time, easing navigation, and exemplifying imperial order the observatory quite literally "made the Empire tick."[37]

Like the observatory, the field has become a space of increasing interest to both geographers and historians of science. As Robert Kohler has influentially argued, the field occupies a curious position between laboratory and landscape.[38] It is a space that exceeds control, troubles any attempt at demarcation, and yet functions as a resource of empirical authority against the fabrications of the laboratory or the abstractions of theory; of empirical authenticity against the purifying practices of the more rarefied spaces of science.[39] This notion of authenticity has often been tightly coupled with discourses of heroic masculinity and self-sacrifice in the name of knowledge.[40] The field can be read as a space onto which imperial and modernist designs were projected, and from which resources, knowledge, and cultural capital were extracted.[41] If lab studies were about demonstrating local specificity, study of the scientific field focused on the uses of the field in cultural, imperial, and global histories.[42] But the field can also be read as a space where gestures of cultural projection and economic extraction could not always be so confidently executed, being read instead as a space of complex encounters, of exchange, and of contestation over modes of knowledge-making and extraction.

Field practices of mapping, surveying, observing, collecting, expeditioning, and even experimenting each have their own spatialities, and the question of place can figure differently into their historical analysis.[43] Yet while much history of field sciences has focused on the horizontal gestures of scientific mobility and landscape surveillance, an emerging interest in the vertical—as a dimension of movement, vision, and practice—"takes us away from human habitation into depths and heights in which no one lives (for long) yet that are vital to global economy and polity."[44]

In disciplines such as meteorology and climatology, looking up involves looking at objects and processes that are irreducibly mobile and lively, rather than objects that are tied to particular places. The "field," then, often becomes a moving assemblage of people, place and practice, rather than a static and well-defined arena of scientific surveillance.[45] So while much recent history of the field sciences has been concerned with the borderlands between laboratory and landscape, the historical study of weather and climate offers the scope to examine the borders between observatory and field, to reconsider the practices of purification and exclusion meant to lend precision and reliability to observation, and to rethink the field as a thing always on the move, and thus always subject to redefinition—both by its actors and by its historians. Mahony's chapter speaks to these borderlands in the case of late colonial agricultural developmentalism and contestations over the place of meteorological observation in an experimental field site whose social, material, and climatic unrulinesses highlight the difficulties involved in taking meteorology out of the observatory and navigating the tensions between regulated experimental space and the multiple meanings, uses, and functions of the field. Anderson (this volume) paints a similar picture of meteorological practice "in the wild," bound up with diverse material infrastructures and cultural discourses pitting technological modernism against the troubling unruliness of oceanic space.

Corporeality has often been missing from, or unevenly present in, our spatial histories of science. We read of bodies excluded from experimental space and of bodies on the move as vehicles of traveling knowledge, but the body as site and object of scientific practice has, understandably, been rather restricted to the history of medicine and the sciences of life and human difference.[46] The effects of climate on alien bodies has been a prominent feature of the imaginative geographies of empire and travel. "Managing the transition of the body through different climates" was a key concern of early European expansion, as imperialists "sought to make tropical climates safe for white settlement."[47] Diverse material cultures and practices for dealing with tropical extremes were developed, with European bodies emerging as both objects and instruments of inquiry in an emerging discourse of "acclimatization," that for many in the nineteenth century was *the* paradigmatic colonial science.[48] Yet efforts to scientifically study the effects of climates on human bodies were never far from efforts to make moral judgments about the inhabitants of faraway places, whether climate was positioned as an *explanans* of racial difference and inferiority or as a cause of bodily, intellectual, and moral decay among those who found themselves dwelling in new atmospheric environments. Powerful "moral climatologies" can be read from histories of climatic thought, wherein the contours of climatic variation were drawn alongside the imagined contours of human difference.[49]

Two chapters in this volume concern the emergence of the body and the bodily atmosphere as a distinctive space of scientific practice, whether through the biometeorological body of early twentieth century Hippocratic revivalism (Livingstone), or within the body of Ellsworth Huntington, enlivened by the ozonated New England air, living proof of his own determinist philosophy (Fleming). The relational spaces of body, atmosphere, and microclimate have emerged in histories of this determinism, and also in studies of the creation of anthropogenic microclimates in architectural practice.[50] Throughout the colonial and Cold War eras, concerns about healthy, liveable climates reached into the homes of the hot and stuffy, the dark and cold, and architectural studios and government planning offices became new sites where meteorological and climatological expertise came into contact with the designers and engineers of space. As Daniel Barber suggests in this volume, it is through architecture and planning that we can perhaps see most clearly how knowledge and imagination of weather and climate have reshaped the spatialities of everyday life, from the design of homes, offices, and factories to the planning of entire towns and cities. In an era of global climate change and new discourse about designer climates at all scales, it is therefore appropriate that the final section of this volume foregrounds unique branches of climatology, organized around the human body and its place in architectural space that emerged through new hybridizations of knowledge, space and practice.[51]

Knowledge-making practices came to inspire particular imaginations of the climates of the world. The experiential body generated geographical imaginations of pathological climates. The ships collating climate observations and the insurers tallying lost lives generated data that created maps of weather and climate risks around the world, with maps and routes then judged and even priced accordingly. The specific contexts, politics, and sites within which North American, European, or Soviet climate science was produced, for epistemic, managerial, or geopolitical ends, shaped important and enduring imaginations of the nature of climate and its changes at a range of spatial scales. Knowledge-making practices came to shape imaginations of the world's climate(s), but these imaginations themselves also shaped and directed practices of making empirical or theoretical knowledges of weather and climate. The notion of "geographical imagination" can help us get at this coproduction of knowledge, practice, and imagination and to render it in new, spatial terms.

THE GEOGRAPHICAL IMAGINATION

Stephen Daniels suggests that the geographical imagination is shaped by and through discourses and practices, encompassing both geography as a discipline and as articulated in sites beyond the discipline.[52] It draws on previous concepts of imagination, particularly the idea of the sociological imagination in reference to C. Wright Mills. For Mills, the sociological imagination was a particular way of looking at and thinking about the world; in other words, it inspired and was a central part of what the discipline of sociology did and contributed to knowledge.[53] Within geography, the term "imagination" has similarly been proposed as an approach to inspire and engage students and even schoolchildren with a mode of inquiry that is explicitly geographical in nature.[54] In other words, to think like a geographer means being aware of the geographically specific nature of spaces and places, knowledge and practices.

We use the notion of geographical imagination in a dual sense—both to describe a particular analytical attention to spatiality that we seek to bring to the history of meteorology and climatology, and to situate these atmospheric knowledges within broader suites of knowledge, experience, and expectation that, historically, have shaped how actors have comprehended the world and how they oriented their actions toward it. Sociologists have recently turned to the notion of imagination to analyse the work performed by widely shared imaginaries of desirable futures in the ordering of social life in the present.[55] Our use of imagination supplements this focus on time and futurity with an emphasis on place and spatiality. It emphasizes that in any given field there will always be multiple imaginaries in circulation, precisely because each imagination will be geographically situated, but some of these are more politically powerful than others, and some will travel further (that is, they are more "mobile" in Latourian terms).[56] We suggest the existence of a multiplicity of geographical imaginations that intersect, interact, work with, and create friction between other imaginations.[57] Indeed, as in the work of Annemarie Mol, it is often the frictions that produce the most interesting insights into how imaginations and practices shape each other.[58] That imaginations are produced in and through experience and practice proves that they are place-based, which means they are not global except in their reach. We therefore need to attend more closely to the sites through which imaginations are produced, whether these are, for our purposes here, ships, insurance offices, or meteorological stations. The geographical nature of imaginations also calls attention to the central importance of exploring imaginations beyond the classic scientific sites of the Global North and to illustrate how some imaginations circulate more freely than others in spaces beyond conventionally defined scientific metropoles. In sum, the geographical imagination—as both a thing to be studied and an analytical stance—enables us to understand how and why some of the knowledges identified in the previous section circulate more readily than others. Interrogating these imaginations is a core contribution of this volume to the history of meteorology and climatology.

One important area where the geographical imagination has been regularly studied is through the historical lens of empire and imperialism. "Imagined geography," for example, was invoked by Edward Said to explore how territories are understood in ways that often re-present knowledge of them as universal rather than partial or situated.[59] To name just three examples in this tradition: Denis Linehan has explored Irish accounts of missionaries in Africa arguing that there existed a particular Irish imagination of Africa as a racialized space, which nonetheless oscillated between imperial and emancipatory ideals through a complex assemblage of ideas, materials, and practices; Georgina Endfield and David Nash likewise show how imaginations of Africa as a pestilential space shaped missionary engagements with it, in bodily practice and discourse; and Diana Davis has shown how French colonial desertification narratives came to shape interventions in environmental policy that marginalized the "destructive" Maghrebi pastoralists.[60] In each case, whether the notion of imagination is explicitly invoked or not, there is a sense that dominant discourses and practices are produced from specific places and knowledges, and despite their frequent presentation as universal are in fact distinctly partial, and based as much on experience and preconceived ideas or theories as on verifiable empirics. As Dee Mack Williams has noted in the case of desertification in Inner Mongolia, Chinese policies to switch local people from nomadic lifestyles to settled agriculture, accompanied by supporting land policy changes, have enhanced the destruction of the remaining areas as herders competed to graze animals on the remaining unclaimed land. With further desertification this was heralded as evidence of the need for stronger land management.[61] Thus, geographical imaginations come to shape practices and these practices in turn further reenforce dominant imaginations. Discourse and practice, narrative and materiality, come together to create, support, and maintain dominant geographical imaginations.

Scholarship has thus drawn attention to the fact that many powerful geographical imaginations are produced from within authoritative centers of calculation, often about "periphery" regions.[62] But this should not lead us to think that imaginations are only produced within these centers or draw on expertise only from the center. As the previous section highlighted, there has been considerable work to expand and decenter our geographies of knowledge to also consider knowledge produced within the Global South, and through processes of cultural encounter and exchange. Imaginations are produced in a great variety of spaces. In the case of Nordic indigenous groups, for instance, the defense of their own geographical imaginations as a vector of resistance to imposed, managerial imaginations provides an opportunity to set out culture and territoriality on their own terms.[63] That being said, the risks of co-option are considerable and these imaginations can come to reenforce an unchanging, pure indigenous essence that perpetuates power dynamics and constrains indigenous peoples to live in ways consistent with certain imaginations (both internal and external) of their lifeworlds. As Bjørn Sletto has pointed out, these imaginations of an essential indigenous environmental ethic have created new forms of eco-governmentality and self-surveillance.[64] Imaginations, even when considered diversely, work within particular power geometries that those imaginations are not fully able to change or to enforce. In other words, we argue that there is a continual tension or friction between different imaginations. The work of Mara Goldman et al. gets closest to this in discussing the multiple ontological politics of drought, with Maasai drawing on different forms of knowledge compared to formally trained scientists.[65] What drought becomes is shaped within different forms of scientific practice but is inspired by—in the case of the Maasai—more geographically specific criteria than in the case of the climate scientists.

Work on geographical imagination has also inspired a focus on creative practices and writing where geographical imaginaries circulate with and through images, poetry, music, and objects.[66] Peter Hulme's work is perhaps best known for exploring the literary geographical imaginations of Cuba that circulate through maps and novels.[67] It is the steamy atmosphere conjured by Joseph Conrad that evokes a particular imagination of tropical Africa as a wild landscape, and it is through specific but very different fourteenth-century lenses that the travelogues of Marco Polo and Ibn Battutah are constructed.[68] In writing about popular news coverage, Gordon Winder and Michael Schmitt claim that the deaths of Mahatma Gandhi and Indira Gandhi were represented through the lens of postwar politics by the *New York Times*, which orientalized India as a space of violence that was rapidly changing under the onslaught of globalization and American diplomatic power.[69]

Literature and the written word have therefore been central to many accounts of imaginations, but words by themselves are not the only aspect to receive analytical attention. Geographers have explored the role that images play in shaping geographical imaginations too, not least the representation of particular climates as ideal for travel or rest.[70] As Kneale and Randalls' chapter highlights, maps of the world divided into risks by insurance companies became easy representational devices that enabled insurers to quickly demarcate the pricing of risk; and as Anderson's chapter shows, and as Edwards has also illustrated, imaginations of "the global" come to be drawn through images and schematic diagrams that imagine climate as a particular kind of spatial scale (see below).[71]

This connection between the imaginative and the visual has inspired a whole series of works in imaginative geographies that draw on ideas of creativity, enthusiasm, and practice to show that imaginations can be reimagined and rematerialized to engender new forms of political and social engagement and foster different ways of being and acting in the world.[72] As Bill Howie and Nick Lewis put it: "If geographical imaginaries are not just socially produced but also socially productive, then this productivity can be both studied and shaped."[73] In this book we highlight how geographical imaginations became productive of particular kinds of relations in the past. Although it is not the specific goal for most of the chapters in this book, we hope that consideration of these relations might also inspire introspection on the role that current imaginations play in contemporary matters of concern such as climate change or air pollution.[74] Historical research into geographical imaginations therefore is not just a passive account of past imaginations but a call to a particular kind of political historical genealogy where we strive to show how social and natural relations are reconfigured in ways that frequently enhance some interests at the expense of others.

One way in which some interests are advanced over others is through the geographical reach of particular imaginations. If all imaginations are local but some circulate more freely than others, then thinking through the spatial scale of imaginations becomes important. What appear to be global imaginations are nested in particular sites of knowledge and experience. This is perhaps not better illustrated than in Sheila Jasanoff's account of an Indian environmentalism that cast the "spaceship earth" image in a rather different light to the triumphalist technological environmental reading that it is frequently given in the United States, where it played into a particular kind of Cold War politics. As Jasanoff puts it: "much work has to be done to make the representations look as if they are the right way of characterizing the world."[75] A global view of the earth from space could equally inspire a technological utopia of the ability to control the planet's systems as much as it might indicate a small, fragile planet.[76] The phrase "think globally, act locally" embeds this geographical imagination in the heart of sustainable living. But by focusing on the global scale in contemporary climate change for instance, perhaps the challenge has been the loss of local connection and the dominance of a particular kind of ponderous global policy negotiation.[77] Likewise, statements of global knowledge or global problems come to shape the possibilities and restrict the potentialities of local knowledge and practices.[78]

Global geographical imaginations can easily translate into power relations that centralize authority and decision-making within global spheres or organizations.[79] Part of what we seek in this book is to highlight the local practices that have legitimated and enabled the generation of widely circulating knowledges and imaginations, without losing sight of the fact that these imaginations are produced from within specific networks and are partial, specific, and multiple. As Francis I, Emperor of Austria, stated: "There is no affair that a priori and according to general principles could be called large or small; matters are only large or small in comparison to and in relation to other things."[80]

Finally, we must remember that geographical imaginations are not only produced—but consumed. A particular representation of the tropics as dangerous space only maintains its validity as a powerful imagination if those reading it actually adopt it. As Sarah Radcliffe has pointed out, we should not make the mistake of thinking that everyone reads texts or images in the same way, as though an explicit geographical imagination can be conveyed linearly from one person to another.[81] Rather, people translate ideas through other imaginations, as situated within their own experience and practices. As Victor Savage has argued, as European travelers experienced the tropics, they began to challenge a universal view of the dangers of the tropics and, rather, rearticulated danger on much finer geographical scales.[82] In an inverse of this process, Endfield's chapter demonstrates how widely circulated imaginations of healthy South African climates were challenged by the bodily experiences of migrants. In the consumption of geographical imaginations of weather and climate, it is important to acknowledge the ways in which imaginations were contested, challenged, or simply ignored. The biography of Ellsworth Huntington is perhaps instructive here, given that his work had relatively less influence within the academic discipline of geography than we might expect (see Fleming, this volume).

We therefore argue that the concept of geographical imagination provides a powerful tool with which to interpret accounts of knowledge-making practices in the history of meteorology and climatology. Knowledge is generated and circulated not just by and between formal scientists but also through networks encompassing field practitioners, ship captains, insurance clerks, and political ideologues. Instances and examples of meteorological knowledge production, we suggest, are always situated within a broader discursive field, where they serve to either construct, reinforce, or challenge dominant imaginations not just of the atmosphere but of spaces of human and nonhuman life, of political domination and contestation, and of technological ambition. Our focus on the coproduction of atmospheric knowledges with broader geographical imaginations helps us interpret how and why popular ideas of weather and climate can outlast even prominent scientific rebuttals of these ideas. Imaginations, in other words, help us understand better the historical transformations in approaches to weather and climate and provide insights of relevance to those struggling with making climate change real today.

OVERVIEW OF THE BOOK

The book is organized into three sections. Section I, "Spaces of Observation," builds on existing work on the histories and geographies of the observatory and field sciences in order to offer a new picture of how weather and climate have been sensed, observed, and understood in diverse spatial settings and with diverse scalar ambitions. Simon Naylor and Matthew Goodman explore the emergence of a network of colonial observatories in the mid-nineteenth century, and the place of those observatories in the development of new regional, national, continental, and imperial geographies of meteorological and climatological knowledge. In examining the cultural and material politics of observational science, the authors situate the observatory as a key site in the furtherance of British imperial progress, civilization, and educational reform. George Adamson furthers their analysis by examining the work of Gilbert Walker, director general of observatories in early twentieth-century India. Adamson shows how Walker's statistical construction of new climatic oscillations was enabled by the kinds of imperial networks described by Naylor and Goodman. Adamson reflects on the implications of these globe-spanning statistical constructions as a new kind of spatial imagination that is difficult to place within conventional historiographies of climatology's progressive globalism. In the next chapter, Katharine Anderson examines how, even as the technological march of the twentieth century seemed to render the entire globe subject to surveillance and control, distinct anxieties circulated around certain atmospheric and oceanic spaces. Interpreting the history of the weather ship as a story of cultural anxieties about technology and the immensity of global space, Anderson situates the evolution of meteorology's spaces of observation within wider imaginaries of the place of human beings within rapidly evolving technological networks and persistently unruly environments. Finally, Ruth Morgan reminds us that, historically, spaces of weather observation are not just products of colonial networks and technological conquest. By examining the knowledge networks through which the Leeuwin Current off the coast of Western Australia has been made known, she shows how local fisher-people, colonial meteorologists, and national and international research bodies have interacted, through various economic, political, and ecological projects, in the production of a distinctive set of imaginations of the Australian climate. Protagonists in each of these chapters worked toward a (re-)scaling of climate by moving from local knowledge networks to an understanding of global climate systems, albeit in a multiple rather than a singular form.

In the chapters of Section II, "Horizons of Expectation," the authors examine how imaginaries of weather and climate have been shaped by different practices of knowing and inhabiting tropical environments. These practices connected scales and brought understandings of climate closer together with cultural, economic, and political ambition. Here, the climate being referenced is not simply that of global climate systems but, rather, a more tangible climate in which local knowledge and experience continued to matter; exceptions and individual variety were as important as understandings of connected, global systems. James Kneale and Samuel Randalls examine how life assurance companies constructed particular imaginative geographies and cartographies of climatic risk, which cannot be understood without reference to contemporary late nineteenth- and early twentieth-century debates about the links between climate, race, and hygiene. Georgina Endfield investigates how the climates of South Africa were mobilized as rhetorical tools to tempt the emigration of British women to the colonies. Although claims about healthful climates—ripe for domestic settlement and imperial improvement—were often dashed, the case provides an important new window onto the gendered politics of climatic knowledge and expectation. Staying in the same region, Meredith McKittrick explores debates around rainmaking in early twentieth-century South Africa, arguing that a racialized discourse of scientific modernity was used to distinguish the "superstitious" practices of black African rainmaking from a surprisingly robust and persistent set of practices and discourses about artificial rainmaking in white farmer and scientific circles. Focusing on the deployment of the notion of "artificiality" as a form of cultural boundary work, McKittrick shows how projects of atmospheric knowledge-making become bound up with broader—and in this case heavily racialized—projects of state-making. Finally, Martin Mahony describes how the meteorological controversies surrounding Britain's infamous postwar "groundnut scheme" in colonial Tanganyika reveal competing conceptions both of climate-society relationships and of what reliable knowledge of tropical climates looked like. A new confident developmentalism saw meteorological expertise sidelined, and when the rains failed to come, efforts were made, as in McKittrick's case, to artificially improve the climate—to save the scheme upon which rested, for many of those involved, the fate of British imperialism in a postwar world. Together, these chapters show the diverse material, epistemic, and spatial practices through which expectations of climate have been fashioned, and the effects of such expectations and imaginations in the world-making practices of colonialism, finance, development, and state-making.

In the final section, "Atmospheric Entanglements," attention turns more concertedly to how relationships between climates and bodies, ecologies, and societies have been conceived in different contexts. The scale of climate emerges as one of connection, as the body or the building becomes an experimental site for enabling and thinking through a broader climatology. Experiences of average weather—the climate of a place—remained relevant while protagonists developed more universal claims about why these climates mattered. David N. Livingstone unravels the links between Hippocratic notions of climate, health, and place, a nascent "biometeorology," and the biopolitics of eugenic thought. Examining the reemergence of the body as a site of climatic inquiry in the early twentieth century, he shows how a set of diverse intellectual conditions converged around new conceptions of the links between climate and health, in a fashion that not only deepened the "pathologisation" of certain places and environments but also has resonance with present-day debates about the links between climate change and future human health. James R. Fleming follows with an exposition of Ellsworth Huntington's ideas about the links between atmospheric ozone and human health and productivity. Digging deeper into the intellectual landscape described by Livingstone through the figure of the controversial Yale geographer, Fleming positions Huntington's revitalized climatic determinism in the context of a curious interwar debate about "biophysics, biocosmics, and biocracy" and provides further analysis of the situated—and yet in some ways enduring—character of this particular variant of the geographical imagination. Finally, in a rather different context, Daniel Barber examines the intersections of modernist architecture and local-scale climate knowledges through the postcolonial lens of Rio de Janeiro. Rather than the body, as with Huntington's concern, it is the building's connection to climate that shapes Barber's story. He looks at how, through a series of design experiments, modernist architectural techniques that were developed in Europe and North America were reworked according to local climatic and sociopolitical conditions. This new climate-centric architecture created new means of encountering climate as something emergent and mediated, a protean feature of manufactured space. Barber's contribution suggests that the study of climate and architecture can enrich our understanding of a particular strand of the geographical imagination, building on the work of Peter Sloterdijk concerned with the evolution of designerly ways of knowing and constructing climatological and atmospheric spaces.[83] The spaces of cities and buildings are imagined and engineered as much as the bodies that inhabit them. All three chapters in this section thus speak to the central importance of exploring climates at all scales, beyond global ones.

Finally, we are joined by Mike Hulme who offers reflections on the forgoing case studies and argues for their relevance to present-day debates about the place of scientific knowledge in societal engagements with the phenomenon of anthropogenic climate change. While notions of climate vary between chapters—in section I being more related to the idea of climate systems and in sections II and III the average weather of a place—geographical knowledges and imaginations produced and enabled connections between the local and the global in multiple and heterogeneous ways. Indeed, the structure of this book resists a historiographical tendency to go from the local and the particular to the universal and the global, preferring to emphasize that place and space matter for the production, circulation, and utilization of all forms of knowledge. This, as Hulme and also Coen point out, is vital to remember and think through in the context of present-day concerns.[84]

PART I

SPACES OF OBSERVATION

1

ATMOSPHERIC EMPIRE

Historical Geographies of Meteorology at the Colonial Observatories

Simon Naylor and Matthew Goodman

ATTEMPTS TO DEVELOP A NEW MODEL SPACE in which the physical sciences could be investigated are at the center of this chapter. In Britain, physical observatories were based on a blueprint provided by astronomy and propelled forward by an obsession with the mapping of the Earth's magnetic field. New or repurposed observatories were established to this end in London and Dublin and across the British Empire from the late 1830s onward. Although often positioned as the poor cousin to the pursuit of terrestrial magnetism, the study of meteorology was a critical component of activities at physical observatories both at home and overseas and were required to conform to the same exacting regulations. Britain's colonial observatories—in Canada, Saint Helena, India, Australia, and South Africa—were designed to produce new meteorological and climate knowledge at regional, national, and continental scales, both through their own observations and their coordination of networks of other sites of observation. This historical geography of instrumental and observatory life shows how these sites were central to the production of meteorological and climate knowledge as well as to other imperial geographical imaginations.

IMAGINING THE OBSERVATORY

The observatory sciences were a loose family of nineteenth-century sciences concerned with both the heavens and the earth, including astronomy, cartography, and surveying, geodesy, tidology, meteorology, statistics, and terrestrial magnetism.[1] All of these insisted on the importance of the observation of nature—in the words of the astronomer John Herschel, the "systematic registry and reduction to fix and realize the fugitive phenomena of the passing moment." The ideal space from which to collect this information was a building or structure that had been built or modified for the purpose, although in practice observatories took many forms. As Aubin, Bigg, and Sibum note in their work on astronomical observatories, certain precision practices *made* the observatory and not the other way around. These were the most crucial components of observatory science. Aubin, Bigg, and Sibum refer to these practices as "observatory techniques," which included the use of precision instruments for making observations and taking measurements; data reduction, tabulation, and conservation; the representation of information; and the management of personnel within and outside the observatory.[2]

The observatory was an experimental site where a wide range of technological devices were developed, tested, calibrated, and put to use. In fact, the observatory helped to further a culture of precision that transformed scientific practices in the nineteenth century. Men of science cultivated the idea of precise instruments, built to exacting standards in metropolitan workshops, "calibrated in centres of excellence and put to work in distant, but well run observatories," which would in turn act as reliable centers of calculation for regional actors. Of course, moving instruments long distances from their place of manufacture to their place of use exerted extreme pressure on what were very fragile networks. Delicate instruments routinely got broken or affected during their journeys. "Instruments were like precious livestock," Schaffer observes, with "every vagary requiring careful husbandry."[3] Their materiality only somehow emerged in these moments of disaster, reminding us that "the definition and character of an instrument is not permanent and fixed but can be altered in many ways, including through repairs, refinements, and reworkings."[4]

Another idea that was deployed to promote observatory science was that such sites were both remote from the world and thoroughly grounded in it. On the one hand, observatories worked by emphasizing their separation from normal life, whether geographically (by being on top of a mountain for instance), physically (by being immune to any effects that might have altered their readings) or socially (by maintaining routines that were not affected by normal life).[5] This said, observatories were also justified on the basis of more worldly concerns. They were routinely positioned and justified as being of utility to expansionist enterprises. As Kapil Raj notes, corporate commerce recognized that the existence and expansion of European overseas trade was dependent on scientific expertise.[6] Of course it wasn't only private companies that invested in science. Observatory science served imperial systems of government. Observatories were justified as being of use to a range of imperial ventures: they aided time-keeping in ports through time balls and the setting of ships' chronometers; they helped improve knowledge of navigation by educating naval officers in practical astronomy. The addition of physical observatories—often in the grounds of the astronomical observatory—promised to extend the usefulness of science to empire by mapping regional magnetic variations, contributing to understanding of the local tides, and improving knowledge of local and regional weather and climate and their relations to trade, health, agriculture, and territorial expansion. Observatories aided knowledge of territories by assisting in surveys and making maps and charts. They were also central to large-scale statistical enterprises. By expanding observatory techniques and networks and taking them into new geographical and cultural territories, "imperialistic projects sometimes overstretched both techniques and networks," but the challenges this overstretch posed often "provided opportunities for overhauling techniques and networks and served to enhance the global prestige of observatory techniques."[7]

Observatories—along with botanical gardens, museums, libraries, and churches—also promised to act as centers of improvement and civilization, which would encourage the development and the refinement of rough-and-ready settler society at the imperial frontier. "Rapidly progressive as our colonies are, and emulous of the civilization of the mother country," opined Herschel on the colonial observatory, "it seems not too much to hope from them, that they should take upon themselves, each according to its means, the establishment and maintenance of such institutions both for their own advantage and improvement, and as their contributions to the science of the world."[8] The Artillery officer and magnetician Edward Sabine agreed with Herschel, arguing that the observatories would help to "arouse, to nourish" natural knowledge in the colonies by helping to form "good observers" out of local residents and by providing a haven for the traveling observer.[9]

THE MAGNETIC CRUSADE

In 1839 the vessels HMS *Erebus* and HMS *Terror*, commanded by Captain James Clark Ross, set off on a four-year voyage to the Antarctic Ocean. The "great scientific object" of Ross's expedition was terrestrial magnetism—to observe the distribution of magnetic influence over the high Southern latitudes along with their changes of form over time. It was hoped that the accurate mapping of such features would lead to the discovery of the causes that engendered "the great features of the magnetic curves, and their general displacements and changes of form," while also helping the navigator at sea.[10] On board the *Erebus* and *Terror* were the staff of several observatories that were to be established on various parcels of British colonial territory en route. The Royal Artillery officer John Henry Lefroy was dropped off on the island of Saint Helena, while another Artillery officer Frederick Eardly-Wilmot (usually known simply as Wilmot) was left to establish a physical observatory in the grounds of the astronomical observatory at the Cape Colony. A detachment of naval personnel was left at Hobart Town, Van Diemen's Land, where a magnetic observatory was established. Another observatory was established at Toronto in Canada. These stations were supported by those operated by the British East India Company at Simla, Madras, Singapore, and Bombay. Observatories at Dublin, Greenwich, and later Kew acted as the standards against which the colonial stations were compared, while Thomas Brisbane established a private magnetic observatory at Makerstoun in Scotland.

The magnetic crusade was sold to the British government by Herschel; Ross and the Admiralty hydrographer Francis Beaufort helped to persuade the Admiralty of the scheme; Sabine ensured the support of the Army; and the magnetician Humphrey Lloyd worked on the details of the fixed observatories and their cost. The Royal Society and the newly formed British Association for the Advancement of Science (BAAS) also played key roles in the project's success. In his address to the 1845 magnetic conference in Cambridge, Herschel argued that "the gigantic problems of meteorology, magnetism, and oceanic movements can only be resolved by a far more extensive geographical distribution of observing stations, and by a steady, persevering, systematic attack."[11] The globalizing tendencies of the British Empire provided the ideal theater and infrastructure on which to prosecute this data-intensive project with global ambitions. The stations in Canada and Van Diemen's Land were judged as approximate to the points of the greatest intensity of the magnetic force in the Northern and Southern hemispheres; Saint Helena as approximate to the point of least intensity on the globe; and the Cape of Good Hope as a station where the changes of the magnetic elements presented features of peculiar interest.

The colonial and domestic observatories that took part in the magnetic crusade shared as their primary concern the observation of magnetic variations, but this was by no means their only preoccupation. Some observatories pursued astronomical research, others collected geodetic and tidal data, and all collected meteorological observations. In relation to the study of meteorology in particular, it was accepted that close relations existed between magnetism and meteorology: that the weather and the earth's magnetic field were subtly related and that both were under the influence of celestial forces, that both were of immediate importance to navigation, and that the atmosphere affected the performance of magnetic instruments such that information on pressure and temperature was needed to effect reductions.[12] The distribution of the observatories across the globe afforded the opportunity to observe the relations between magnetic variations and climate, and investigation into the diverse climates and weather conditions of British imperial territory was deemed of interest in its own right. As Herschel put it: "We depend for our bread of life and every comfort on [the globe's] climate and seasons, on its winds and waters."[13] On these grounds, an observatory in Canada promised to reveal the limits imposed by the northern climate on agricultural expansion, while the Cape observatory promised something similar with regard to the southern African climate. Observations across India and the Bay of Bengal hoped to reveal the laws of storms and of the monsoon.

It is unsurprising that the publicly funded colonial observatories were devoted to utilitarian ends and guided by the principle of service, whether to Dublin and Greenwich, to European science, or to British overseas trade.[14] Despite this, public financial backing for the observatories was for a limited time only. As it was, the sums provided to run them were not enough to meet their full costs. The Toronto Observatory, for instance, frequently had to borrow money at the end of each year to be able to carry on its work. Government funding was cut in the early 1850s, with Sabine conceding that sufficient data collection had been completed, such that "the laws themselves are not likely to be subverted or contradicted by a larger series of observations."[15] In his review of the accomplishments of the colonial observatories, he argued that five years of hourly observations was sufficient to obtain the mean values of the magnetic and meteorological elements and their diurnal, annual, and secular variations, "as well as the peculiarities of climate bearing on the health and industrial occupations of man." This was a convenient figure for Sabine because five years of observations printed in full detail would occupy two quarto volumes.[16] Most of the various observatories continued in some form after funding was withdrawn and were usually sustained by local authorities and volunteers.

TRAINING IN EXACT PRACTICE

In the spring of 1837 Humphrey Lloyd persuaded the University of Dublin to establish a magnetic and meteorological observatory in the grounds of Trinity College. Observations began in November 1838, in time for the commencement of Britain's magnetic crusade. Although Greenwich established its own observatory at around the same time, Lloyd's observatory at Dublin established its place at the epicenter of Britain's network of observatories. Sabine summed up the nature of the observatory's relationship to the colonial sites when he observed to Lloyd, "What a fine family of children the Dublin Observatory will have!"[17]

Indeed, Lloyd used Dublin as a model for other physical observatories within its ambit, which he set out in his *Account of the Magnetical Observatory of Dublin* in 1842.[18] In this account he laid out a detailed plan of the work and the ends of his own observatory and those like it. His thoughts echoed James Forbes's ideas on the necessary functions of the ideal meteorological observatory, which Forbes laid out in his report to the BAAS in 1840, and those of Herschel on the ideal physical observatory, thoughts Herschel shared with key correspondents.

Dublin and its subordinates were to operate and keep in constant repair a set of standard philosophical instruments, including barometers and thermometers, for which constant errors had been determined and to which local instrument makers, observers, and travelers could compare their own devices. Second, the observatories were to contribute to the establishment of laws of phenomena in nature. In a letter to Beaufort in 1835 Herschel claimed that observatories should provide "regular observations of local phenomena of a variable or temporary nature" as well as "the deduction and establishment of their laws of periodicity and local coefficients." Chief among these would be the laws of magnetic intensity and direction, laws of meteorology "in all its extents," and laws of the tides.[19] Third, the observatories should help to fix secular constants, such as mean annual temperature or mean annual pressure at sea level. Herschel demanded accurate determinations of local data that were "invariable, or subject only to very slow (secular) variation," such as magnetic force, air temperature, atmospheric pressure, and sea level.[20]

This final requirement bound all three functions together. The determination of such "normal data" with sufficient precision, James Forbes argued, was "incompatible with any but an official system of registration, which shall be conducted for very many years on exactly the *same system*, with instruments of the same kind, with unremitting attention not only to the *fidelity of the observations*, but to the *perfect repair and comparability of the instruments*."[21] Finding laws and fixing constants required this exacting regimen to be applied in multiple localities, "by observations strictly simultaneous, made according to the same instrumental methods, and with the same instrumental means."[22] As Elizabeth Musselman reminds us, the observatory sciences "required the most extreme levels of timeliness, reliability, and sobriety—the kind of moral and physical purity that only missionaries and astronomers were just crazy enough to demand."[23] Meteorologists too: according to Lloyd, meteorological science depended on these principles of method and cooperation more than any other science.[24] The fact that the observatory sciences required both labor *and* precision to effect their ends and the fact that the two were so difficult to reconcile were at the heart of the challenge for crazy meteorologists with global ambitions.

The only sure way of developing a global terrestrial physics was to provide well-funded stations at home that could verify instruments and train observers prior to their dispatch to state-funded colonial observatories. For Wilmot, Charles Riddell, Lefroy, and other colonial observatory superintendents, Dublin operated as a school of exact practice, where they were sent to receive instructions in observatory techniques. Once observers were in post, Lloyd continued to advise them through a steady flow of correspondence, even if Dublin's own marginal position in British communications networks proved to be a niggling problem. Wilmot at the Cape complained, for instance, that Dublin was out of the way for East India Company ships. He also passed on Lloyd's letters to Lefroy at the Saint Helena Observatory, given that letters were "so very uncertain" in reaching that island.[25] In fact, writing to Herschel in 1842, Wilmot proposed to return home, "where I shall supply more of my own individual wants in the matters of observation, information etc. by one week spent at Greenwich or Dublin etc. than by all the communication of a 12 month."[26] For those times when letters were not forthcoming and observers could not leave their posts, instruction manuals had to suffice. Detailed instructions and correspondence could legitimately train observers on the fly. Diane Josefowicz has noted the high degree of faith that the British—particularly Herschel—placed on written guidance. Herschel, the editor of the Admiralty's 1849 *Manual of Scientific Inquiry*, placed great emphasis on clear and efficient communication and on the ability of guides, questionnaires, and skeleton forms to "regularize and coordinate data collection and communication."[27]

Herschel's advice to meteorological travelers and enthusiasts in his 1835 *Instructions for Making and Registering Meteorological Observations in Southern Africa* demonstrates the value he placed on written instruction as the least-cost path to data accumulation. He spent a great deal of time laying out the basic principles of common observational practice, including the best way to transport instruments, the use of stated and regular hours of observation and registry, the importance of the continuity of the observations, the value of the careful training and supervision of a family member to ensure this, the use of the instruments' own scales, and so on. As much, if not more, emphasis was placed on reduction techniques than it was on observation and archival practice. Herschel provided very careful instructions in the reduction of various observations, including zero point errors, height about sea level, the temperature of the barometer's mercury, and the change of level of the mercurial surface in the barometer's cistern.[28] In turn, poor reductions were worse than no reductions at all—unreduced observations tethered to knowledge of the relevant errors was a much preferred outcome.

One way or another, the data *had* to be able to move. The proper application of corrections allowed data to travel and be compared with data from elsewhere much more efficiently. As a science that studied processes occurring over large distances, meteorology had to spread its observations over "the greatest extent of territory, and the greatest variety of local and geographical position."[29] The reward was the establishment of secular constants and various laws governing weather and climate. For instance, the systematic development of laws to explain periodic changes in the atmosphere promised probable conjecture as to the general course of seasons to come and better preparation for singular events such as gales and droughts. This dual emphasis on data analysis and on the drawing of relations between observed phenomena and physical causes were, Gregory Good notes, two of the biggest preoccupations in the physical sciences during this period.[30]

INSTRUMENTS ON THE MOVE

Before physical data could move from the colonial observatories back to London and Dublin, instruments had to be moved in the other direction—packaged up in workshops in England and carried by ships, hands, on horse, in canoe, and by wagon over oceans, roads, and deserts. The movement of precision instruments was a perilous yet essential aspect of any instrument's life cycle. According to the Royal Society's *Report of the Committee of Physics, including Meteorology, on the Objects of Scientific Inquiry in those Sciences*, the barometers sent out to all observatories established by her Majesty's government and at those in India had first been "independently graduated and compared with the standard of the Royal Society." The report went on to explain the relatively simple process by which a barometer may be compared against a standardized barometer and itself made standard. "By this means," it was stated, "the zero of one standard may be transported over all the world, and that of others compared with it ascertained." The magnetic and meteorological crusade was built on the ideal of observations made by the same instruments operating on the same method all over the world. It was, in theory, to be a scheme conducted simultaneously in its temporal, material, and methodological facets. However, to transport a standard barometer around the globe "with perfect effect" required the utmost care in the transport of the intermediate or "compared" barometer and was by no means "an operation either of trifling import or of hurried or negligent performance," because "some of the greatest questions in meteorology depend on its due execution."[31] The problem of shipping out barometers and other meteorological instruments was a critical issue in the success of the enterprise.

Meteorological instruments were routinely described as "fragile" by those who were charged with their transportation and their operation, and these instruments generally suffered more during transit than their magnetic counterparts.[32] When Riddell's instruments arrived at the Toronto Observatory without too much in the way of breakages, he wrote that "considering their perils on the voyage and 500 miles inland journey, they [the instruments] have *escaped* wonderfully well."[33] Riddell's choice of verb here is illuminating. Riddell understood, as much as members of the Royal Society's Committee of Physics (formerly the Committee of Physics and Meteorology) did, that instruments on the move were vulnerable and subject to a number of possible contingencies through which they could not easily pass but, rather, that they had to "escape" from, implying the need for not a little luck and good fortune.

The instruments traveled alongside observatory personnel on the *Erebus* and *Terror*. The *Erebus* was outfitted to contend with Antarctic ice, and less attention was paid to how best to store the instruments, materials, or even the people, who traveled on board. The ship's holds were hot and damp spaces for fragile meteorological instrumentation. The catalogue of afflictions caused by the transmission of instruments in such an environment was long and proved extremely "disheartening" for Wilmot, as he headed for his post at the Cape.[34] Every iron and steel article arrived "rusty" and "completely spoilt," Wilmot wrote to Lloyd.[35] The standard barometer was broken; the portable barometer "so nearly *stewed* that the ivory ring which [was] marked with the various corrections, was *burst open*"; the ivory scales of the "beautiful wet bulb" thermometer were bent and the screws burst out "by the warping"; all glued cases were melted out and open; and the "sliding boards of the anemometer table shrunk."[36] Several of the magnetic articles that had been sent were found to be similarly bent, warped, rusted, and broken. At the Cape it was largely Wilmot's job to recalibrate the instruments and make them work within the observatory setting. He insisted on having a workshop made alongside the construction of the main observatory building, so that he might repair his instruments following their shipment to the colony.[37]

Lieutenant Lefroy traveled on the *Erebus*'s sister ship, HMS *Terror*. While these two ships were remarkably similar (bomb vessels fitted to withstand Antarctic ice), Lefroy's instruments did not suffer as badly as Wilmot's. The only serious injuries among Lefroy's instruments were those that befell the standard barometer, which leaked some mercury, and the minimum-registering thermometer for terrestrial radiation, which broke.[38] Another of Lefroy's barometers, made by Newman, was also put out of action by a loss of mercury, but this was because it was transported over the rough roads of Pico Ruivo, one of many short stops the Ross party made on their way to Antarctica, and not when it was in transit on the ship.[39] As Lefroy noted in a letter to Sabine after this incident, "could Mr

Newman [the instrument maker] see the roads over which it had been carried such a catastrophe would I imagine in nowise surprise him."[40]

Following a recommendation by the Royal Society the complement of colonial observatories was bolstered with a new station at Aden, then a British settlement. The observatory, which was to be administered by the East India Company, was to be equipped in the same manner as all the other observatories operating on the scheme—its instruments and their adjustment were to mirror the apparatus and methods employed by Lloyd's Dublin observatory. To reach Aden, this equipment had to undergo a total of "*five* transhipments and a passage of four days across the desert."[41] It was not an experience "conducive to the preservation of delicate instruments," reported Henry Yule, a geographer and Central Asian scholar who had joined the East India Company in 1837.[42] In 1840 he was appointed to the Bengal Engineers and ordered to India for service in the Khasia Hills. However, Yule's journey to India first included a stop at Aden to report on the water supply and to "deliver a set of meteorological and magnetic instruments for starting an observatory there."[43]

Several of the most important magnetic instruments arrived "in pieces . . . adrift in the boxes," two maximum registering thermometers were broken, and a small number of additional equipment and magnetic bars were spoiled, but it was the damage incurred by the three barometers, all of which arrived "broken at the upper end," that was the matter of deepest regret to Yule, as it was these cases to which his care was "most particularly directed on route."[44] The "experience of this vexation," Yule wrote in a letter affixed to a report by a committee of Bengal Engineers on the state of the magnetic and meteorological instruments conducted after the articles had arrived in Aden, had convinced him that "never again" should it be ventured to "transport *filled* Barometers . . . in cumbrous cases which can only be carried *horizontally*." He had shown the damaged barometers and their packaging to "Mr D'Abadie [*sic*] a French gentleman . . . well known to scientific men at home for his attention to these subjects [who] said at once that he would not warrant Barometers for 100 yards if packed in that manner." Yule further drew attention to the "utter insufficiency of *glued* fastenings and veneering for the cases and frames of instruments sent to this climate."[45]

The loss of the barometers was a source of great regret to Sabine. The magnetic bars that had been spoiled by rust could have been cleansed and the magnetic instruments that had been damaged made right again had a "competent person" been on the spot in Aden to effect such repairs. With "the exception of the breakage of the Barometers," Sabine wrote to Herschel, these were injuries "of no great consequence."[46] However, although other instruments could be repaired, there was nothing to be done about the broken barometers, except to send out a replacement. This time Sabine demanded that the barometer should be "in a leather case (six guineas) which the officer should himself carry in his hand, or sling on his back, in transhipments, or in land traverses."[47] Although it did not seem that Yule had been supplied with printed instructions to this effect, Sabine's guidance conformed to Herschel's *Instructions for Making and Registering Meteorological Observations*, where travelers were urged to carry their barometers upright, inverted and in hand, or strapped "obliquely across the shoulder of a horseman," so as to break the shocks inflicted by rough roads.[48] Shipping the barometer was an involved and intimate process; the individual in charge had to be conscious of it and alive to its needs at all times during transit and to maintain a close attachment throughout.

Both Yule's and Wilmot's accounts remind us that the main challenge of shipping instruments was not simply a matter of the time spent on board a ship. It also involved instruments' packaging too. In transit aboard ships the supporting arrangement of frames, boxes, and cases chosen to hold the instruments mattered almost as much, if not more, than the instruments they supported. When such items failed they admitted their contents to the local environment of the holds of the ship: spaces hot and damp where meteorological instruments were stewed and bent and warped. The ship could be a floating observatory for those instruments kept above board and in use; for those instruments below deck the experience could be quite contrasting. As the Aden incident also demonstrates, shipment could mean transhipment—unloading and reloading—with the addition of different modes of transport. In all of this, the position of an instrument like the barometer mattered. It had to be, or ought to have been, carried vertically and carried personally. After all, a barometer sent from England was not only a material instrument of measurement. It was an intermediate or compared barometer too, which had been standardized and graduated at the Royal Society and which could be used to further the chain of standardized barometers across the globe. The shipment of such a barometer was a link in this chain, and often a weak one. If in these moments a barometer underwent irreparable change, the whole process had to be reset, and the chain restarted in London at the Royal Society.

OBSERVATORY LIFE

Instrument troubles did not end when the objects reached their final destinations, either. Observatories had to be assembled, and the construction work placed further stresses on the delicate instruments. At the Cape it took eleven months for the physical observatory to be built. The neighboring astronomical observatory did not possess any "unoccupied room in which the instruments might be temporarily established," and Wilmot was unwilling to work outdoors in the day, because "Mr Maclear [director of the astronomical observatory] once had a large telescope blown over by a sudden gale," or at night, because it was deemed "unsafe."[49] The British Admiralty, on whose land the observatory was built, refused to pay for security to ward off looters.[50] As a result, observations did not begin until April 1841. In the meantime, Wilmot wrote to Lefroy, complaining "My instruments groan in their cases at their long confinement, but what can I do?"[51]

The completed observatory was made up of a number of structures—what Wilmot described as "8 or 10 queer looking houses & sheds for instruments."[52] A Venetian screen was placed around the veranda on the north side of the building to house the thermometers, which could be partially opened depending on the wind. The standard Newman barometer sat in the main building. A two-room cottage was provided for Wilmot's use, but he gave it to one of his assistants who had a family. Instead, Wilmot lived in the wind tower. This was a circular building that supported A. Follett Osler's newly invented anemometer, a version of which was also in use at Dublin, as well as a self-recording rain gauge.[53] However, Wilmot's choice of accommodation proved ill-advised. Writing to Lloyd he complained: "The anemometer is a thorough beast—and I wish with all my heart it was blown into the sea. The pressure part is not manageable with truth—for the results depend upon a variety of causes—and as it only blows from SW and NW the direction is not of such value. Perhaps I am rather prejudiced against him as his noise prevents many a nights' sleep in the Tower. But seriously . . . it is quite useless now in its main object" (see figure 1.1).[54] A few months later Wilmot wrote to Herschel about Osler's anemometer, complaining of its "crankiness" with regard to the measurement of wind force and stating that "it will bring little credit upon any who have care of it."[55]

For a man who confessed to knowing little about the physical sciences upon his arrival at the Cape, and who admitted that he fully intended to leave magnetism and meteorology behind when he returned to his regiment, Wilmot's commitment as an observer in Africa was sustained by his Anglican ideals of self-denial, stoicism, and duty. To his mind, both the church and the observatory were sites from which the wilderness would be cultivated.[56] However, both were impoverished. The Cape's Anglican church, finished in 1834, had only one colonial chaplain and little in the way of funding, while the Afrikaners refused to give up their language and way of life despite efforts to anglicize them.[57] Meanwhile, beastly, groaning, cranky instruments conspired with drunk soldiers and tardy building work to retard Wilmot's ability to bring his flock under control and fulfil his duty as a scientific serviceman.

Wilmot's observatory at the Cape was not the only station that struggled to meet the instrumental standards set in Dublin and London—if anything,

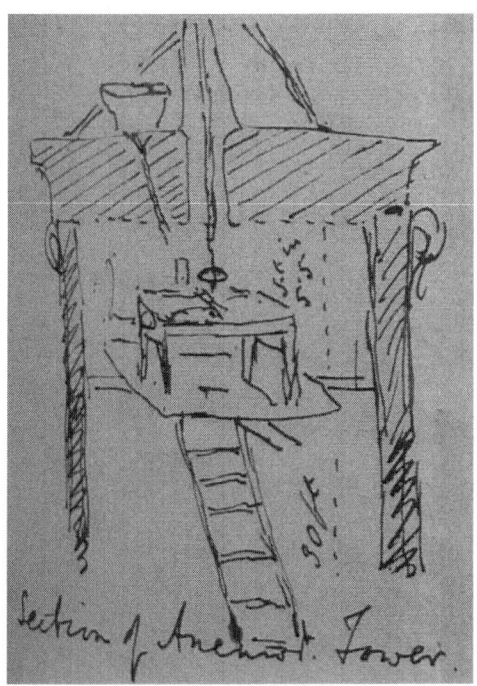

FIGURE 1.1. Wilmot's sketch of the Cape Observatory's anemometer tower. Letter from Wilmot to Lloyd, August 7, 1841, MS/119/II/83, Archives of the Royal Society, London.

matters were much worse in Bombay. The Bombay Observatory was one of four East India Company observatories. Similar to the Cape, it was established in the grounds of an extant astronomical observatory on the small island of Colaba, just south of the Island of Bombay. Meteorological and magnetic observations began in December 1841, in a building that had been hastily erected on the plan supplied by Lloyd, using the instruments that had been intended for the observatory at Aden.[58] In 1834 Arthur Bedford Orlebar had been appointed superintendent of the observatory, a post he held in conjunction with his professorship of mathematics and astronomy at Elphinstone College, located a few miles north of Colaba Island in the main part of the city. When Orlebar retired to Europe during a period of illness in 1842, he was replaced by George Buist, the editor of the *Bombay Times*, and secretary of the Bombay Geographical Society.[59] Buist sent his meteorological observations to Sabine, who used them as the basis of his paper on the diurnal variation of barometric pressure, where he praised Buist and his work at Colaba as of "particular value."[60] In the late 1840s the post of observatory superintendent became allied to the office of the hydrographer. William Montriou, a commander in the Indian Navy, replaced Orlebar in 1847. Montriou was in turn replaced by Lieutenant Edward Fergusson in 1851, and he held the post until 1863.[61] Both men visited Kew Observatory for training.

Concerns about the operations of the Colaba Observatory eventually induced the Bombay government to form a committee of inquiry to investigate its performance, which reported the findings in January 1865.[62] The committee began its assessment by listing the problems with the observatory's instruments, their maintenance and use. Assuming that all of the instruments had begun their life in good order, the committee concluded that poor maintenance standards had been kept and at times there seemed to have been no maintenance at all. For instance, the wind gauge was reported to have "never been well suited for registering sudden changes of wind" with a poor record of force (*Report of the Committee of Inquiry*, 9). The readings from the barometers were pronounced untrustworthy because of "very abundant oxidation" (11) of the mercury in the cisterns as well as the accumulation of dust and cobwebs. Those from the moist bulb thermometers were suspect because of their exposure to variations of the wind, while the cloth covering the bulbs was sometimes covered in salt, which it was assumed would influence the evaporation and "affect the accuracy of the indications to a very considerable extent" (11). The observatory was also deemed to have adopted a problematic observation regime, which went counter to that of all other British colonial observatories: Colaba had adopted the Göttingen day for its magnetic observations and the local day for the meteorological observations. It was therefore difficult to relate one set to the other. Moreover, committee members presumed, confusion over Göttingen and Bombay time had led Professor Orlebar erroneously to assume that the dry-bulb thermometer readings were faulty. He had therefore decided not to publish them, and so the data was lost.

Not that committee members were complaining about the lack of observations—quite the opposite. Huge numbers of observations were published, when they could have easily been condensed. However, where mean values had been provided, it was said that "the mere arithmetical operations of addition, subtraction, and division have been but carelessly performed" (21). They also criticized the Colaba Observatory for its problematic locality. Two eighteen-pounder iron guns sat 220 yards from the observatory and fired shots to announce the arrival of ships into the harbor, which had noticeable effects on the observatory's various delicate instruments. The observatory was also felt to be "isolated" from the Island of Bombay so that it was "by no means adapted for directly ascertaining its general meteorological phenomena" (21). One of the immediate actions of the committee was to appoint a new superintendent at Colaba, Charles Chambers, who did a lot to improve the prospects of the Colaba Observatory, especially in terms of its research into terrestrial magnetism. The observatory also benefited from the establishment of the India Meteorological Department in 1875, which brought all meteorological work in the country under the control of Henry F. Blanford, the first meteorological reporter to the government of India.[63]

REGIONAL SCIENCE

By the end of the century the Bombay observatory was one of a large number of stations across the subcontinent that fed meteorological information to the Imperial Meteorological Office at Calcutta.[64] A similar sort of fate was in store for the Toronto Observatory, although it ended up at the center of its network. This was the case even though, at the outset, Toronto had nothing to compare to the networks and infrastructure already in place in India, or indeed the Cape, prior to the magnetic crusade. The original intention was to place Canada's magnetic and meteorological observatory in Montreal, but an initial survey revealed the presence of magnetic rocks. The site was transferred to Toronto in 1839, where some land was granted by Upper Canada College and a small wooden hut was built.

In the 1840s Toronto was something of a frontier town, and Charles Riddell claimed to find little local interest in science. In fact, Good argues that Riddell and his replacement, Lefroy, were disappointed at the state of Canadian science and so spent time cultivating relations with American personnel and societies.[65] When financial support was threatened in 1850, Herschel wrote to the president of the Royal Society to argue that, if support was continued, the observatory would surely become a "local centre of reference" for magnetic and meteorological research in North America—"the national observatory, the centre of diffusion of astronomical and of all exact scientific inquiry, and the zero point of a future trigonometrical survey."[66] The observatory did receive three more years' funding, after which time the newly named University of Toronto took over the observatory, with financial support from the provincial government. A more permanent stone observatory was built in 1855, the same year in which the university established a chair of meteorology, appointing George Kingston—previously head of Quebec's Naval College—to the position and as director of the observatory (see figure 1.2).[67]

The Toronto observatory had achieved some success in its early years. For instance, it was the first of the colonial stations to publish any of its results. At the BAAS meeting in York in 1844, Sabine presented Toronto's 1841–1842 meteorological observations and praised them for their "completeness and fullness."[68] Toronto acted as base station for the magnetic and meteorological survey of British North American territory, led by Lefroy.[69] It also bore out the prediction and promise made by Herschel in 1850. In his report to the Honorary Secretary of State for the Provinces in 1870, Kingston stated that "I have been much engaged

FIGURE 1.2. Toronto Observatory (1857), photograph by Gagen and Fraser, 1885, Acc: E 3-34B, reproduced courtesy of Toronto Public Library.

during the past year in an endeavour to extend the operations of the Toronto Observatory by making it a centre for collecting atmospheric and other physical data from different parts of the Dominion, and also by promoting improvement in the mode of observing and registering practiced by private observers and by encouraging the establishment of new stations." By 1870 Kingston was receiving monthly returns from eighteen stations in Ontario, six in Quebec, seven in Nova Scotia, and two in New Brunswick.[70]

Many of Kingston's observers were connected to the railway and steamship companies and did not receive financial support. However, the government did fund the observatories at Quebec and Saint John, and Queen's College at Kingston and McGill College at Montreal received allowances. The Quebec and Saint John observatories were primarily time observatories, for shipping, but their interests expanded. In 1856, for instance, Lieutenant Ashe at Quebec wrote to the Legislative Council of the Province of Canada to ask for funds to extend the observatory's capabilities. Ashe argued the case for studies of longitude, astronomy, and meteorology. He asked for funds to provide better meteorological instruments, "as this science is likely to give much valuable information to the agriculturalist, and to promote the prosperity of the country."[71] Kingston managed to persuade the minister of marine and fisheries in the new Canadian government that a network of stations was invaluable to the country. In 1871 five thousand dollars were provided to support the network, with the ultimate end being the provision of storm warnings. This was effectively the beginning of the Canadian Meteorological Service, with the Toronto Observatory as its central office.[72] In 1897 the magnetic work was moved to Agincourt, northeast of the city, because of magnetic interference caused by the electric streetcars, while the original site was refitted to better fulfil its meteorological and time obligations.[73]

A further means of developing a wide view of the weather was through the establishment of meteorological stations in senior county grammar schools across the province. Lefroy put forward a proposal to Egerton Ryerson, the chief superintendent of education for Upper Canada, which was included in Ryerson's 1853 Grammar School Act. It was suggested that improved weather and climate knowledge would aid settlement and agriculture and help to test theories in the physical sciences; that the stations' operation would foster habits of observation and an attention to nature in the schoolchildren; and that a province-wide network of stations would help to develop a sense of Canadian community.[74] Each participating school was supplied with a barometer, a Daniel's hygrometer, an air thermometer, a rain gauge, a wind vane, and standardized forms for daily observations and monthly abstracts.[75]

Complaints were made about the quality of the American instruments, and so instruments were sourced from Britain.[76] While on a visit to London in 1855, Ryerson discovered that Lefroy was also in town and solicited his advice on the subject of meteorological instruments. On Lefroy's advice Ryerson purchased the equipment from Negretti and Zambra, who arranged for the instruments to be tested at Kew and shipped to North America through Negretti's brother in New York.[77] Ryerson expressed the hope that the grammar school network of stations, "when once established, will be more complete than that in any other part of [North] America."[78] The Toronto Observatory took on the responsibility of receiving, arranging, and publishing the observations and of comparing the schools' instruments with its own standards. The project began in 1858, although two years later only sixteen out of the thirty-one counties with senior grammar schools had purchased instruments, and most submitted abstracts intermittently.

CONCLUSION

In his introduction to the third published volume of magnetic and meteorological observations produced at Toronto, Sabine took the opportunity to highlight the more general achievements of what he called the "experiment of Colonial Observatories." What had been achieved above all was "the union of detailed knowledge . . . with the opportunity of generalization, and consequent insight, afforded by results admitting of strict comparison and combination, obtained from well-selected stations at such distant points of the globe, and by a uniform system of observation."[79] In this chapter we have considered the geographical imaginaries that were central to this observatory experiment. The model observatory was one such imaginary. Personified by Dublin, and later Kew, this was a space apart from the world, where social and physical forces were heavily regulated. This model was then translated to other "well-selected stations" across the British Empire through the transfer of architectural, social, observational, and computational principles, most crucially through the training of staff prior to embarkation and through the movement of compared standard instruments, such that a "uniform system of observation" could be effected. The management of data flows from these "distant points of the globe" back to Sabine's computational hub at the Royal Arsenal at Woolwich produced another geographical imaginary: that of a quasi-global physical record.[80] For instance, writing at the height of the crusade in 1845, Sabine combined and compared the observations from Bombay, Toronto, Greenwich, and from the *Erebus* and *Terror* in the Antarctic Ocean, to claim that meteorology was now one step closer to a clearer understanding "than we have hitherto possessed on those great aërial currents . . . and a field of research appears to be thus opened by which our knowledge of both the persistent and the periodical disturbances of the equilibrium of the atmosphere may be greatly extended."[81]

The translation of this model space from metropole to imperial province routinely overstretched techniques and networks. Observatory data supported Sabine's claims as to the development of magnetism and meteorology, even while observatory practices were under scrutiny. Instruments and their cases literally became unstuck when exposed to environmental and social conditions far from home. At the end of long and uncertain communication networks, lonely observers also became unstuck as they struggled to reconcile imperial and scientific imaginaries. However, such challenges sometimes provided opportunities for overhauling techniques and networks. In both Canada and India for instance, struggling observatories eventually helped to build new meteorological networks at a range of scales, and to refine instrumental and observational techniques that were better suited to climates very different to those found across the British Isles. In conclusion then, colonial observatories were zero-points for testing out acts of imperial and scientific imagination, where meteorologists fantasized about bringing equilibrium to turbulent atmospheres.

2

IMPERIAL OSCILLATIONS

Gilbert Walker and the Construction of the Southern Oscillation

George Adamson

THERE HAS BEEN A RECENT CALL within geography, anthropology, and science and technology studies (STS) literature to examine the spatial imaginaries created by discourses of climate. The primary motivation for this call has been anthropogenic climate change. In *Weathered* Mike Hulme drew a distinction between the distant, politics-infused, and socially constructed global "climate" and the local, embodied "weather" that is experienced directly and connected to place.[1] However, this binary is too simple to fully explain atmospheric science, which incorporates intermediate scales of analysis such as synoptic or regional-scale systems. One particular dimension is the spatially connected statistical world of climatic oscillations and teleconnections. These climate "modes" are not weather, but neither are they a unified climate. The framing of a statistically interconnected climate constructed through this form of atmospheric inquiry permeates a number of imaginaries. These climate modes bound global weather in particular ways and present local disasters as elements of an interconnected climate.

The oscillatory climate mode that has the largest place in the public imagination is the El Niño Southern Oscillation (ENSO), a coupled ocean-atmosphere phenomenon that affects the equatorial Pacific Ocean. Unlike many modes (for example, the North Atlantic Oscillation) the underlying physical mechanism that causes ENSO is generally considered to be reasonably well understood. This is the El Niño phenomenon, a quasi-periodic movement of warm water from the west to the east Pacific. The atmospheric component of ENSO is referred to as the Southern Oscillation, a "seesaw" pressure relationship between the east and west equatorial Pacific.

Previous work undertaken to understand the imaginaries associated with ENSO has primarily focused on the oceanic El Niño/La Niña component of ENSO. The anthropologists Kenneth Broad and Ben Orlove, for example, have described how the 1997–1998 El Niño event in Peru was used as a conduit for various imaginaries of globalization, presenting Peru both as a participant in a global technoscientific network and also as the passive victim of malign foreign forces. The cultural analyst Marita Sturken has explored the El Niño in the imagination of Californian males who followed the event on the Weather Channel: El Niño was presented as a conspiracy narrative, the "über explanation for all events." Through this event Californian males constructed their identity as a uniquely resilient people living in a dangerous location at the edge of the Western world.[2]

Other researchers have explored the imaginaries associated with the names El Niño and La Niña. In the English-speaking world these terms give ENSO a "Latin exoticism," allowing the phenomenon to be Othered. In southern California in 1997 the threat of El Niño storms was consciously paralleled with the threat of illegal Central American immigration. The personality of El Niño has also been gendered, with the term "La Niña" applied to the opposite extreme of ENSO, a name given by the South African climatologist S. George Philander in 1985 to reflect its status as El Niño's "consort," the "lesser" of the pair.[3] The environmental historian Julia Miller has explored the politics of this artificially gendered duality, with El Niño presented as an embodiment of masculine destruction. La Niña is expressly described as the less destructive extreme in ENSO, therefore comprising a Derridian hierarchical dualism. This gendering of the phenomenon may have implications for the way that communities prepare for hazards associated with El Niño or La Niña.[4]

ENSO can be framed in multiple ways. Depending on the definition adopted, the phenomenon can be described as extremes, deviations from a norm, a smooth and quasi-periodic oscillation, a dynamic ocean-atmosphere process, or a spatially aggregated set of teleconnections. Most previous work on ENSO imaginaries has focused on the framing of El Niño as an extreme event or deviation. The focus in this chapter is on the ways of understanding relationships in world weather created by the spatial correlations that make the Southern Oscillation (and other global climate oscillations) visible.[5] The visibility of the Southern Oscillation is contingent upon specific sets of statistical practices that bound the Earth's climate through degrees of correlation, that is, the extent to which

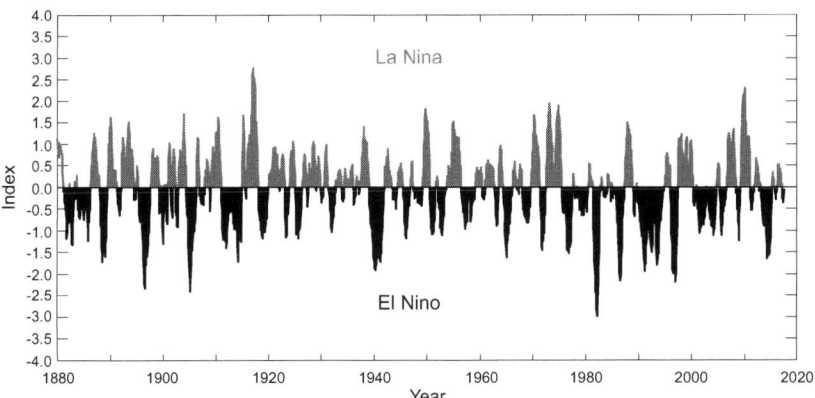

FIGURE 2.1. Time series of the Southern Oscillation Index from 1880 to 2018, based on CRU data. Drawn by Cath D'Alton in UCL Department of Geography Drawing Office.

weather patterns in certain parts are related to weather patterns in other parts. In the case of ENSO this is the relationship with the central Pacific.

The correlation patterns associated with the Southern Oscillation form specific spatial imaginaries. The Southern Oscillation also generates temporal imaginaries, with weather patterns viewed as a constantly varying fluctuation or oscillation (see figure 2.1, which is the time series of the Southern Oscillation Index, determined through a standard measure of the Southern Oscillation Index developed by A. J. Troup in 1965: the difference in pressure between Darwin and Tahiti).[6] Negative values are indicative of El Niño conditions and positive values indicative of La Niña conditions. This imaginary places El Niño and La Niña in an unbroken cycle and shows ENSO as a relatively smooth oscillation. Both of these sets of imaginaries—the spatial correlation and the temporal fluctuation—bound the four-dimensional ENSO into two-dimensional space, although the nature of the imaginaries created by the two sets of images is quite different.

In this chapter we explore the emergence of these spatial and temporal imaginaries through the work that led to the construction of the Southern Oscillation. We focus on the career and research of Gilbert Walker, a British mathematician and meteorologist who worked as director general of observatories in India from 1903 to 1923. Walker was a colonial scientist working within a particular set of contexts, and understanding these contexts is vital to appreciate the role that climate oscillations have played in society. We will discuss his personal experiences during the work toward the creation of the Southern Oscillation, the epistemic traditions in which he was working, and the practices that he devised while working toward the eventual formalization of the Southern Oscillation. We will also explore some of the legacy of the Southern Oscillation as Walker defined it. This chapter is therefore part of an ongoing project to explore the (historical) geographies of ENSO knowledge production and circulation.

GILBERT WALKER

Gilbert Walker (figure 2.2) was born on June 14, 1868. He received a scholarship to St. Paul's School in 1881 and a scholarship in mathematics to Trinity College Cambridge in 1885. His academic career in mathematics was impressive: he was named Barnes Scholar in 1887, Senior Wrangler in 1889, and elected Fellow of Trinity College in 1891. He became a lecturer in 1895.[7] His research focused on pure and applied mathematics and particularly the mathematics of projectile flight, an interest spawned by a visit to Australia in the late 1880s and his fascination with the boomerang, which earned him the nickname "Boomerang Walker."[8] He also focused on the mathematics of waveforms and electromagnetism, and a paper on aberration won him Cambridge University's Adams Prize in 1899 together with Joseph Larmor.[9]

In 1903 Walker was appointed assistant to the meteorological reporter in India. In 1904 he became the third head of the Indian Meteorological Department (IMD) and the second director general of observatories, succeeding Sir John Eliot, who had himself succeeded Henry Blanford.[10] Walker had no formal background in meteorology and preceded his journey to India with a tour of observatories in the United States, where he met and subsequently corresponded with a number of American meteorologists and astronomers. While in America he demonstrated the unusual flight of the boomerang to several of his hosts, founding "boomerang clubs" at the US Department of Agriculture Weather Bureau in Washington, DC, and the Yerkes Observatory in Williams Bay, Wisconsin.[11] His interest in boomerangs continued while he was in India, and he would throw them on the Annandale racecourse in Simla to the amusement of the viceroy.[12]

As head of the IMD Walker brought in a number of British scientific assistants, whom he allowed to conduct their own research alongside their duties at the IMD.[13] Four of these employees—J. H. Field, John Patterson, Charles Normand, and George Simpson—later became directors of meteorological services (in India, Canada and the United Kingdom). Walker corresponded regularly with Simpson, a researcher on atmospheric electricity who worked on James Scott's Antarctic expedition in 1910–1912. Under Walker's directorship the

FIGURE 2.2. Gilbert Walker. From Walker (1997) "Pen Portraits of Presidents."

number of officer posts in the IMD that were occupied by Indians rose from three (of thirteen) in 1920 to ten in 1923. He maintained a large team of Indian clerks, including one meteorologist, Rai Bahadur Hem Raj, who was promoted to the position of imperial meteorologist in 1910 in recognition of twenty years of service.[14] Walker and Hem Raj coauthored a paper on the "Cold Weather

Storms of Northern India" in 1913.[15] Little is known of his other clerks. In 1913 Walker wrote a recommendation for the "boy-mathematician" "genius" Srinivasa Ramanujan, who did not work for the IMD.[16]

Walker conducted substantial research activities while he was in India, including painstaking mathematical research that would lead to the codification of the Southern Oscillation. This was published in Walker's own journal the *Memoirs of the Indian Meteorological Department* from 1909, and then in the *Quarterly Journal of the Royal Meteorological Society* on his return to the United Kingdom. He published nine editions of "Correlations in Seasonal Variations of Weather" between 1909 and 1924, followed by six editions of "World Weather," forming a continuum of research on weather correlations. Based on his observations in India, he also published three papers on the mathematics of tropical bird flight during the 1920s and wrote the *Encyclopaedia Britannica* entry on "natural flight" in 1929.[17]

Walker suffered poor health on at least two occasions. From 1890 to 1892 he spent three winters in convalescing in Switzerland where he became an expert ice skater.[18] Sometime in 1911 or 1912 he again suffered a "breakdown in health" and was forced to spend time away from India, causing Simpson to end his tenure in Antarctica early in order to return to India.[19] Walker returned to India in May 1912.[20] There is little evidence that Walker associated his poor health with climatic conditions in India, indeed he seems to have found the climate agreeable.[21] After his return from India in 1924, Walker spent the rest of his working life as professor of meteorology at Imperial College (then the Imperial College of Science and Technology) where he undertook research on the cellular structure of unstable fluids and the formation of clouds and cyclones.[22] He retired in 1934 but continued to publish until 1947, acting as editor of the *Quarterly Journal of the Royal Meteorological Society* from 1925 to 1941. He was awarded a Royal Medal by the Royal Society in 1924 and the Symons Gold Medal by the Royal Meteorological Society in 1943, and he was knighted in 1924. He died on November 4, 1958.

WALKER'S WORK

It is often stated that Walker's selection as successor to John Eliot is surprising.[23] This somewhat misrepresents the nature of Walker's role in India and the status of meteorological research at the turn of the nineteenth and twentieth centuries. The director general of observatories was in charge of observatories with three remits: meteorological, astronomical, and geomagnetic observation (as well as, in some cases, tidal and geodesic observation). Many of the observatories were primarily set up for geomagnetic observation with meteorological instruments introduced later; correspondingly, meteorologists during the nineteenth century were often primarily astronomers. Meteorology, geophysics, and astronomy were thus regularly researched interchangeably using the same or similar theories.[24] Walker had a strong background in electromagnetism and came highly commended in this field, having won the Adams Prize in 1899 because of his work on the aberration of light—research that the senior assistant to the Greenwich Observatory, Frank Watson Dyson, described as being "of as much interest of the Astronomer as the Physicist or Mathematician."[25] He was elected Fellow of the Royal Society in 1904 on the back of this work and received an employment reference from Lord Kelvin.[26] He therefore had a highly successful career in the disciplines that contributed to "observatory science" (see Naylor and Goodman in this volume), even if this did not include meteorology.

Walker's early research in India focused on electromagnetism. He gave a number of lectures in Calcutta in 1908 under the title *Outlines of the Theory of Electromagnetism*.[27] Some of his earliest administrative responsibilities as director general of observatories were to settle arguments among magnetic, meteorological, and astronomical observers, as is evident in a letter from Louis Agricola Bauer and the Department of International Research in Terrestrial Magnetism in Washington: "Is it not possible in consultation with all parties concerned to make an annual division of expenses? For example, so much for astronomical work, so much for meteorological work, etc., so that the astronomers, meteorologists and magneticians will know precisely how much to count on and be made responsible for limiting their expenditures and researches within the amount assigned. This is the way we do it in the Coast and Geodetic Survey where also varied interests are represented."[28] The three-way nature of Walker's administrative responsibilities is reflected in the scientists he visited in America before arriving in India: the meteorologist Cleveland Abbe, the astronomers George Ellery Hale and Edward Pickering, and the geophysicist Louis Agricola Bauer. He also visited Gustav Hellmann at the Preußischen Meteorologischen Institut (the Prussian Meteorological Institute) in Berlin on his way to India.[29]

Walker's other primary responsibility as head of the Indian Meteorological Department was to devise a long-term average forecast for the Indian monsoon, which had been the principal goal of the Indian Meteorological Department since its establishment in 1875. In 1886 Henry Blanford had established a regression-based formula for predicting average monsoon rainfall over the subcontinent, and from 1887 this had included data from beyond India, with Blanford inferring in 1887 that droughts in India were associated with pressure systems over Mauritius, Australia, and Central Asia.[30] John Eliot extended the forecast to incorporate the strength of the trade winds over the Indian Ocean, Nile flood data, and pressure systems over the southern Indian Ocean, South America,

and South Africa. These forecasts had failed around the turn of the twentieth century, neglecting to predict the drought of 1899 that led to a devastating
famine.[31]

Walker's work to predict the monsoon was influenced by these regression-based forecasts and also by a search for a terrestrial footprint of astronomical
cycles that had been prevalent within meteorology since the late nineteenth
century. This tradition emerged from the joint practice of meteorological, astronomical, and geomagnetic research and was led by two particular scholars
later named by Walker as being among his influences: Sir Norman Lockyer and
William James Stuart (WJS) Lockyer, a team of father and son who were both
astronomers by training, with Norman Lockyer best known for his pioneering
work using spectroscopy ("spectroheliography") to examine variations in solar
activity.[32] The elder Lockyer's meteorological work primarily focused on the
actions of sunspots on weather patterns. This was influenced by the discovery
in 1843 of the eleven-year sunspot cycle, which Lockyer considered the "Saros
of meteorology" after the Saros cycle that was used to predict solar and lunar
eclipses. Norman Lockyer's philosophy regarding meteorological research was
laid down in his 1872 monograph *Meteorology of the Future*, where he stated that:
"Surely in Meteorology, as in Astronomy, the thing to hunt down is a cycle, and
if that is not to be found in the temperate zone, then go to the frigid zones or the
torrid zones to look for it, and if found, then above all things, and in whatever
manner, lay hold of, study it, record it, and see what it means"[33]

Lockyer's work included a report in 1879 to the IMD on the links between
the sunspot cycle and southern Indian rainfall and the cyclic nature of famines
in Madras. This was well received in London, but it was rejected for monsoon
forecasting.[34] Later, working with his son, Lockyer sought to uncover pressure
relationships in world weather and to determine cycles within those relationships
through harmonic analysis, principally to ascertain the role of sunspots in global
weather patterns. This included in 1906 the discovery of a "barometric see-saw of
short duration (about 3.8 years) occurring between India and Cordoba."[35]

Another important influence on Walker was the Swedish meteorologist
Hugo Hildebrand Hildebrandsson, who in 1897 published work on atmospheric
"centres of action." This involved plotting pressure data from sixty-eight meteorological stations around the world for the period from 1875 to 1884.[36] Among
other trends, this demonstrated an opposition of pressure between Sydney and
Buenos Aires. Although the relationship was only established by observation of
plotted curves, Walker later claimed that this was the first observation of the
Southern Oscillation, stating that Hildebrandsson was the first to visualize "the
far-reaching character of the subject."[37]

Walker also corresponded regularly with the British meteorologist William Napier Shaw, who in 1905 published a paper on the potential relationship between rainfall in northwest Europe and the trade winds over Saint Helena.[38] It is likely that this work on weather relationships between geographically unconnected regions of the world was an influence on Walker's later approach. It is also likely that he was influenced by the directors of observatories whom he met in the United States in 1903. For example, Cleveland Abbe—an astronomer before he was a meteorologist—wrote the following regarding Walker's departure from New York: "Our weather between Washington and New York is very apt to repeat itself in seven-day periods for a little while and then start off on another period. To-day, therefore, I may venture on a long-range prediction, to the effect that on Wednesday April 8th, at New York, you will sail out of the harbour with pleasant weather."[39]

Walker's work to understand the variability of the Indian monsoon, therefore, derived from correlation-based studies of relationships in "world weather" in order to derive regression-based forecasts. He did not follow the development of dynamical meteorology that was being developed by Abbe and Vilhelm Bjerknes among others, despite Abbe, in a letter of 1902, encouraging Walker to turn his attention to dynamical meteorology.[40] Instead, Walker held that "the relations between weather over the earth are so complex that it seems useless to try to derive them from theoretical considerations; and the only hope at present is that of ascertaining the facts and of arranging them in such a way that interpretation shall be possible."[41] His research did eventually turn to dynamical meteorology on his return to the United Kingdom where he undertook laboratory-based analysis on the formation of clouds.[42] While he was in India, however, the pragmatic considerations of forecasting a system as large and diffuse as the Indian monsoon apparently drove Walker toward a statistical approach, his obligations as a colonial meteorologist and forecaster trumping his instincts.

Three decisions made by Walker early during his time in India were fundamental to his work. First, an understanding of world weather and hence forecasting the Indian monsoon required an extension of focus beyond the Indian Ocean to incorporate the entire world, including relationships with a lag of several months (here he did take a lead from Abbe).[43] Second, correlation analyses could not be undertaken with individual meteorological stations but, rather, with "centres of action" where meteorological conditions were closely correlated and affected regions around them. This concept of centers of action had been introduced by Teisserenc de Bort in 1881 and adopted by Hildebrandsson, but Walker extended the definition to regions that are seasonally important as well as annually and to include rainfall and temperature.[44] He also introduced the definition of "passive" centers of action, which are homogenous but affected by other regions rather than affecting them. The seventeen centers of action that Walker ultimately isolated included Iceland, various parts of Australia, South America, the Indian peninsular, and Java.[45]

Walker's third decision, and arguably his most important contribution to the search for relationships in world weather, was to explore new mathematics of the significance of correlations and periodicities.[46] This was in response to a criticism of the Lockyers' earlier work; namely, that the likelihood of finding strong relationships or patterns increases when large numbers of relationships are selected. Walker was the first meteorologist to apply the Pearson coefficient to relationships between sets of recorded meteorological data.[47] He also calculated the "probable error" of a relationship so as to determine the likelihood of the relationship being derived through chance, correctly surmising that if one relationship was artificially selected from a number of relationships the chance of a trend being ascertained would increase with the number of relationships compared. This criterion was also applied to harmonic analysis. Here he was building particularly on the work of Arthur Schuster (a friend with whom he corresponded regularly) who had published a method for determining the reliability of periodicities with an a priori specified frequency.[48] Walker presented a mathematical "criterion for reliability" in his third paper in the *Memoirs of the Indian Meteorological Department* in 1914 under the title "Correlation in Seasonal Variations of Weather III: On the Criterion for the Reality of Relationships or Periodicities."[49] He considered a relationship reliable when the probability of it having occurred by chance was less than or equal to one-twentieth (equivalent to the 95 percent significance value commonly used in statistics today).

Walker used his criterion to examine the reality of apparent relationships in world weather. One particular area that Walker applied his criterion to was the influence of sunspots. In "Correlation in Seasonal Variations of Weather III" he calculated correlations between sunspots and temperature, rainfall, and pressure in different parts of the world.[50] In "Correlation in Seasonal Variations of Weather VIII" he added relationships with the number of cyclones in different regions and cloudiness in India. He ultimately concluded that the sun "plays a subsidiary part as a control of weather," basing his arguments on a mixture of empirical analysis and the absence of a physical explanation.[51]

> In general it appears that for explaining the weather abnormalities of the seasons the variations of the solar radiation are inadequate, and we must seek the reasons in the previous distribution of seasonal features over the earth. In this conclusion we disagree with the widely prevailing idea that

such abnormalities are an immediate consequence of changes in the heat given out by the sun. As a definite example I know nothing in solar physics to explain without reference to previous terrestrial conditions the contrast between the biggest Indian monsoon on record in 1917 and the monsoon in 1918 that has only been surpassed in scantiness twice in the last sixty years.[52]

The philosopher Christopher Pincock argues that the Lockyers were one major casualty of this analysis.[53] Indeed Walker seems to have had somewhat of a frosty relationship with the Lockyers. In a letter of 1903 George Hale remarked to Walker that "I am glad your visit to Lockyer turned out to be more satisfactory than you had anticipated." In a postscript of a letter to Walker in 1909 Arthur Schuster stated, "you need no longer be afraid of Lockyer in the Observatory Committee his influence is nil." However, there is no evidence that Walker sought to directly contradict the Lockyers, and the mentions of them in his papers are generally respectful. Indeed, he named the Lockyers as among "the previous workers to whom I owe most," alongside Hildebrandsson and Felix Maria von Exner, C. Braak, Bjørn Helland-Hansen and Fridtjof Nansen, and Robert Cockburn Mossman.[54]

THE SOUTHERN OSCILLATION

Between 1914 and 1923 Walker utilized his criterion for reliability to search for statistically robust correlations in world meteorological data. This involved the undertaking of large quantities of calculations in Simla, aided by a surfeit of Indian clerks who were otherwise underemployed because of the unavailability of British supervisors during the First World War.[55] He also applied his criterion to analysis of "oppositions" or "see-saws" of pressure between different regions of the world, some of which had previously been suggested by the Lockyers and others. In doing so he was able to provide a mathematical justification for a "well known" opposition of pressure between Iceland and the Azores.[56] He also corroborated a "familiar" fluctuation in the North Pacific, the discovery of which he attributed to a relationship between pressure in the North Atlantic and weather over Europe that had been postulated by Niels Hoffmeyer in 1878. In 1924 he named these relationships the North Atlantic Oscillation (NAO) and the North Pacific Oscillation (NPO).[57] This word "oscillation" was taken from a paper by his precursor as director of observatories John Eliot in 1895 on "certain oscillatory changes of pressure of long period and short period in India," although this was the first time it had been used for such large-scale variability.[58]

The relationship that Walker considered most significant was a "swaying of pressure on a big scale backwards and forwards between the Pacific Ocean and the Indian Ocean."[59] This "swaying" he identified between two clusters of meteorological stations. The first cluster included "the high pressure centres in the north and south Pacific (San Francisco, Tokio [*sic*] in summer, Honolulu in spring, summer and autumn, Samoa, S. America)." The other included "the centres of relatively low pressure in the region of the Indian Ocean (Cairo, N. W. India, Port Darwin, Mauritius, S. E. Australia and the Cape)."[60] In 1924 he codified this relationship as the Southern Oscillation (SO).[61] Walker credited Hildebrandsson and the Lockyers for the first discovery of this relationship, although both had represented the relationships only graphically with no statistical relationships.[62] It is unclear whether Walker was inspired by his contemporaries to apply his criterion to these relationships or whether he arrived on the relationships by chance and gave the credit due to his good nature. However, in demonstrating the relationship mathematically Walker gave the relationships what Pincock refers to as a "superior epistemic position," as well as names.[63]

In 1932 Walker, together with Edward Bliss, published indices for the three oscillations (NAO, NPO, and SO).[64] These were derived through correlation and some relatively arbitrary mathematics whereby data points that co-varied less strongly were assigned a value of 0.7, explained as follows:

> We might divide the centres of the oscillation into two groups, those in the first group tending to have positive departures when those in the second are below normal; and we might further suppose that a, b, c represent the proportional departures of three strongly marked members of the first group, and d, e two less marked members of that group; and that f, g, and h, i, j stand for strongly and less strongly marked members of the second group. Then we might take $(a + b + c - f - g)$ as a first approximation for the oscillation as a whole. On correlating this first approximation with the data of the individual centres we might find that while a, c, and g have coefficients of .78, .75 and −.82 respectively, b has only .64 and must therefore go into the less strongly marked or B group, and perhaps the coefficient of i shows that it ought now to go into the more strongly marked or A group. So as a second approximation we try $(a + c - f - g - i) + .7 (b + d + e - h - j)$, and after one or two more experiments we find a formula consistent with its relationships.

The formula for the Southern Oscillation in austral winter (June to August) was then defined as:

> (Santiago pressure) + (Honolulu pressure) + (India rain) + (Nile flood) + 0.7 (Manila pressure) − (Batavia pressure) − (Cairo pressure) − (Madras pressure) − 0.7 (Darwin pressure) − 0.7 (Chile rain)

For the austral summer the formula derived was:

> (Samoa pressure) + (Northeast Australia rain) + 0.7 (Charleston pressure)
> + 0.7 (New Zealand temperature) + 0.7 (Java rain) + 0.7 (Hawaii rain) +
> 0.7 (South Africa rain) − (Darwin pressure) − (Manila pressure) − (Batavia
> pressure) − (Southwest Canada temperature) − (Samoa Temperature) − 0.7
> (Northwest India pressure) − 0.7 (Cape Town pressure) − 0.7 (Batavia
> temperature) − 0.7 (Brisbane temperature) − 0.7 (Mauritius temperature) −
> 0.7 (South American rain)

From these indices Walker calculated correlations with the Southern Oscillation and meteorological parameters (rainfall, temperature, and pressure) from all of the observatories he had data available for. From this he created spatial correlation charts for the Southern Oscillation (see figures 2.3 and 2.4). Walker also produced charts of pressure lagged by six months (not presented here). Later he applied harmonic analysis to his Southern Oscillation Index, demonstrating a period of around three years.[65]

Walker's charts show many of the relationships that are now commonly associated with the Southern Oscillation, including strongly positive associations with rainfall over India, Australia, southern and eastern Africa, and parts of Southeast Asia (i.e., higher rainfall during what we now consider La Niña conditions), and negative relationships over the equatorial Pacific and South America. Likewise, pressure in June–August displayed a strong positive relationship with the Southern Oscillation in a band across the central Pacific. It is interesting that Walker did not identify a similar relationship for December–February, although the chart for relationships of pressure in that region with the Southern Oscillation six months previously (not shown) does show this. Pressure relationships in the Indian Ocean and around Australia/Southeast Asia are negative.

Walker produced several hypotheses to explain the underlying mechanisms of the Southern Oscillation. For all three oscillations he attributed the mechanism to "an increase in the general circulation."[66] For the Southern Oscillation he first attributed the change in the circulation to unusually heavy ice flows in Antarctica leading to high pressure in South America; South America being as an "active" station within the Southern Oscillation during the Southern winter.[67] However he had dropped this hypothesis by 1932.[68] Port Darwin was noted to be a particularly active location, but Walker was not able to suggest any mechanism that explained its importance. Regarding sunspots, Walker noted that they were thought to offer "some influence" on the Southern Oscillation, yet "solar activity cannot be the cause of the SO."[69] In "World Weather V" (1932) he and Bliss recorded that the strong negative coefficients with temperature off

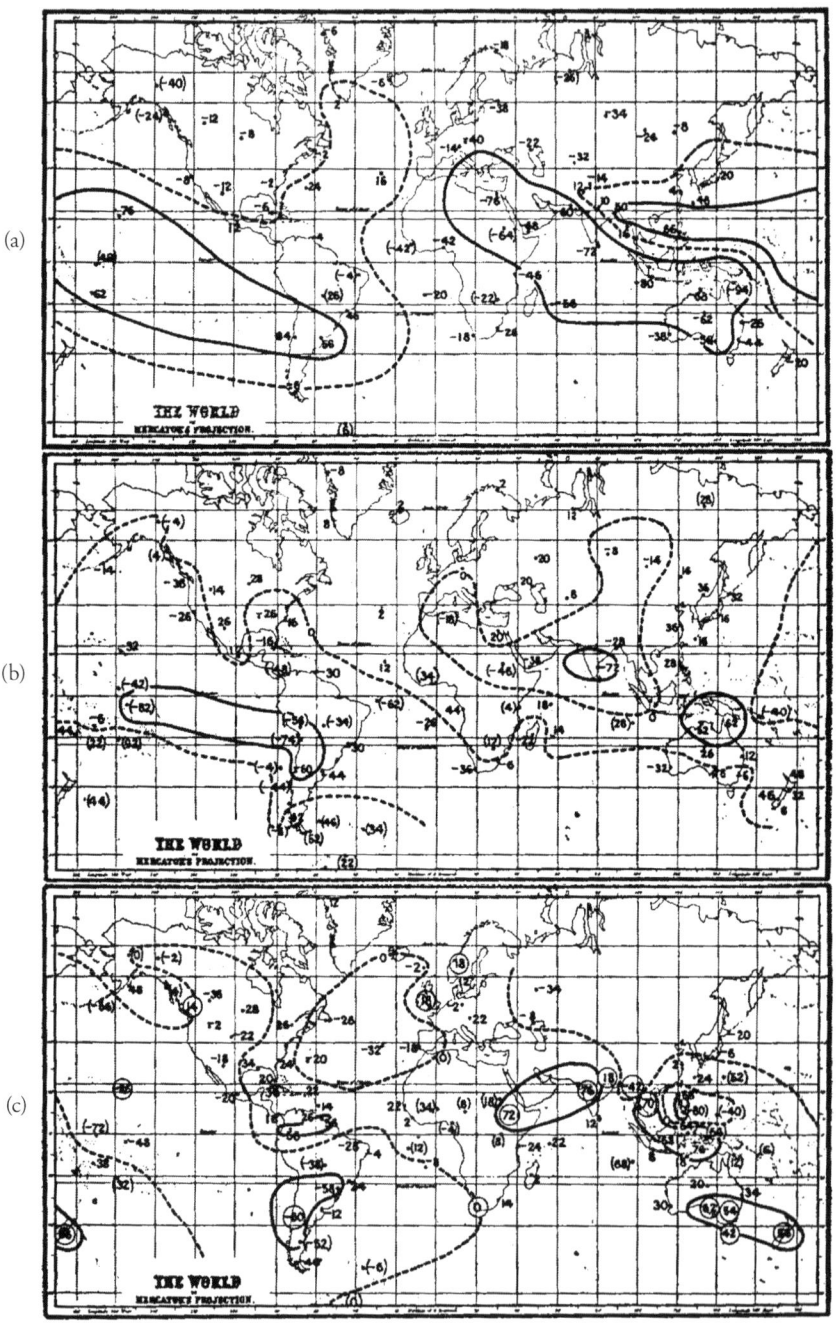

(a)

(b)

(c)

FIGURE 2.3. Spatial correlations of Walker's Southern Oscillation index for June to August with (a) pressure, (b) temperature, (c) rainfall.

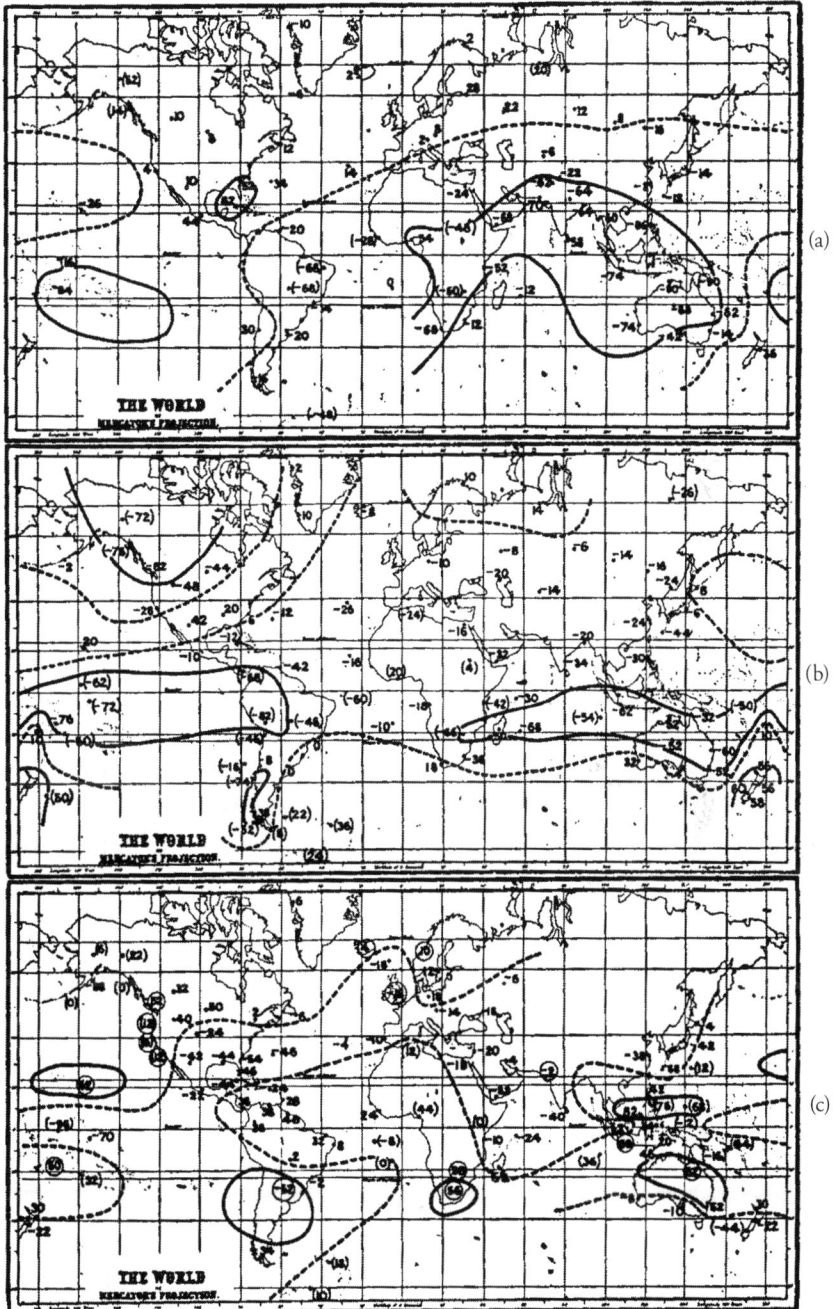

FIGURE 2.4. Spatial correlations of Walker's Southern Oscillation index for December to February with (a) pressure, (b) temperature, (c) rainfall.

the coast of South America could be related to "the cold Humboldt current and [to] the relatively cool SE and ESE winds blowing from over it, so that an increased circulation might furnish the explanation."[70] That is, the strength of the easterlies deriving from the Humboldt Current may explain the mechanism of the Southern Oscillation (an "increased circulation" being associated with a positive Southern Oscillation and hence what are now considered La Niña events). Walker thus came close to suggesting the mechanism of ENSO as it is currently understood, although he did not follow this line of inquiry.

Walker saw that the lack of a physical mechanism should not be an impediment to using statistical methods. He viewed the Southern Oscillation index as a useful measure of "foreshadowing" seasonal weather patterns in different parts of the world, a term he invented to account for the lower level of precision in these techniques compared to forecasting.[71] To this suitability of this approach he invoked his own experience in developing a statistical forecast of the monsoon in 1908, which was not based on an underlying understanding of the monsoon. This had provided a "closeness of fit" of around 0.5, a similarity he illustrated with a time series in World Weather.[72]

DISCUSSION

Philipp Lehmann has argued that the first decades of the twentieth century represented a shift in the science of the weather, as meteorology and climatology were increasingly split off from one another.[73] Studies of the weather developed along the lines of the dynamical meteorology being developed by Bjerknes. Climatology continued the study of cycles popularized by Eduard Brückner and others. This latter tradition both incorporated understandings of anthropogenic influences on the climate through deforestation and also used climate to explain human history and racial difference. Climatology was also increasingly associated with eugenics, as David Livingstone has demonstrated elsewhere in this volume, with Ellsworth Huntington becoming somewhat of a parody of this tradition. Lehmann argues that the climatic tradition was scrutinized by dynamical approaches through the first half of the twentieth century and was found lacking, leading to a (temporary) demise during the middle of the twentieth century that was aided by climatology's unpleasant political associations.

It is difficult to fit Gilbert Walker into this simple dichotomy. He avoided dynamical meteorology as a tool for forecasting and instead followed a statistical approach, producing what Abbe disparagingly called "climatological forecasts" (an endeavor that Abbe described as "wandering away from the higher intellectual light").[74] Nevertheless, Walker was a mathematician by training with advanced understanding of physical processes and was closer to the dynamical meteorologists in his personal relationships. He corresponded regularly with Abbe, Exner, and Napier Shaw and showed no evidence of sharing the political views of Huntington and his contemporaries. Walker also did eventually embrace dynamical meteorology through his work on convection and cloud formation, and he acknowledged Arthur Schuster in 1943 for helping him recognize "the physical rather than the mathematical side of meteorology."[75]

Walker generally avoided use of the term "climate" in both his research papers and published correspondences. He apparently did not have much interest in theories of anthropogenic climate change. The American climatologist and eugenicist Robert DeCourcy Ward wrote him on "the influence of forests upon rainfall in India" in December 1904.[76] Walker's response is not recorded but, given the lack of any discussion of anthropogenic influences on climate in Walker's writings, it can be assumed to have been in the negative. He was also dismissive of the cultural, physiological, and economic implications of climatic cycles that interested Huntington, William F. Petersen, Clarence Mills, and others (see Livingstone, this volume). There is no evidence that he based this dismissal on political grounds; rather, his criticisms related to the weak mathematics that characterized economic climatology. For example, in 1922 the economist William H. Beveridge published a paper on "Wheat Prices and Rainfall in Western Europe" in the *Journal of the Royal Statistical Society*, in which he claimed to find a link between wheat prices from 1500 to 1869 and the sunspot cycle. Walker wrote a letter to *Nature* to demonstrate the weak mathematics in Beveridge's paper, which clearly irritated Beveridge.[77]

Criticizing periodicities that did not use the criterion of reliability became a fixation in the later years of his career. In 1945 Walker stated that he had "spent much time and energy in attempting to diminish faith in very doubtful periodicities."[78] For example, in an editorial for the 1936 volume of the *Quarterly Journal of the Royal Meteorological Society* Walker reproduced a comment in *Earthquakes and Mountains* by Harold Jeffreys, which stated that "it is to be wished that editors of journals would make it an absolute rule not to publish papers on periodicities if these criteria are not applied; if the results are not significant they are worthless, and if they are significant opinion is prejudiced against them in advance." Walker replied: "If, as is probably true, ninety-five per cent of the periods announced are non-existent, it will save costs in printing, as well as valuable time spent in reading, if such a rule is enforced."[79] As well as helping to popularize the idea of climate cycles, Walker should perhaps then also be credited with helping to instigate the movement away from them.

Rather than Walker's place within an intellectual schism in meteorology and climatology in the early twentieth century, it may be better to classify Walker through the tradition that he emerged from, and the context within which he was working. In regard to the former, Walker's approach to meteorology was shaped primarily by his training in electromagnetism and mathematics and through his administrative and intellectual responsibility to all of the fields that came under his jurisdiction as director of observatories. Walker's approach married harmonic analysis in physics with meteorology, carrying on a tradition that was at least half a century old but giving it new mathematical rigor. It is interesting that Walker chose the word "oscillation" to describe the climate modes he uncovered; although the Lockyers recognized the introduction of this term to tropical meteorology by John Eliot, it is a word they used sparingly, preferring "periods" or "variations."[80] Huntington and Napier Shaw used "pulses."[81] Walker himself originally used the term "swayings," and it is certainly not at all clear from the time series of the Southern Oscillation presented in "World Weather V" that the phenomenon can be described as a smooth "oscillation."[82]

His use of the term may be partly explained by his long correspondence with Schuster who was a pioneer of harmonic analysis in meteorology, most significantly applying the analysis to sunspot data in 1906 and validating the eleven-year Schwabe cycle.[83] Like Walker, Schuster's background was in physics and his primary contribution to meteorology through mathematics. Walker apparently saw himself as building on Schuster's work, and he acknowledged him when receiving the Symons Medal in 1934.[84] His influence is evidenced in his note in response to Beveridge's paper:

> Before Schuster's papers on the periodogram it was customary for a period to be accepted as real provided that it had an amplitude comparable with that of the original figures under analysis; and he revolutionised the treatment of the subject by showing that if the squares of the intensities of the various periodic terms are plotted in a periodogram . . . then the probability that a particular ordinate will equal or exceed ka is e^{-k}. . . . Sir William Beveridge . . . goes a stage further, and after picking out the largest from a number of intensities he applies the same criterion as if no selection had occurred. . . . It is however clear that if we have a hundred intensities whose average is decided from a number of random figures then the probable value of the largest of these chance intensities will not be a but will be considerably greater, and it is only when the largest amplitude actually derived materially exceeds the theoretical chance value thus obtained that reality can be inferred. . . . As pointed out in a paper "On the criterion for the reality of relationships or periodicities" in the Indian Meteorological Memoirs (Vol. XXI, No.9, 1914) the same principle is valid when discussing relationships.[85]

Walker, like Schuster, worked from an epistemology in which mathematics was primary. His approach to meteorology was to apply the mathematical rigor he had used in studying electromagnetics and the physics of motion, in this case through the mathematics of statistics. Within this context, the attribution of the term "oscillation" (suggesting a misleading regularity) is understandable.

Walker's approach to meteorology can be classified as the collection of data on weather variability and the processing of these data to reveal relationships over large spatial and temporal scales that were not previously visible. Through this he sought to uncover patterns that might aid in the seasonal prediction of weather. There is little evidence that Walker saw himself as a tool of colonial power. He fought for the independence of the IMD as a separate entity from the colonial observatory committee, and under his directorship unprecedented numbers of Indians were promoted to officer positions within the IMD.[86] Yet his work was influenced by a particularly imperial imaginary. Much has been written about the way that attempts to formalize the natural and social world through data gathering enabled colonial power, with statistical surveys such as William Henry Sykes's *Statistic of the Deccan* allowing the colonial government to understand and therefore control territories recently acquired.[87] These data-driven practices constructed a particular British vision of the subcontinent: colonial science and statistical practice drew together disparate landscapes, environments, communities, and religions to construct a unified territory that was made visible through technoscience. This territory therefore required technoscientific governance, providing a justification for—as well as an aid to—the British Raj.

Katharine Anderson has described the role of meteorology within this colonial scientific enterprise. Control of the Indian atmosphere provided a synecdoche for governance of the subcontinent. Meteorology provided "a timely example of underlying law and practical command."[88] Imperial meteorology and medicine could also validate colonial rule by diminishing the effects of the famines, epidemics, and cyclones that had been a recurrent aspect of life on the subcontinent, as well as naturalizing such events and thus partially removing them from the domain of direct governmental responsibility.[89] Observatories had a particular function within this endeavor (see Naylor and Goodman, this volume). Observatories were deliberately located in places where they could be beacons of imperial expansion and colonial superiority. They could act as "centers of improvement and civilization" to "encourage the development and the refinement of rough-and-ready settler society at the imperial frontier." These were spaces "apart from the world, where social and physical forces were heavily regulated."[90] Within them (male) observers could embody the imaginary of the ideal colonial scientist and administrator: an intelligent, precise, and unemotional individual who governed through supreme patience and intellect, which was an example to his native subordinates.[91] Inside these spaces the meteorologist could fantasize about "bringing equilibrium to turbulent atmospheres," just as the district collector, also located in spaces away from the Indian metropole, fantasized about bringing order to a chaotic territory through clearheaded British governance.[92] The colonial hill stations and "summer capitals" of the Raj, such as Simla in which Walker was located, served similar purposes.[93]

Walker's archives reveal little evidence that he paid particular attention to these considerations. In fact, it would be fair to say that he primarily considered himself a scientist who happened to be working in India, rather than an imperial scientist. Yet the specific geographies of colonial meteorology enabled his work. Walker's examination of "world weather" relied almost entirely on data that were collected from observatories in imperial centers or in European colonies. Each of the seventeen meteorological "centres of action" that he isolated were located within the territories of Great Britain, Russia, Portugal, and the United States, with the exception of South America (comprising former colonies) and Iceland (from 1918 in an Act of Union with Denmark). Walker's complicated Southern Oscillation Index includes data from twenty-six locations, each of which were located in extant or former colonies at the time the indices were derived. The maintenance of these expensive observatories around the world and the sharing of data globally—which allowed Walker to undertake his analysis—was also arguably enabled only by the collaboration between a relatively small number of imperial powers that began with the Vienna Congress of 1873. The proliferation of meteorological data that led to the codification of the Southern Oscillation was the outcome of a particular manifestation of imperial power, itself building on an imaginary of the interaction between colonial science and governance.[94] Furthermore, it was only through the availability of large amounts of cheap Indian labor that Walker was able to undertake the mass of calculations required to uncover the oscillation.

This brings the argument back to the spatial imaginaries of climate at the opening of this chapter. Walker's approach to meteorology (which would now probably be considered "big data") created a way of understanding the atmosphere that was separate from the lived experience of weather and that relied on mathematical calculations to uncover and understand. The correlation charts that Walker produced in "World Weather V" (figures 2.3 and 2.4) construct two particular imaginaries. The first is that weather is spatially interconnected across the globe and that these relationships are only loosely related to terrestrial topography. Walker's climate is both globally integrated through correlations and cosmopolitan insofar as these relationships have no association with national borders. The Southern Oscillation Index itself was derived from data points across the world with no particular relationship to the central Pacific as the Southern Oscillation is understood today. The correlation patterns are therefore self-referential and do not bound weather parameters to any specific region. Walker's charts thus represent a manifestation of the meteorologist Douglas Archibald's earlier description of weather as recognizing "neither political nor superficially physical divisions of the land" and "require[ing] to be studied on the largest possible scale."[95]

Yet, as Anderson has shown in relation to Archibald's statement, the political neutrality of the Southern Oscillation is at best partial. Walker's maps were derived from data collected in observatories, the formal "havens for the travelling [European] observer," that often shared little commonalities with the imperial cities in which they were located, let alone the colonized countries that hosted them.[96] The new way of understanding the atmosphere as revealed by the charts relies entirely on colonial spaces of science. Walker's climate is globally interrelated, but reliant for its construction on observatories for the generation of data, and the "ramshackle" headquarters of the IMD in Simla for the analysis.[97] Thus, these imperial spaces were presented as central to an understanding of weather across the entire global atmosphere, without any need for an appreciation of physical processes. Likewise, the architecture of European imperialism was necessary both for their creation and for their interpretation in understanding and predicting weather. The significance of this imaginary for the subsequent history of statistical climatology is an interesting avenue for future analysis.

CONCLUSIONS

At the time of his death Walker's legacy was presented as one of failure. The statistician Ronald Fisher stated to the Royal Meteorological Society in April 1922 that "no new meteorological fact has been discovered by means of correlation coefficients; certainly up to the present no practical forecasts had been obtained from correlation coefficients."[98] The meteorologist William Dines, reviewing Walker's assorted contributions to the *Memoirs of the Indian Meteorological Department*, concluded that "correlation is of very little use for the purpose of forecasting unless the coefficients concerned are very high."[99] Walker's only notable defender was Hendrik Berlage Jr., head of meteorology at the Koninklijk Magnetisch en Meteorologisch Observatorium te Batavia. Berlage published a paper in 1934 that linked the periodicity of the Southern Oscillation to rainfall over Indonesia and suggested utility for forecasting, but this enthusiasm was somewhat of an anomaly.[100]

It is perhaps the association of his work on cycles with controversial Huntingtonian climatology that was responsible for the rejection of Walker's meteorology during the middle decades of the twentieth century, despite his lack of engagement with this tradition. Walker himself lamented "an attitude of scepticism [in England] in such matters that is of the greatest value as an antidote to rashness; but it is, in my view, excessive when applied to conclusions based on an adequate number of years such as forty or fifty."[101] Walker's obituary by his Imperial colleague Percival Sheppard in the *Quarterly Journal of the Royal Meteorological Society* included a statement that has been often quoted: "Walker's hope was presumably not only to unearth relations useful for forecasting but to discover sufficient and sufficiently important relations to provide a productive starting point for a theory of world weather. It hardly appears to be working out like that. We still do not have a quantitative theory of the general circulation but such progress as has been made in this study since World War II derives from different beginnings. Nevertheless, as theory advances, it may well be profitable to look back and see whether Walker's findings give a lead for further advances."[102]

Contemporary commentators, however, missed how close Walker came to identifying the mechanisms of what would later be called ENSO. As well as recognizing a possible role for the Pacific easterlies and the Humboldt Current, and postulating a particularly important role for pressure over Darwin, Walker also twice suggested that ocean circulations and temperatures may be responsible for the mechanism.[103] In an address to the Royal Meteorological Society in 1927 he suggested that relationships between regional pressure and the Southern Oscillation may be explained by "variations in temperature of a local current [or] variations in activity of the general oceanic circulation." In the same paper he demonstrated a relationship between temperature over Samoa, the Nile floods, and pressure over Port Darwin, all now considered important ENSO teleconnection regions.[104] Likewise, in the Smithsonian Annual Report for 1935 he recommended "a search for an explanation in terms of slowly changing features, such as ocean temperatures."[105]

It was not many years after Walker died that his approach was validated. The year of his death, 1958, was also the International Geophysical Year. During this period oceanic expeditions generated direct observations of the 1957–1958 El Niño event, kick-starting a new drive to understand El Niño.[106] The period saw several new papers by Berlage on the Southern Oscillation, including the development of a simplified index.[107] The culmination of this work was the linking of the Southern Oscillation with sea surface temperatures in the west-central Pacific in 1969 by Jacob Bjerknes and the postulation of what would later be called ENSO.[108] Bjerknes's discovery of a mechanism for the Southern Oscillation legitimized Walker's approach to finding relationships first and mechanisms second. Since then statistical climatology has become a fundamental part of climatology, reinforced by the development of climate models that allow mechanisms for statistical relationships to be determined experimentally.

It is in the context of this proliferation of statistical climatology and development of several new oscillations over recent decades (including the Pacific Decadal Oscillation, the Indian Ocean Dipole, the Atlantic Multidecadal Oscillation, etc.) that the imaginaries associated with Walker's work must be reviewed. Despite being rejected in his lifetime, since his death Walker's approach has become paradigmatic. It is therefore particularly important that the context in which it was developed is understood. Broadly speaking, the Southern Oscillation can be construed as emerging from the confluence of two imaginaries. The first came from Walker's background, an epistemology that gave primacy to robust mathematics over all other forms of knowledge production and thus elevated the persona of the empirical (male) scientist above those whose knowledge was interwoven with emotional or political considerations. This was cultivated during his formative career in late nineteenth-century Cambridge and exported by Walker to India.[109] The second was more overtly imperial: the disciplining of the Indian atmosphere through the particular practices of colonial observatory science. The integration of these imaginaries by a man who believed that "what is wanted in life is ability to apply principles to the actual cases that arise" precipitated the construction of the Southern Oscillation.[110]

Several areas emerge from this analysis that would merit closer attention. Walker's work was reconsidered in the context of scientific diplomacy during the Cold War. Greg Cushman has shown how oceanic El Niño research during this period was harnessed to build soft-power relationships with Latin American countries and create a culture of reliance on US technoscience.[111] The similarity to the knowledge-governance practiced by the colonial government in British India is striking. Likewise, it is notable that the one researcher to enthusiastically utilize Walker's work during his lifetime was Hendrik Berlage Jr., director of the Batavia Observatorium and essentially Walker's direct contemporary within the Dutch East Indies. This suggests an intriguing relationship between colonial governmentality and statistical formulations of the atmosphere and poses questions regarding center–periphery relations in the context of the legitimacy of colonial science. Further work is therefore required to understand how the imaginaries bound up with the Southern Oscillation traveled, and the contexts within which they eventually gained legitimacy.

Accounts of the history of climatology have highlighted the importance of colonial meteorology in the globalization of climate during the nineteenth and twentieth centuries. Imperial concerns expanded the meteorological-geographical imagination of the European centers at the same time that colonial territories provided new field sites for the installation of instruments to extend meteorological knowledge across space and into the sky.[112] Walker's work both complements and complicates these accounts. The Southern Oscillation produced an imaginary of a globally interconnected atmosphere, but the interrelatedness is incomplete and large regions are excluded through lack of data or lack of relevance. The Southern Oscillation itself is global in its reach, yet it is given agency as a discrete feature rather than as a component of a holistic climate. Thus, Walker's work provides new opportunities within which to examine the distinctive nature of colonial meteorology in the peripheries and to provide nuance to our understanding of climatic and meteorological science during the twentieth century.

3

THE WEATHER SHIP

Networks, Disasters, and Imaginaries after 1945

Katharine Anderson

THE NORTH ATLANTIC OCEAN STATIONS (NAOS) network was developed by the International Civil Aviation Organization (ICAO) in the immediate aftermath of the Second World War. The ten member states of ICAO's North Atlantic regional group (Belgium, Canada, France, Ireland, the Netherlands, Norway, Portugal, Sweden, the United Kingdom, and the United States) agreed to station a number of ships to take upper-air and surface observations, relay radio messages from aircraft and merchant shipping, and serve as navigation beacons that could help aircraft fix their positions during the crossing. For meteorologists promoting the plan in 1946, NAOS was the realization of an old dream. It marked the opening of a new era of global scientific cooperation. "[We] purred like a hundred cats," remembered one delegate, Sverre Petterssen, a Norwegian meteorologist who had come to work in wartime Britain as head of the upper-air research forecasting group at Dunstable.[1]

In truth, the early years of NAOS saw more sparks than purring. Looking at the network's history in its first decade shows scientific, political, and commercial interests in precarious balance. Postwar austerity made the expensive network difficult to implement, with its funding short-term and subject to budgetary pressures. Competing national interests in commercial aviation troubled negotiations. The development of jet aircraft, which flew at a higher level than the first generation of transatlantic commercial aircraft, brought other questions.

Would jets simply fly "above the weather," ending the need for detailed forecasting for aviation? While meteorologists like Petterssen were firmly committed to an ocean meteorological network, bureaucrats, airline executives, and politicians demanded repeated justification of the value and purpose of an ocean network.

The frictions surrounding NAOS reflected the imagined futures of the atmosphere and the oceans that were in play after 1945. Behind these debates over the weather ships lay broader ideas about the oceans and the atmosphere. Unlike the idea of ships as agents in motion across featureless space, the weather ship was significant because it was *not* en route: it was stationed in its designated ocean square. Through this contrast, the network called attention to a three-dimensional world with different kinds of mobility: the rolling grey surface of the North Atlantic, the vertical layers of moving air above this surface, and the technologies that flowed through these environments. Aircraft moved above the ships, sounding balloons drifted off them, and radar and radio signals pulsed in and out. The weather ship was a marker of both changing technologies and international structures.

It pointed as well to questions about whether and how human eyes and hands were essential elements of these new conditions. Weather ships seized the public imagination in part because oceans and ships were already iconic sites of human observation. Viewed from the deck of a ship, the ocean had been a place where the sharp-eyed individual could grasp the grand dimensions of nature and discover new worlds. If its scale dwarfed the human observer, the ocean was also a surface for heroic action, a site where humans survived only by their strength, wit, and judgment. In many ways, then, the weather ship slid easily into established patterns of knowledge, power, and place. At the same time, changing technologies and international infrastructures of NAOS challenged such conventional geographies of knowledge. The weather ship connected the maritime tradition to new technical systems and to new political and economic imperatives. Bridging visions of past and future, it prompted questions about human sensibilities and a human scale of action in a large technoscientific network.[2]

In this chapter I approach the history of NAOS and its place in the geographical imagination from three different perspectives. First, a review of the network's difficult first decade shows how contemporaries envisioned meteorological science and the work of the weather ships. The early years of NAOS are the most revealing, beginning with the negotiations immediately after the war through its first crisis in 1953–1954, when the Americans abruptly threatened to leave the network at the height of the Cold War. Second, an examination of the public reputation of the weather ships shows that the humanitarian role of the network was a tenacious and illuminating aspect of its history. A final section situates the weather ship within a wider oceanic imaginary, examining three works that explored the relationship of ships, technology, and the bodies of human observers. From different genres and national contexts, these evocative representations speak to the place of the weather ships in the contemporary imagination. They illustrate the possibilities and anxieties of the networked oceans of the mid-twentieth century.

BUILDING A METEOROLOGICAL NETWORK AT SEA FROM 1946 TO 1954

By 1945 the organized exchange of scientific observations from ships at sea already had an established history. In the nineteenth century the American naval officer Matthew Fontaine Maury had argued that knowledge of both the oceans and the atmosphere could be transformed by making "every ship a floating observatory."[3] Responding to this appeal, an international meeting in Brussels in 1853 standardized a weather log to be returned to their respective national offices or observatories. The exchange of ships' logs thus predated wider meteorological coordination within the International Meteorological Organization (IMO), founded two decades later in 1873. Ships renewed their importance as meteorological agents in the early twentieth century. With the emergence of radio, ships could transmit observations directly to land rather than hand in logbooks at the end of the voyage. The French meteorological service carried out a series of voyages in the *Jacques Cartier* from 1922 to 1925, envisioning this ship as a true floating observatory that could link North America and the Pacific to Paris.[4] A few years later, the IMO adopted another, less centralized model, coordinating a system of voluntary ship observers along the world's increasingly busy ocean routes. From 1929 participating nations undertook to enlist a number of observing ships in proportion to their share of world shipping tonnage.[5] Over the next decade the pace of development of aviation added further impetus to meteorological networks. By the late 1930s the Germans, French, British, and Americans had all experimented with Atlantic weather ships to support aviation.[6] In 1937–1938, for example, meteorologist Frank Entwhistle carried out a program of upper-air soundings on several Atlantic crossings on behalf of the British Meteorological Office and Imperial Airways. Because of the demands of aviators, Entwhistle optimistically noted, "we are thus gradually reaching a position where a network of observations will be in existence over the whole globe and material for a study of synoptic meteorology will be available on a scale hitherto unprecedented."[7]

At the end of the Second World War, Entwhistle's sense of the inevitable relationship between meteorology and aviation reappeared within the framework of the International Civil Aviation Organization. This new association replaced the International Commission for Aerial Navigation (ICAN), an interwar organization that by this time was closely associated with the failures of Versailles and the League of Nations. The new ICAO took shape over eighteen months, beginning with the Chicago Convention signed by fifty-two nations in December 1944 and followed by a provisional planning group (known as PICAO) that operated through 1945 and 1946. This group divided the globe into ten air navigation regions and created regulations and principles for aviation worldwide. These included mapping standards for aeronautical charts, licensing aircraft and pilots, establishing passenger and cargo landing agreements, and developing regulations for customs, quarantine, air traffic control, accident investigation, and meteorological information. The ICAO arrangements preceded the formation of the United Nations, but in 1947, at the first formal general assembly of the United Nations, ICAO became one of its Specialized Agencies, like UNESCO (1946), World Health Organization (1948), and the World Meteorological Organization (1950).[8]

This last raises an obvious question. Why didn't ICAO leave meteorological projects like the weather ships to the World Meteorological Organization (WMO)? Perhaps in part because its predecessor, ICAN, had found international scientific organizations too slow to respond to the needs of aviation, ICAO was unwilling to defer or delay its meteorological interests.[9] The ICAO especially wanted to ensure immediate international regulation of ocean observations. Britain and Canada were anxious to prevent the complete American control of air commerce; other nations in continental Europe similarly had their eye on the future of long-distance routes. A list of scheduled monthly transatlantic flights in 1947 by ten major airlines gives a snapshot of the rapid development of civil aviation (as well as a glimpse of the dominant position of the United States): United Airline had 61 monthly flights, Pan-American Airline had 385, American Overseas Airline had 235, Air France had 35, Sabena had 17, Koninklijke Luchvaart Maatschappij (KLM) had 25, British Overseas Airways Corporation (BOAC) had 70, and TransCanada Airline had 87, for a total of 942 monthly flights.[10] The rapid postwar development of such an intensive schedule makes it easy to imagine how contemporaries saw the potential value of ocean observations. Moreover, the ICAO member states already understood that ship-based observations on the North Atlantic route were critical because of a clear demonstration of their effectiveness in the last months of the war. During the return of aircraft to the United States from Europe in 1945, there had been more than twenty weather ships positioned in the North Atlantic, providing support for two different transatlantic routes: the crossing via the Azores and the more northerly route via Gander in Newfoundland to Foynes in Ireland.[11] The issue in 1946 was simply how to convert a vital system immediately to international

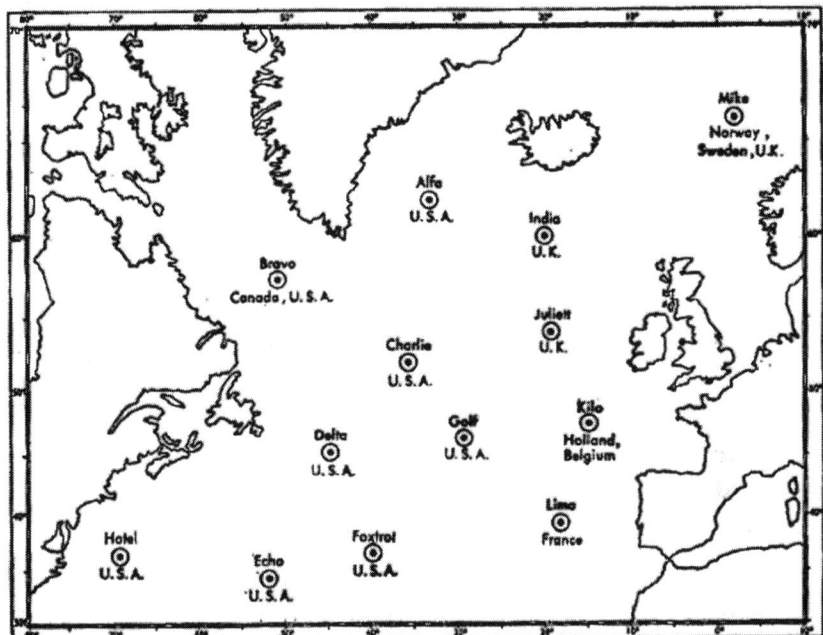

FIGURE 3.1. A chart of the Atlantic showing the positions and national responsibilities of stations in the 1946 agreement. Source: Downes, "History of the British Ocean Weather Ships." British Crown copyright, Met Office. Used with permission.

regulation. An ocean observation network in this region accordingly became a priority for ICAO.

The agreement reached in London on September 25, 1946, negotiated positions for thirteen ocean stations to be staffed by an estimated twenty-five ships (see figure 3.1). The agreement was to take force by July 1, 1947.[12] It allocated national responsibilities for these ships (either directly or indirectly, as financial contributions) on the basis on the proportion of traffic carried by airlines of the country, just as the observing network of the IMO had done with shipping in 1929. Since fully 75 percent of the passengers traveling in the second half of 1946 were carried by airline companies of the United States, the number of the stations assigned to United States was correspondingly large. Seven of the initial thirteen stations were American, with an additional eighth station shared between the United States and Canada.

If NAOS took shape fairly readily on paper, however, it was an entirely different matter to get the ships to sea and keep them there. The thirteen stations of the 1946 agreement, seen by meteorologists as an "irreducible minimum" for the service, in fact never existed operationally. Economic, administrative, and diplo-matic obstacles proliferated. For example, the British Meteorological Office—despite its strong commitment to the NAOS proposals—faced many challenges converting the four corvettes it had hurriedly purchased from decommissioned naval stock. There were further complications involved in training both observers and seamen for what was seen as a low-status and uncomfortable service. The last of Britain's four weather ships agreed to in 1946 made its first tour of duty only in February 1948.[13] Belgium and Netherlands opened their shared station briefly before the end of 1947 but then suspended observations in the second half of 1948 because of problems with the vessels. Sweden, Norway, and France only began to operate their stations in 1948. Of the seven solo stations that the United States agreed to in 1946, only two were in operation by late 1948. Some stations proved especially difficult. In the case of Station B, northeast of Labrador, Canada and the United States had initially agreed to start their shared operations in July 1947. When it emerged that the United States officials planned to delay their start to September, Canadian naval authorities (much to the frustration of Canadian meteorologists) also began to slow down preparations. Rotation schedules for the ships conflicted, and each nation's estimate of operating costs for the station were widely different (for one thing, the Americans estimated that each station needed three ships, while the Canadians thought a rotation of two was sufficient). With such divergent estimates, any agreement on monetary compensation in lieu of staffing Station B seemed unlikely to succeed. Perhaps in light of these issues of cost and coordination, Canada early on sought to trade its Atlantic stake in Station B for a Pacific station for which it would be solely responsible. Canadian authorities also argued, persistently but unsuccessfully, that its contributions to the ICAO should be reduced to reflect the costs of its land-based operations (such as Gander or Goose Bay), since other nations such as Iceland, Denmark (for posts in the Faroe Islands and Greenland), and Portugal (for posts in the Azores) could count theirs in NAOS negotiations.[14]

Beyond complicated arguments over the expense of NAOS, the weather ships network also navigated with great difficulty among the shoals of postwar American internal politics. Advocates of ICAO and NAOS in the United States had lauded the weather ships as "essential not only to the safety of transoceanic air operations, but also to preserve our leadership in international air commerce," but there were plenty of opponents as well.[15] Tensions between the US Coast Guard and the Navy, which had shared responsibilities for ocean weather duties between 1940 to 1946, made matters worse. On the one hand, the Navy disliked the idea of the Coast Guard turning into a second "blue-water" navy, a vivid possibility given the number of ships required for the projected Atlantic and

Pacific stations. On the other hand, the Navy had no interest in taking on the routine of weather observation duties itself, on either ocean. For its part, the Coast Guard realized that staffing the seven and a half Atlantic stations alone would require almost the entirety of its heavy fleet, and it resisted weather duties for ICAO until more resources were confirmed.[16]

Meanwhile, other factors fed the interservice rivalry and concerns about costs that defined debates about NAOS in the United States. The technological uncertainties facing the aviation industry was one consideration. Experiments with unmanned instrument buoys that could be moored in increasingly deep water suggested that an alternative to ships would emerge quickly. Perhaps a system of reconnaissance flights offered a cheaper solution than weather ships? More dramatically, perhaps the future of trans-ocean flying lay with the development of large mid-ocean communication towers or even artificial islands, complete with stations for refueling.[17] Finally, there were repeated suggestions that, since the new jet propulsion aircraft flew higher, above the most turbulent conditions, oceanic weather observations would become steadily less important to aviation in the near future. Beyond the technological issues, the larger context of aid to postwar European reconstruction also contributed to the American reluctance to commit to NAOS. Owen Brewster, senator from Maine and chair of the hearings on the weather ships, commented to Delbert Little, assistant chief of operations at the Weather Bureau, "you apparently intimate that we are considered a piker in international affairs because we are not maintaining our five [further] ships . . . I hardly think the world is entitled to suggest we are pikers right now."[18]

The reluctance to fund weather ships after 1945 in the United States, like the Canadian administrative problems with Station B, could be multiplied by similar examples in France, Norway, and elsewhere. They were typical of what one official called "a decided lag and a considerable degree of procrastination" in ICAO more generally.[19] In 1949 NAOS members agreed to cut the number of stations from thirteen to ten, reducing the US responsibility. They also negotiated financial contributions from more European nations not formally part of the ICAO North Atlantic region but who possessed airlines that had begun to use the Atlantic routes, like Switzerland and Cuba. At the end of this second three-year term from 1949 to 1952, however, the agreement ran into fresh difficulties. With the Republican Dwight Eisenhower newly elected as president of the United States, resistance to the ocean stations intensified once more. American airline companies increasingly objected to the idea that they should pay for a meteorological network of general value to all, and their resentment grew rather than lessened with the growth of the aviation industry.[20] Dragging its feet on the renewal of NAOS, the United States requested a survey of members' views on the weather ships in July

1953. Then, on October 21 (apparently without consultation with its own Weather Bureau), the United States abruptly notified ICAO of its intention to withdraw from NAOS altogether: "The services provided by the Ocean Stations in the North Atlantic are not required by the United States . . . that the benefits derived by the United States are no longer commensurate with the cost."[21]

This decision stirred international outrage on both scientific and political grounds, especially from allies as important to American Cold War policies of containment as Canada and Britain. A flurry of consultations ensued as both individual states and international bodies like the ICAO developed their response to Eisenhower's move. A characteristic report by an aeronautical engineer, H. Roxbee Cox, to the British Aeronautical Research Council, for example, argued that "the value of the ships was tied to the network principle." Losing stations would turn out to be a cost to airlines: without forecasting of headwinds and with optimal route selection no longer possible, aircraft would need greater fuel reserves and so would be able to carry fewer passengers. Moreover, Cox noted, "future and promising lines of meteorological forecasting, such as by using digital computers, would be severely handicapped."[22] An ICAO working paper agreed with this emphasis, pointing out that "the network value of the NAOS programme decreases gradually as the number of stations decreases and it is not possible to say at which stage the network value is lost entirely . . . [with stations becoming] isolated observation posts."[23]

The Canadians were particularly frustrated. In a pointed allusion to a parallel sphere of cooperation (the American pressure on Canada to accelerate the development of military bases in the Arctic), one acerbic editorial in Ottawa commented that the sudden and unilateral withdrawal set a poor precedent. "International cooperation must be based on consultation."[24] Over the next month Canadian meteorological authorities, who felt Eisenhower's advisors were "looking at this matter through political [i.e. domestic economy] spectacles only," produced detailed estimates of the economic costs of less accurate forecasting. They began calculating how to charge for weather forecasts from Gander and Dorval to fund the ocean stations and further suggested (in a response described approvingly as "rather Machiavellian" by a British representative in Ottawa) that Gander authorities could henceforth require a 20 percent fuel margin for error for planes leaving the airport on Atlantic journeys that were now more meteorologically uncertain and dangerous.[25]

With its own Weather Bureau continuing to stress the essential scientific value of ocean stations, with the meteorological requirements of jet aviation still uncertain, and with alternative technological fixes still more theoretical than practical, the US government backed down. In February 1954 a new NAOS agreement established nine stations. It also created a new structure for assessing "non-aeronautical benefits," that is, the wider scientific value of observations from the Atlantic, which disproportionately affected European meteorological forecasting to the east rather than continental United States to the west.[26] Yet although the tone of NAOS negotiations moderated after this crisis in 1953–1954, the underlying concerns never went away. NAOS remained constantly on the defensive about expenses. "All the weather ships put together cost only slightly more to operate in a year than the purchase price of . . . one of the jetliners crossing the North Atlantic seven miles above the changing sea," ICAO reminded the public in 1960.[27] In 1972 ICAO agreed to shift most NAOS responsibilities to the WMO's World Weather Watch program. By 1974 the United States had entirely pulled out of its stations, and the 1975 NAOS agreement reduced the number of stations to four (with station "Charlie" operated by the Soviet Union, "Lima" by Britain, "Mike" by Norway and the Netherlands, and "Romeo" by France). Buoy to satellite communication links were established in the 1970s, and the NAOS agreement formally expired in 1986.[28]

The sketch here of the early history of the weather ship network reveals the close connections among scientific, economic, and political concerns with the atmosphere and the oceans in the decades after 1945. From the geographical perspective of the current volume, however, NAOS is especially interesting because it marked a reconception of the North Atlantic as a region rather than a route. As an article in *Popular Mechanics* put it, the weather ships meant the decline of the age of "ocean blanks."[29] With aviation and telecommunications expanding, the Atlantic Ocean acquired a new vertical dimension, but also new places—the ICAO zones, the alphabetized ocean stations, the range of direction-finding transmitters, the layers of air recorded by the sounding balloons, the standard flying height of different types of aircraft, the routes that shifted around headwinds. Older landmarks acquired new familiarity: airports in Newfoundland, Ireland, and the Azores and observatories in Iceland, Greenland, and the Faroe Islands (see figure 3.2).

The Canadian response to the United States in 1953 had made this new sense of the region especially clear by insisting that the North Atlantic was the equivalent of the continent's interior: "if the ocean station network was disbanded we would be faced with a meteorological problem similar to that which would exist in Canada if we had no radiosonde stations away from the coast."[30] In this sense, the term "network principle" that was used to defend the system in the crisis of 1953 is a telling one for spatial reasons as well as diplomatic ones. Despite its origins in the spread of commercial aviation, NAOS was not tied to the idea of paths, criss-crossing the air above the ocean in the same manner that ships'

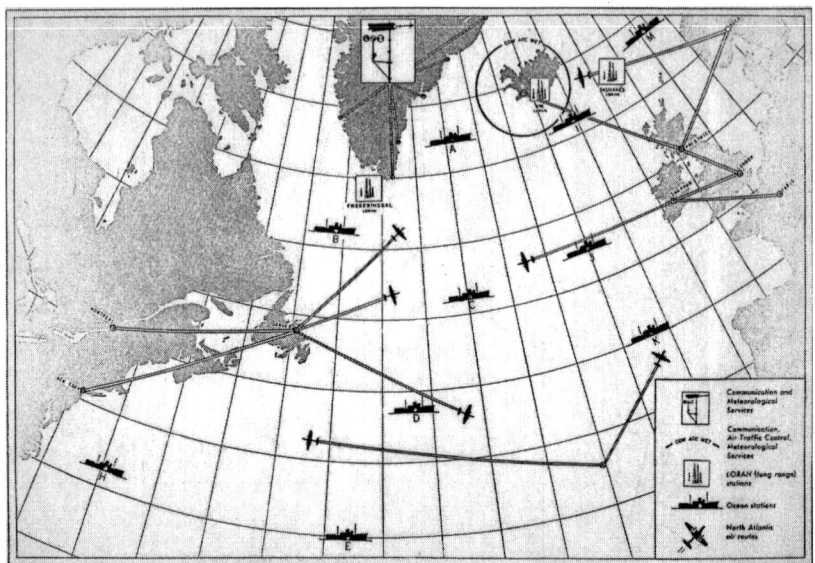

FIGURE 3.2. A chart indicating the reduced network of ten stations, with air routes and communications marked, shows the growing complexity of the Atlantic as a region. "Chart with LORAN," METEOROLOGY: Stations and Observatories (Code B, 50/5): North Atlantic Ocean Weather Ships Scheme: UK policy on follow-up to 1949 agreement. Item 6, AIR 2/12199, TNA.

tracks had on its surface for centuries past. Instead, the network defined a space of physical phenomena that needed to be understood synoptically, on a large scale and in three dimensions (see figure. 3.3). Weather ships gave the North Atlantic a new shape and identity.

At the same time, the ships also represented the connections between and among regions globally. Calling attention to linked physical phenomena, they fostered a new understanding of the relationship of ocean and air and of the procession of weather systems from North America to western Europe. But beyond these physical features, the ships signaled changing economic and political systems as well, both cooperative and competitive. While a system for meteorological observations of the North Atlantic was important practically for the expansion of commercial aviation, it was also important symbolically, as a model of international cooperation, with shared responsibilities and shared access. The network itself—and the transatlantic aviation enterprise that it served—recorded the shifting center of political gravity from Europe to North America.

Yet its identity as a region underlined complicated new questions. If the

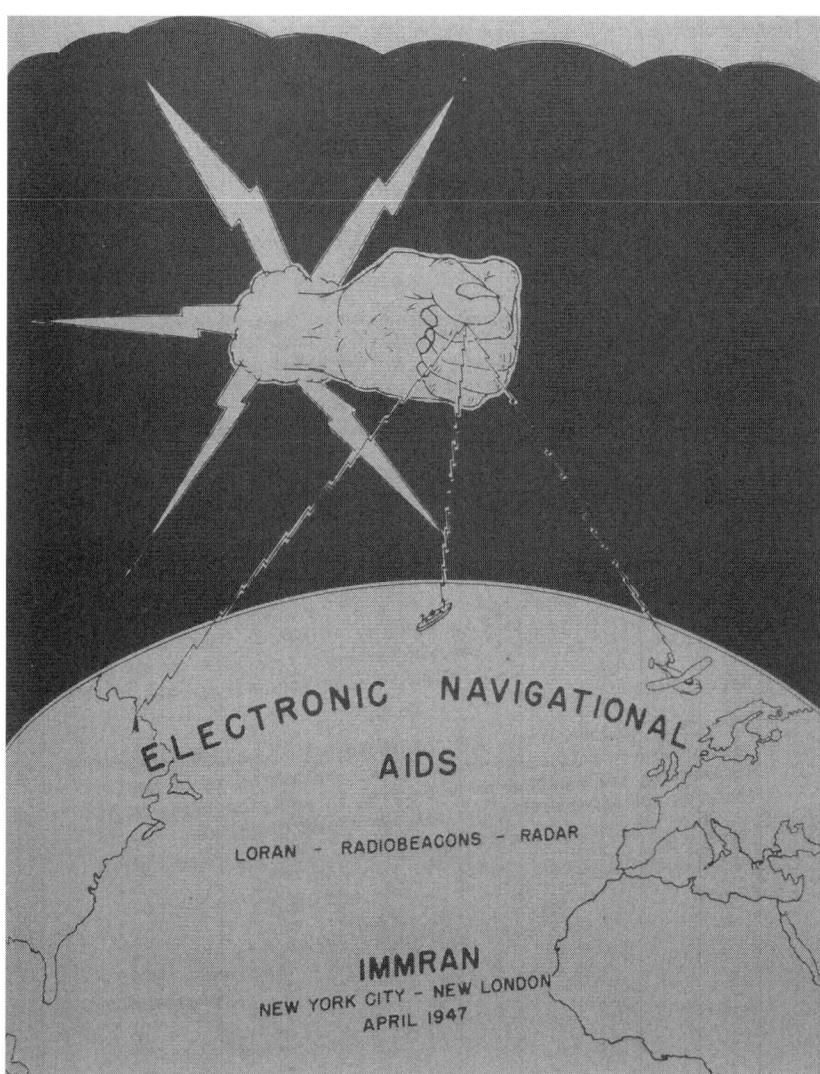

Figure 3.3. The three-dimensional space of aircraft, ships, and radio over the ocean-atmosphere. [Title page of document 43 showing ocean, ship, and loran station], *International Meeting on Marine Radio Aids to Navigation, Proceedings and Related Documents* (New York, 1947), 457.

North Atlantic was best understood as a coherent space, how were its vertical layers connected? How few distributed observations would capture its essential features?[31] How to understand the different meaning of the region when traveling eastward than when traveling westward? How to translate the region into ideas about users or measurements of the value of knowledge? How to allocate the distinction between direct forecasting benefits and the value of understanding the oceanic atmosphere more generally? By posing such questions, the NAOS network also raised the even more fundamental issue of who should answer them. Scientists or industry experts? Governments and politicians or international bodies? These wide-ranging challenges of knowledge, technology, and their publics took shape generally in the decade after 1945, as we have seen, but they emerged as well in distinctive form in the search-and-rescue (SAR) function of the weather ship. Investigating the SAR role of weather ships—rarely practiced, but often invoked—shows that the cultural representation of oceans and atmospheres is as much a part of NAOS history as its scientific ambitions or political negotiations. Such histories of representation repay exploration, especially for vast environments like the oceans and the atmosphere whose scale and inaccessibility make them difficult to know and to experience.[32]

THE WEATHER SHIP AND THE SAFETY OF TRANSATLANTIC FLIGHT

From its inception the ICAO had emphasized the multiple functions of weather ships, listing navigational aid and oceanographic research as well as meteorological observations. Yet these activities did not play an important part in justifying the network.[33] Weather ships made plankton observations, launched drift bottles to observe currents, and studied bird migration, ocean temperatures, and terrestrial magnetism.[34] In the 1950s vessels also began to monitor levels of "atmospheric pollution," an euphemism for atomic fallout.[35] But such research activities, while they often got a mention, were always viewed as opportunistic benefits. In contrast, the heroic role of weather ships as rescue outposts seized the public imagination and became a part of the network's identity.

In several respects, contemporary preoccupation with the SAR function of the network seems surprising. First of all, the role was clearly exceptional. Second, it contradicted the weather ship's primary purpose, to monitor and record as a stable point in a world of moving phenomena and signals: a rescue was the one reason a ship could move "off station." Third, for perhaps obvious reasons of reputation, aviation companies themselves rejected the characterization of the weather ships as essential rescue units. In the diplomatic alarms following Eisenhower's threat to withdraw from the network in late 1953, for instance, a BOAC representative asked the air minister to use more care in his public statements. He reminded the politician bluntly that "the use of such vague expressions as 'adverse effects' or 'consequences which we fear' in the context of transatlantic civil aviation should be avoided."[36] In the face of these considerations, the constant repetition of a search-and-rescue rationale for the ships is striking.

Why did SAR acquire this importance? A comparison helps understand how this function emerged as a kind of trump card in discussions of the network. As an exceptional and potential benefit, SAR could be compared to another aspect of the network, its military significance. The history of weather operations in the Second World War had made the strategic relevance of meteorology rather obvious, and as the Cold War intensified in 1952 and 1953, large NATO exercises made it clear once again that the Atlantic was more than just a region for civil aviation.[37] Yet SAR also overshadowed this other potential justification.[38] In at least one case, meteorologists strenuously resisted bending the network to serve military exercises in the Atlantic and did so by invoking safety. Asked to endorse the temporary relocation of Station "I" in spring 1953 to support critical military exercises, the Canadian director Andrew Thomson reminded his minister that the stations were chosen for effective meteorological coverage, not Air Force support. Thomson noted, "there are, at any one time, over one thousand civilian passengers airborne over the North Atlantic ocean. There can be no question that the movement of Ocean Weather Station India to the position required by the RCAF reduces the accuracy of our forecasts for the whole of the North Atlantic. It is suggested that the needs of this body of civilians should not be accorded second place."[39] Safety concerns, then, broadened the value of knowledge about the oceans. As Fiorella LaGuardia, then mayor of New York City and a key figure in developing civil aviation in the United States, famously commented in the opening negotiations for ICAO in Chicago in 1944: "Everybody is against bad weather."[40] Invoking "civilian" safety conferred authority on meteorologists as the arbiter of the environment and its phenomena.

References to safety gave meteorologists this authority because of a particular transatlantic history of scientific cooperation. In 1914, in the wake of the Titanic, the International Convention for Safety of Life at Sea (SOLAS) had identified the humanitarian role of ships in two ways. First, it set procedures for communication—that is, for priority of distress signals and the sharing of urgent weather or navigation information. These regulations included detailed provisions not only for radio equipment but also for the activities of a "certified watcher." The qualifications of the operator and the requisite duty hours were matters of intense negotiation in the first SOLAS Convention, negotiations that stressed the vital necessity of connecting the instrument to a trained ear.[41] Second, the convention led directly to ice patrols to monitor the presence of icebergs in shipping lanes. The ice patrols, conducted by the US Coast Guard, were funded by international subventions. The records circulated to maritime traffic in company with weather statistics and navigational updates.[42] The SOLAS Convention was renewed in 1929 by eighteen nations, including most of those who became members of the ICAO North Atlantic regional group after the war. Like SOLAS a generation earlier, the NAOS agreements responded to new technological conditions in the Atlantic and proposed the internationalism of the maritime community as a fundamental aspect of these conditions.

The strong association of weather ships with the ideals of SOLAS intensified following one dramatic event: the mid-ocean ditching of the *Bermuda Sky Queen* in October 1947. The *Bermuda Sky Queen* was one of twelve "Flying Clippers," or Boeing 314s, originally made for Pan American Airline between 1938 and 1941 (see figure 3.4). Turned over to the US Navy in 1942, the *Bermuda Sky Queen* was then sold again in 1947 to American International Airways for commercial use. In its fatal October journey, en route from Foynes to Gander, the *Bermuda Sky Queen* faced strong headwinds. The pilot, Captain Charles Martin, realizing that he had misjudged his fuel and would not make land, turned the airplane back east to head for Weather Station "C" or Charlie, five hundred miles from Newfoundland.[43] He successfully landed beside the USCG Cutter *Bibb* despite seas of thirty to forty feet. Passengers and crew tried to wait out the gale in a slowly leaking aircraft, its life rafts too small to inflate or launch in the high seas. Eventually, fearing that the trapped passengers would force open a door and "lead a lemming procession" into the sea, the *Bibb*'s crew risked the approach in its own lifeboat. In several attempts, the sixty-two passengers and seven crew members were successfully transferred to the vessel. The *Bermuda Sky Queen* was sunk, and the *Bibb* returned to Boston in triumph, met by reporters who recorded the emotional tributes of passengers for the newsreels.[44] Later, an aviation inquiry dammed (and fined) Martin for the decision to fly, but the rescued passengers petitioned to praise Martin's "major achievement" of the landing at Station Charlie.[45]

The crash landing of the *Bermuda Sky Queen* was the quintessential shipwreck of the aviation age.[46] The incident was in reality exceptional: one evaluation of NAOS in 1953 called the SAR function of weather ships "a psychological aspect" only.[47] Nevertheless, rescue—and the *Bermuda Sky Queen* rescue in particular—shaped discussions of the program throughout its first decades to the point of caricature. Two months after the *Bermuda Sky Queen* disaster, a Canadian vessel, HMCS *St Stephen*, set off to Station B, accompanied by a journalist. He described the routine work of meteorological observations dramatically. The meteorologist "had to be tied down with a rope around his waist in order to keep from being blown overboard during launching of a [six-foot diameter] huge

FIGURE 3.4. The Boeing 314 pictured here is the *Yankee Clipper* (1939). The Boeing 314 *Bermuda Sky Queen* was built two years later in 1941; at the time of the 1947 ditching, it was owned by American Airways. Source: Harris & Ewing, photographer (public domain), via Wikimedia.

balloon." But the meteorologist and the ship returned safely to harbor several weeks later, having carried out "two of three basic purposes"—that is, collecting meteorological observations and performing as a navigational beacon for aircraft. The ship's third purpose as an "air-sea rescue unit" was not called for, except to succor "a weary sea gull" who made "a crash landing on the St. Stephen's deck." The victim "was placed in sick bay for two days" where "under the careful ministrations of the ship's personnel, it rapidly regained strength." On the third day, the gull was enlisted into a photo session with the ship's commander and "deciding this was carrying matters too far, the patient took wing and disappeared." The journalist's description, with its three stages of crash-rescue-publicity, was a deliberate echo of the *Bermuda Sky Queen* affair.[48]

This popular account indicates that SAR function of the weather ship was both less and more than it seemed—impractical and exceptional but symbolic enough to caricature. On a long transatlantic flight, the presence of weather ships clearly reassured the pilots and the public alike.[49] It linked the network and ICAO to an older internationalist tradition represented by SOLAS, a history that

WORLD METEOROLOGICAL ORGANIZATION

WMO BULLETIN

APRIL 1954   VOL. III NO. 2

FIGURE 3.5. The front cover of the *World Meteorological Organization Bulletin* 3 (April 1954) placed search and rescue functions of the weather ship in the spotlight with this photograph of lifeboat and crew in the foreground and the Dutch weather ship in the background. Used with permission of the World Meteorological Organization.

underlined the adaptation of technologies to humanitarian purposes alongside commercial ones. It also offered a way to assert scientific control of the network, in resistance to military interests or the influence of aviation companies. But the so-called psychological effect extended to subtler considerations as well. The repetitive focus on safety was a means of highlighting a human presence within the large systems involved in the network. The tenacious concern with "safety" dramatized a human scale of action, in contrast to large technological, geopolitical, and physical systems. The WMO literally placed this in the spotlight in their celebration of the weather ship following the tricky renewal the NAOS in 1954 (see figure 3.5).[50] Weather ships, like the "certified watchers" of the SOLAS Convention, operated as a reminder of the presence of human bodies within these environments. Instantiating the maritime tradition of the vigilant observer at sea, they remind us that the global ambitions associated with knowing climate and weather also encoded a sense of human vulnerability to vast natural and technical forces.

WEATHER, SCALE AND HUMAN SENSIBILITIES

The significance of the marine observer can be traced further by turning away from the history of NAOS itself to the larger oceanic and technoscientific imaginaries in which the weather ship was embedded. Such imaginaries spell out the genealogies and anticipations, the expectations and uncertainties, of a new world order.[51] Consider now three different genres and three different points in time: an illustration of the human nervous system that the science popularizer Fritz Kahn published in 1939; E. J. Pratt's 1947 poem about wartime North Atlantic, *Behind the Log*; and finally, Rudyard Kipling's 1905 fantasy of the Atlantic airship. These vivid accounts place NAOS within a wider cultural history of thinking about oceanic spaces. All three representations emphasized a new command of oceans, illustrating how the distance between continents was collapsing. At the same time, however, they highlighted the strains involved in positioning human experience within the physical system of ocean-atmosphere and the equally complex system of airplanes, signals, and weather data.

In the first example, a remarkable medical illustration titled "The Speed of Thought—Has Been Surpassed by Technology!" coupled ocean and atmosphere to the human nervous system. The image was part of the publishing enterprise of the Weimar era science popularizer Fritz Kahn (1888–1968). Kahn was known particularly for the five-volume serialized work *Das Leben des Menschen* (1922–1931), which famously depicted the body as an industrial system. In 1933 Kahn, a Jew, was forced to close his Berlin medical practice. He moved to Palestine, France, and Spain before finally leaving Europe for refuge in New York

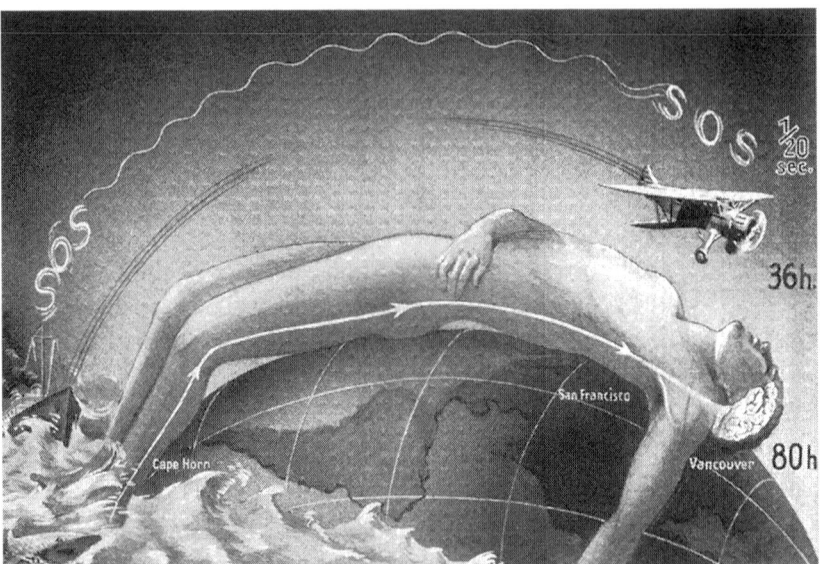

FIGURE 3.6. "The Speed of Thought," in Fritz Kahn, *Man in Structure and Function*, 2 vols. (New York: Alfred Knopf, 1943) 1:472. From author's collection. Used with kind permission of Dror Kahn.

from 1941 to 1949. He spent the last two decades of his life in Switzerland and Denmark.[52] "Speed of Thought" was one of many striking uses of global imagery in his work. Kahn reworked and added to his medical and physiological works throughout his life, and his work had depended on a group of illustrators, often uncredited, making the origins and creator of this image hard to identify exactly. It appeared in Switzerland in his *Der Mensch Gesund und Krank* in 1939 and then in the United States as *Man in Structure and Function* in 1943.[53]

"Speed of Thought" shows a nude male figure draped across a globe (see figure 3.6). The globe, etched with lines of longitude and latitude, identified only three points: Cape Horn, San Francisco, and Vancouver. The body, lying in a position suggesting unconsciousness, or even drowning, is placed mid-Pacific, with his toes reaching across Antarctica and the head just above Vancouver, one arm extended toward the viewer across the North American continent and dangling into the Atlantic. His other arm leads toward a Europe that is beyond the frame of the image. This literally global framework is juxtaposed with reminders of the vertical dimensions of ocean and air. Waves pound the feet and the left hand, while a shark with open mouth lurks near the feet. In the ocean near the shark is a ship; in the air over the head is an airplane. Connecting the two are the undulating lines of a radioed SOS signal. With the image, Kahn sought to capture a physiological fact, the speed of the nerve impulse, by comparing it to speed of an aircraft and radio signal. The caption reads, "A nerve current travels at the rate of 250 kilometres per hour. A person reaching from Cape Horn to Alaska would first feel a shark's bite after 80 hours, and it would take just as long for the volitional impulse to retract the foot to pass from the brain to the toes."[54] Interestingly, the comparison replaced the example of a harpooned whale that was commonly used in earlier scientific popularizations to convey the speed of the nervous impulse and thus reversed the position of the human at sea from hunter to victim.[55]

Kahn's message about scale and speed of the nervous system used oceans, aviation, and signals as central, but ambiguous, metaphors. Besides his medical work Kahn was interested in a wide range of other scientific subjects. He had planned studies of the Arctic and the Sahara in 1932, and in the same year he started work on a natural history of Palestine—all of which suggest an interest in extreme or precarious environments.[56] "Speed of Thought" is another case of geographic extremes. The whole of the Pacific is the yardstick by which the length of body was measured and understood, while the Atlantic serves the same purpose visually for the other spatial dimension of the body, the arm span.[57] The ocean and atmosphere together surround the body, with the radio signal and the flight path of the aircraft emphasizing the air encircling the subject. The ship, the airplane, and the radio in some ways triumph over these spaces—or so the image suggests: the SOS shortwave signal and aircraft both easily "surpass" the speed of the neural impulse. Yet the power of these technologies is eerily unresolved. The aircraft hovers over the helpless figure without making contact, and the signal emerging from the ship's apparatus reaches no answering apparatus at the other end. In the meantime, the ocean waves continue to menace the ship, with the shark waiting nearby. Kahn used to dramatic effect a cliché of the aviation age—our changing experiences of global distance. Yet with this iconic frame of ocean and rescue, he calls attention to the forces of nature without in fact displaying a human mastery of them.

My second imaginary is E. J. Pratt's *Behind the Log* (1947), a long narrative poem written to commemorate the Battle of the Atlantic that is a brilliantly ambivalent commentary on oceans and human observation (see figure 3.7). Pratt (1882–1964), the leading Canadian poet of his time, was born in Newfoundland, the son of a Methodist minister, and was himself ordained in 1913. He pursued doctoral studies in philosophy and theology at Victoria College, University of Toronto, there developing a deep interest in Wilhelm Wundt's psychology. He was a psychology laboratory demonstrator for several years before obtaining a

Figure 3.7. Figure of the mariner. Illustrated by Grant MacDonald in Pratt, *Behind the Log*. From author's collection.

position as a lecturer in English literature in 1920.[58] The 1947 poem reflected Pratt's long-standing interests in the changing social and technological experiences of maritime communities. In *Behind the Log*, he recounted a dangerous convoy voyage from Sydney, Nova Scotia, to Liverpool lasting eleven days, September 9–20, 1941: the group of sixty-five vessels lost sixteen ships to submarine attacks over three nights.[59] Commissioned to write the poem in early 1945, Pratt spent the summer months based in Halifax and aboard HMCS *Skeena* and *Haida*, where he conferred with officers and radar technicians to get what he called a naval education. Overwhelmed with "data, data, data," Pratt later called it "the hardest poem I ever wrote."[60]

Behind the Log is an exploration of human community, picturing the ocean and the technologies that penetrated it as the spaces within which this community took shape.[61] Pratt opens this theme in the manner that the title of the poem would lead us to expect, with the log itself:

> There is a language in a naval log
> That rams the grammar down a layman's throat,
> Where words unwind in paragraphs and lines
> In chapters. Volumes lie in graphs and codes,
> Recording with an algebraic care
> The ways of storms, their lairs, habits and paths[.] (lines 1–6)

The poem turns immediately to the preliminary conference on shore in Sydney, Nova Scotia, as the convoy prepares to depart. In a complicated dialogue section, Pratt designs the voices of different captains to pull at the ear in different languages, dialects, and rhythms. The reference to the challenge of internationalism was deliberate. Pratt, who as a schoolboy had watched Marconi set up the first transatlantic radio signal in St. John's, Newfoundland, had already used radio signals as a metaphor for human community in an earlier poem of September 1939.[62] In *Behind the Log*, his return to this idea sees the seriocomic exchanges of convoy bureaucracy set up the problems of communication and cooperation. The test of community is the poem's theme: strained during the prolonged crisis of submarine attack, communications and thus community manage to recover as the ships of the convoy doggedly work together to locate and attack submarines and to rescue survivors from the damaged ships.

In a similar fashion to Kahn's image, then, Pratt's poem speaks of ocean, human bodies, and the nervous system in conjunction with risk, mortality, and rescue—the problems of survival and human society. The poem's interest in signals and codes marks a concern running through all Pratt's work: the limits of a mechanistic view of human consciousness. Conferring with the illustrator,

Grant MacDonald, a former naval war artist, Pratt insisted that MacDonald cut the detail on one image in order to isolate the asdic (sonar) operator. That is, Pratt wanted the illustration to focus on the act of listening to show this modern technology as the extension of human sensibilities.[63] In the poem itself, Pratt likewise stresses how bodies connect to the signal and the spaces through which it moves. Just as the seaman of old "called to the rocks" with a sounding-lead, he wrote, so the modern mariners call

> To steel gliding beneath the sea; they pierce
> Horizons for the surface hulls: they ping
> The sky for the plane's fuselage . . .
> But though the radio burst and vacuum tubes
> And electronic beams were miracles
> Of yesterday, dismissing cloud and rain
> And darkness as illusions of the sense,
> Yet always there to watch the colours, note
> The V break in the beam's straight line, to hear
> The echoes, feel the pain, are eyes, ears, nerves:
> Always remains the guess within the judgment
> To jump the fire perfection of the physics
> And smell mortality behind the log. (lines 42–44, 46–55)

The focus on acts of observation recalled Pratt's background in the psychology laboratory. The mariner was Everyman, heightened into heroism by the need to literally strain his "eyes, ears, nerves" to deal with wind, rain, and waves. "[C]loud and rain and darkness" present an environment to penetrate with technology, but, Pratt argues, we are still bound to a world of bodies, senses, and judgment.

My third example, Rudyard Kipling's short story "With the Night Mail," poses a chronological oddity for this trio of ocean imaginaries. In terms of their publication dates Pratt and Kahn are more closely associated with the era of the NAOS weather ships, while Kipling's story is set further into both the past *and* the future: first published in 1905, it presents an imagined future of 2047.[64] Yet it is also the example that most explicitly invokes the weather ship. At the time he wrote "Night Mail," Kipling (1865–1936) was at the height of his reputation as a journalist, poet, and short story writer who represented the narrative voice of the British Empire. The story followed an airship, a mail packet, on a single transatlantic crossing in a designated fast lane in the air. One of the story's striking features is its account of the new geography of the air age, describing atmospheric forces in vivid and specific terms. Clouds—whose nomenclature was a scientific debate of Kipling's day—were disparaged as mere "fluff" (117). Instead, Kipling emphasizes the variety of winds, electrical and magnetic conditions.

FIGURE 3.8. The weather ship on her post. H. C. Seppings Wright, "She passed slowly beneath us." Illustration for Rudyard Kipling, "With the Night Mail, from Windsor Magazine, 2047, A.D.," *Windsor Magazine* 23 (December 1905): 62.

These included "a wulli-wa [williwaw] down below that has knocked us into umbrella-frames" (123), an equally dangerous "electric" wind (131), and other "blow-outs, vortices, laterals, etc." (128–29). With the winds taking up horizontal and vertical space, the atmosphere has a striking geographical presence in Kipling's story. But the more interesting features of the story for our theme of human observers are its descriptions of the officers and the ship itself.

The story was partly a portrait of a pair of pragmatic officer engineers, Purnall and Hodgson.[65] Like Pratt's sailors, they are characterized by their sensory acuity: "each has the brooding sheathed glance characteristic of eagles and aeronauts . . . that fathomless abstraction of eyes habitually turned through naked space" (112). But the ship itself is another character and, like the airship officers, a symbol of the future. In the opening passage, Kipling describes the mail room command port as a nerve center in which indicator arrows pulse to register the movement of the mail on a chart, "degree by geographical degree" (112). Later in the story, vivid descriptions of the ship's parts in motion make the vessel a living thing.[66]

As the tumultuous transatlantic journey continues, the apparently routine mail-packet reveals more and more dimensions. It is "a shutterless observatory; a life-boat station; a salvage tug; a court of ultimate-appeal-cum-meteorological bureau for three hundred miles in all directions" (135). Ultimately, the weather ship symbolizes regulation on a planetary scale, serving the eerily named Aerial Board of Control (ABC). "Her black hull, double conning tower and ever-ready sling [to rescue sinking airships] represent all that remains to the planet of that odd old word authority. She is responsible only to . . . that semi-elected, semi-nominated body of a few score person of both sexes [which] controls this planet. Transportation is Civilisation" (135). For Kipling, the mail-packet airship simultaneously embodies communication, a reassuring mid-Atlantic rescue post, and a sinister international techno-authority (see figure 3.8).[67] Kipling was writing at the beginning of the age of aviation and radio, but the story suggests recognizable features of the NAOS system half a century later, down to the alphabetic code of ocean stations and the sideline of search and rescue. Picturing a new technological system that transcended older geographical and political boundaries, Kipling's airship stories gave new materiality and force to the atmosphere as well as to questions of international regulation.

CHANGING GEOGRAPHICAL EXPERIENCE

What can this cluster of imaginaries tell us about NAOS and the weather ship in the mid-twentieth century? These three instances—a sketch of the nervous system, a poem of the modern sailor, and a tale for the air age—are all different, each one fascinating in its own right. Yet collectively, they point to a shared imaginary of oceans, mariners, and technological change. They indicate first how the technologies of this period combined, and how they made the ocean a vertical as well as a horizontal space. The continuum of "ocean-atmosphere" was reproduced in the technological triangle of aviation, radio, and the vessels on the surface in all three of these examples. They also each emphasized a picture of knowledge made possible by signals and information exchange. Finally, they gave an ambivalent account of the human within the system. The vast natural forces represented by the ocean setting and its weather are only partially, not wholly, mastered. These imaginaries help identify the technological anxieties of the postwar era. They suggest a context for, or a sharper edge to, the contemporary insistence that the NAOS ships had several purposes—that is, that they were not just weather stations, not just navigational beacons, not just rescue posts.

In their first decade after the Second World War, the weather ships exposed a changing geographical experience. The network provides an explicit case that shows how the environment was regulated and recharted after 1945. But the ocean and atmosphere were also reimagined. The North Atlantic region became more than a gap to bridge within existing networks of weather information. Contemporaries understood that the winds and weather over the oceans were forces that shaped the future development of aviation, economic prosperity, and international relations. The weather ship network then allows us to glimpse how our changing understanding of regions and global spaces emerged in mid-century. The idea of filling in the vast blank spaces of the ocean with meteorological observers was, and remains, a familiar one: it stretched from the nineteenth century and continues beyond NAOS to the automated weather buoy and satellite era. Weather ships, buoys, and eventually satellites made possible a new blanket of information, and NAOS in this sense does look like a key intermediate step toward the global meshes of contemporary data worlds.[68] The ship was never a singular agent: NAOS stood for ocean and atmosphere, ship and aircraft, sounding balloons, radio and radar. Their combination of functions was part of a distinctive sense of the complexity of the large-scale technological ensemble.

Above all, the story of the weather ships emphasizes how such physical, political, and cultural geographies emerged in tight combination. Contemporaries were preoccupied with the human element of the weather ship: its heroic observers, its role in search and rescue, the importance of the signal reaching from mid-ocean to pilot. Questions about human sensibilities seem to lie at the heart of these concerns. Weather ships then share in the wider history of the scientific observer and remind us how, in weather and climate knowledge especially, observation adapted an essentially maritime tradition of vigilance in the face of nature's power and uncertainty. The responses to the *Bermuda Sky Queen* ditching and the imaginaries of ocean and atmosphere illustrate a heightened sense of anxiety. The anxiety was not just about the dangers of flying but was, more broadly, related to technological, political, and economic regimes. The NAOS network both formed and reflected a new, more intensely regulated world order.

In this history I have underlined how the idea of a geographical imagination adds to our understanding of the environment. It is not simply that some spaces or phenomena were (and remain) hard to understand because of their scale. The realist appeal of the very large or the very deep may well shape our understanding of natural environments in seductive and misleading ways (as Sam Randalls and Martin Mahony remind us in their introduction to this volume). But the atmosphere and the ocean resist their own history in other ways as well. It is harder to see traces of human ideas and experiences in the trackless spaces of the atmosphere and the oceans than it is on the land. It is also harder to examine ideas and experiences because the challenges of accessing these environments tend to foreground technological innovation, with new instruments (or aircraft or submarines) as the driving forces of change. The place of weather ships as part of the story of the air age is a striking case in point. NAOS is unfamiliar to us now in part because it involved ships and the ocean surface—the worlds that aviation appeared to supplant. Adopting geographical imagination as our tool of analysis makes the influence of such subtle forces of technological and environmental determinism part of our subject. It directs our critical attention to the patterns of human experience and understanding, both as they changed and as they resisted change. To make sense of our own urgent contemporary debates, accounts of environmental knowledge need capacious methods of inquiry that can capture the combined influences of material conditions, intellectual practices, and cultural exchange.

4

LOOKING FOR THE LEEUWIN

An Environmental History of the Leeuwin Current

Ruth A. Morgan

"WHERE IS THE LEEUWIN CURRENT?" asked Australian oceanographer George Cresswell in 1991. A decade earlier, Cresswell had named the shallow stream of warm low-salinity water after a Dutch East India Company vessel (the *Lioness*) that had charted the area in the early seventeenth century. When he mused as to its location, Cresswell was asking his readers to turn their gaze from a stretch of sea between Java and Antarctica, to the western maritime fringe of the Australian continent. There, he explained, was the Leeuwin, flowing at the surface from near the Northwest Cape down to Cape Leeuwin, before curling eastward to Tasmania (see figure 4.1). Strongest in the austral autumn and winter, it is the longest coastal current system in the world, skirting the western and southern coastline over a distance of fifty-five hundred kilometers, or the distance between London and New York.[1]

The search for the Leeuwin is a project that commenced nearly two centuries before Cresswell christened the current. European awareness of its existence dates at least to Matthew Flinders's circumnavigation of Australia between 1801 and 1803. Around Cape Leeuwin, the most southwesterly point of the mainland, Flinders noted the presence of an "eastward current along this part of the South Coast" that flowed from the Cape past King George Sound (Albany). Subsequent investigations of the eastern Indian Ocean confirmed the existence of a relatively warm stream off the continental shelf. Broad and shallow near the Northwest

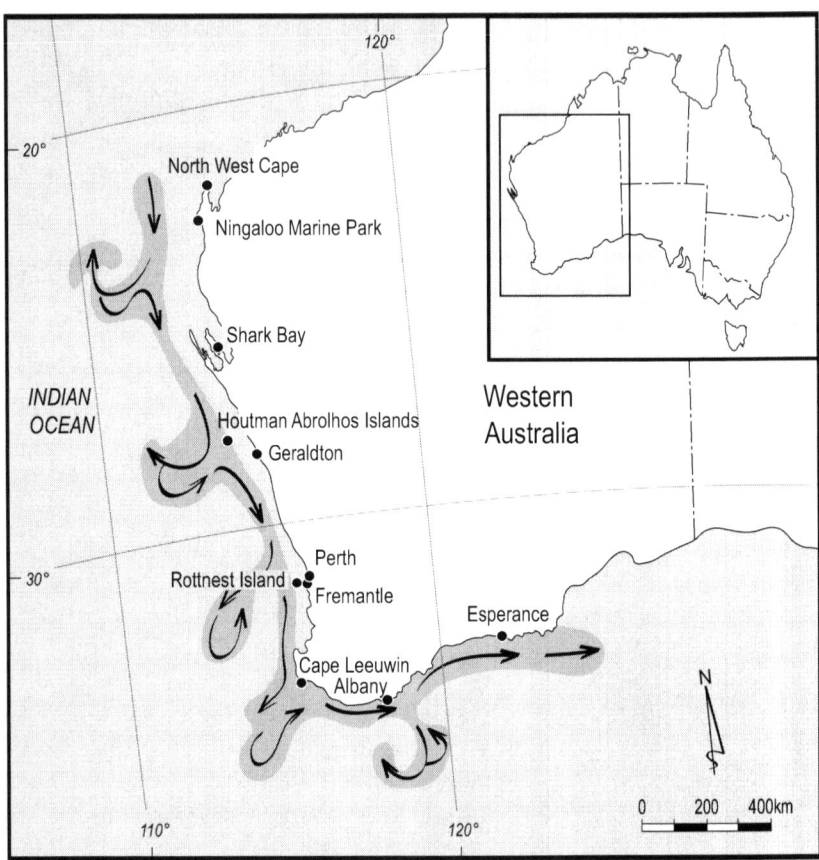

Figure 4.1. The Leeuwin Current. Courtesy of Kara Rasmanis, Monash University.

Cape, this stream of tropical water narrows and deepens on its journey down the coast.[2] This poleward flow distinguishes it from other eastern boundary ocean currents such as the Peru or Humboldt Current, the Californian Current, the Benguela Current, and the Canary Current. Such currents skirt the west coasts of the Americas and Africa as they flow *toward* the equator.[3]

The direction of the Leeuwin has consequences for both the land and the sea. Its poleward flow contributes to the relatively high rainfall of Western Australia, compared to similar latitudes in western South America and southern Africa where the Atacama and Namib deserts lie. In these subtropical regions, rainfall increases away from the coast, whereas Western Australia becomes drier further inland.[4] Meanwhile, the Leeuwin's warm, tropical flow combined with the lack of nutrients from Australia's ancient soils and a low but variable rainfall ensures that the coastal waters of Western Australia support a high diversity of marine species but have more modest concentrations of fish than other eastern ocean currents. Elsewhere, northward-flowing eastern boundary currents carry cold, nutrient-rich waters that support large finfish populations such as sardine (*Sardinops sagax*) and anchovy (*Engraulis encrasicolus*), whereas the Leeuwin favors the production of bottom-dwelling invertebrate species.[5]

Among those that thrive there is the western rock lobster (*Panulirus cygnus*), which lies at the heart of one of the most commercially valuable single-species fisheries in Australia.[6] Spanning a thousand kilometers between Cape Leeuwin and Shark Bay, its catch is closely related to the fluctuations in the strength of the Leeuwin Current.[7] Human encounters with the western rock lobster possibly date to 1500 BP, as Aboriginal rock art from the Dampier Archipelago and Pilbara suggests.[8] Today, the western rock lobster attracts both commercial and recreational fishers, whose activities are closely regulated by the Western Australian government. Each year the commercial fishery involves about 250 licensed vessels and catches about 6,000 tons of western rock lobster by submerging wooden traps (called "pots") in rocky or weedy areas for periods of up to several days.[9] This fishery was the first in the world to be certified to the Marine Stewardship Council Fisheries Standard (in 2000) and the only fishery to be recertified a fourth successive time (in 2017) for its "environmentally sustainable" practices.[10]

Looking for the Leeuwin as it skims the Australian coast draws attention to this littoral as a space for entangled ecologies, livelihoods, and imaginaries. In the shallows of the East Indian Ocean and its Australian shoreline, global oceanic and atmospheric processes combine to produce what is a local phenomenon far removed from national and international political and scientific centers of action. The current's peripheral existence beyond the endeavors of Cold War internationalism encouraged cooperation between fishers and marine scientists focused on the optimal harvesting of coastal resources. These relationships helped cast particular imaginaries of an otherwise opaque ocean current that connected a local fishery to wider international concerns about fish stocks and climate variability. What follows is a triptych of imaginaries—as a fluid frontier, a fisheries index, and a climate mechanism—that, studied together, reveal the eddies of the "material and mental worlds" of the Leeuwin.[11]

LOOKING FOR THE LEEUWIN

Scientific investigation of the Leeuwin Current has been largely an Australian endeavor. This interest can be attributed to the current's close proximity to the extensive Australian coastline; the historically limited resources available to scientific agencies in Indonesia, in whose southern waters the current originates;

and to the political friction between these two nations since the end of the Second World War.[12] Elsewhere, as Helen Rozwadowski, Jacob Hamblin, and Gregory Cushman have shown, the study of ocean phenomena has historically warranted transnational efforts that cross cultural, political, and scientific borders.[13] The very interconnectedness of oceanic and atmospheric processes prompted international scientific collaboration and data collection across vast geographical areas from the nineteenth century. These endeavors later benefited from the geopolitical ambitions of the Cold War era, which produced the "knowledge infrastructure" that made these processes comprehensible.[14]

But the motivations for their study were also rooted in local concerns—after all, the manifestation of such phenomena is experienced at the local level. There, ecologies and cultures, the human and more-than-human, are intimately entangled. The variability of currents, as Cushman and Kristin Wintersteen have shown in the case of the Humboldt Current, can effect enormous ecological change, with far-reaching consequences.[15] Ocean currents—with their boundaries marked by a particular temperature, salinity, and flow, for example—forge ecosystems in which, as environmental scientist Stan Ulanksi explains, "living organisms—plants, animals and microbes—function in conjunction with nonliving components of their environment."[16] As these ecosystems change, so too do the ways in which people experience, imagine, and understand the ocean.

The atmospheric-oceanic phenomenon of the El Niño–Southern Oscillation (ENSO) is a key driver of the Australian continent's boom-bust climate, but historical studies of its effects have been largely terrestrially focused.[17] Despite its vast coastline and marine territories (the third largest in the world), Australia's environmental histories and histories of science have been largely landlocked, focusing on the sciences of exploration and development, on agriculture and pastoralism, the suburbs and the desert.[18] There is a growing interest, however, in the environmental histories of the eastern Indian Ocean and the western Australian coast. These studies have explored the socioeconomic impacts of tropical cyclones; coastal fisheries and fishing cultures; and the science and politics of natural heritage conservation.[19] In this chapter I integrate marine environmental histories into histories of climate and climate science by centering on the Leeuwin Current and the fortunes of the western rock lobster fishery.

Studies of particular fisheries have tended to concentrate on addressing what marine biologist Daniel Pauly describes as the problem of "shifting baselines"—the difficulty of establishing a static reference point for evaluating ecological change.[20] These concerns have stimulated studies of Western Australian fisheries along the coast, using oral histories, scientific studies, and memoirs to trace and interpret narratives of ecological change, projects that collected data from interview subjects with this objective in mind.[21] By contrast, in this chapter we sample interviews from two studies: the Oral History of the Australian Fishing Industry Project, undertaken in the 1980s, and the recently completed doctoral research of Jenny Shaw, who studied the responses of people in an Australian fishing community to anthropogenic climate change.[22] Rather than focus on the population of western rock lobsters, interviewees' accounts of the fishery in the interviews are situated in much larger narratives of work, family, and identity. The Leeuwin Current is entirely absent in these interviews. Used in conjunction with scientific studies, the interviews offer insight into the ways that the management of this fishery helped to render a particular ocean current visible.

The visibility of oceanic phenomena—or the lack thereof—has been a key concern among marine environmental historians.[23] This conundrum of visibility extends to the capacity for "knowability" and certainty that arises through direct human encounter. Compared to most terrestrial settings, the ocean is a largely inhospitable environment for sustained human experience, a setting akin to what the literary scholar Siobhan Carroll describes as a "natural atopia." Although these spaces might be physically accessed by humans, their nature is such that they resist permanent settlement or bodily habitation.[24] Prolonged human contact with the atopia of the ocean, therefore, is a highly mediated experience, whether through technologies or imaginaries that produce both particular and partial representations or "versions" of the marine world.

For historians and geographers, their terrestrial bias can produce depictions of the ocean as a flat static space. In this representation the ocean is an unchanging space of transit and connection between terrestrial places, a space that is mapped using terrestrial devices and populated only at the surface. What this perspective overlooks, geographers Philip Steinberg and Kimberley Peters argue, is the fluidity of the ocean as a dynamic, liquid, three-dimensional physical entity.[25] They advocate a "wet ontology" that conceives of the ocean not only as a facilitator of movement but also as a "space that is *constituted by* and *constitutive of* movement."[26] This study of an ocean current and the imaginaries that have represented its flow reflect this concern for the mobilities that are inherent in the underwater world.

Yet a vertical perspective does not confine this analysis only to the deep. Rather, it encourages looking up to the atmosphere, which, like the ocean, evades direct visual and sensorial experience. As Stan Ulanski reminds us, these are connected systems: "Wind drives the ocean's great gyres; the sea, in turn, drives the atmospheric heat engine. The circle remains unbroken."[27] This connected system of atmosphere and ocean may be inhospitable for humans, but neither atopia has escaped the effects of human activity.[28] Their connectedness and shared human history resonate with the call of historian Alison Bashford for "terraqueous history," an approach for the Anthropocene that accounts for "human impact on atmospheres, hydrospheres, biospheres, and geospheres [that] together constitute the terraqueous globe under the heavens."[29] In this chapter I identify "terraqueous meeting places" between land and sea in the material world of coastal waters, fishers, and fishing vessels, as well as in the geographical imaginaries of opportunity, prediction, and planetary change. From these meeting places imaginaries of the Leeuwin Current developed as a circulation not only of liquid and lobsters, but also of capital and climate.

THE LEEUWIN CURRENT AS FRONTIER

In 1972 the scientific journal *Nature* published an article on thermal anomalies in the eastern Indian Ocean. Its author, the Western Australian geographer Joseph Gentilli, described the presence of "'rafts' of warmer water that remains warmer for a longer time" along the Western Australian coastline. These rafts, he suggested, accounted for the occurrence of higher sea surface temperatures than might be expected for the area north of 36° during the winter months as well as their "lag" into the spring. Although ships' records had earlier suggested the existence of such phenomena, their presence, Gentilli marvelled, "have not been described before."[30] His characterization of the Leeuwin Current in terms of warm rafts imagined the eastern Indian Ocean as a scientific and commercial frontier, ripe for exploration, discovery, and exploitation.

Such an imaginary was a local manifestation of a global scientific undertaking to render the Indian Ocean visible—the International Indian Ocean Expedition (1959–1965). The data amassed during this project provided the basis for Gentilli's research into the climates of (Western) Australia. There, in his adopted home, the state government's developmentalist call to clear a million acres of land a year reflected a pioneering ethos to postwar land settlement that encouraged population growth, the expansion of agricultural areas, and the harvesting of natural resources. Gentilli, who had arrived in Western Australia in 1939 as a refugee of Mussolini's Italy, embraced this ethos in his early work. In 1941 he had declared, "Western Australia's key problem is how to increase her population; nearly all other problems are mere corollaries of this one great question."[31] In this continental west at least, the frontier remained open and its expansion widely valorized after the Second World War.

For Australian scientists in particular, the Leeuwin Current became imagined as a frontier for scientific research in the interests of resource development. In the Australian context, a frontier imaginary resonates with a history of settler colonial contestation and violence. In this imaginary, the frontier is a static, solid, and advancing frontier that misrepresents the complexity and porousness of colonial interactions. To counter this narrative, historian Luke Godwin proposes a "fluid frontier" that allows for a "more dynamic view of [the] frontier" with "temporal and geographical dimensions."[32] This notion of a fluid frontier presents an opportunity to navigate the mobility and dynamism that materially characterize marine spaces and human interactions with them over time. As environmental historians and historical geographers have shown, the association of the ocean with the frontier has been a common and persistent theme in maritime scientific and economic endeavors. Drawing on terrestrial rhetoric, this frontier imaginary envisioned the ocean and its depths as virgin territory similarly awaiting discovery, exploration, and exploitation.[33] In terms of Carroll's schema, an imagined fluid frontier of the ocean conforms to the permanently unsettled condition of a natural atopia, which "await[s] neither improvement, nor inevitable wide-scale settlement, nor seamless incorporation into the domestic space of the nation."[34] Once the atopia of the ocean is rendered visible and accessible by advances in science and technology, it becomes possible to imagine penetrating and closing this fluid frontier—to explore, possess, and make legible ocean space.

Gentilli's choice of the term "rafts" arose from conversations with his friend Dominic Serventy. A zoologist and keen ornithologist, Serventy had learned of "the warm water and the tropical species in it" through discussions with fishers in the Houtman Abrolhos (28.66°S, 113.81°E) during the 1940s (see figure 4.1).[35] Serventy's interest in seabirds and his work establishing the new Western Australian fisheries laboratory of the Council for Scientific and Industrial Research (CSIR) led him to explore the state's waters, including the Abrolhos, an archipelago of 122 islands and coral reefs located some sixty kilometers west of Geraldton.[36] There, he led a research program on the life history of the rock lobster (then classified *Panulirus lingipes*) and the state of the lobster fishery (see figure 4.2).[37] To better understand these so-called rafts, which seemed to account for the unexpected presence of tropical marine and avian species, Serventy and his colleagues undertook their own studies of the coastal currents, because a current's flow is one of the more important factors influencing species' availability and movement. The instruments they used were rudimentary—they dropped postcards (sealed in plastic envelopes) and empty bottles into the ocean between the mainland and the Abrolhos and Rottnest Islands (32.00°S, 115.51°E). Once these makeshift drift-cards and bottles were collected and returned to the researchers (for a modest reward), their destinations revealed the presence of a southward flow in winter and a northward flow in the summer.[38]

Fishers had also furnished their observations of the nature of coastal currents to an earlier generation of visiting naturalists. In 1913 and 1915, the zoologist

FIGURE 4.2. Fishing for western rock lobster at the Abrolhos Islands, 1949. Sourced from the collections of the State Library of Western Australia (003343D) and reproduced with the permission of the Library Board of Western Australia.

William Dakin led two Percy Sladen Trust expeditions to the Abrolhos Islands during his tenure as foundation chair of biology at the University of Western Australia (1913–1920). In his report, Dakin admitted: "I have had to depend upon information culled from the fishermen who are out at all seasons of the year." They told him about a "southerly current" during winter that for Dakin seemed "quite in favour of the tropical current theory."[39]

The theory to which he referred was the hypothesis of English naturalist William Saville-Kent, who had been the Western Australian colony's commissioner of fisheries between1893 and 1895. On a tour of the Abrolhos in the winter of 1894, which was the first scientific expedition to the archipelago, Saville-Kent was surprised to find in the temperate waters off the west coast fish that he had earlier encountered in the tropical climes of the Torres Strait and the Great Barrier Reef. To account for their presence, he suggested the existence of "an ocean current setting in from the equatorial area of the Indian Ocean [that] penetrates as far south as this island group without impinging on the adjacent mainland." Such a "warm stream" would carry these tropical species "to such congenial conditions and environments as exist in and among the Houtman's Abrolhos."[40]

The Abrolhos that Dakin encountered during the Great War (1914–1918) were "quite uninhabited." His measure of habitation was likely the presence of a permanent population of Anglo residents. Malay guano workers visited during the summer months, he noted, while the fishing luggers were "very frequent visitors," bringing mostly Italian and Scandinavian fishers from the ports of Fremantle or nearby Geraldton. The daily and seasonal rhythms of their labors were tuned both to the creatures they sought and to the conditions at sea. As Dakin observed, "They may fish in the deeper waters around the islands, or send one or two small boats among the reefs, whilst keeping the lugger off during the day. At night, the numerous sheltered regions between the islets are at hand." As sites for resource extraction, the islands belonged to a fluid frontier—a rugged and hostile maritime world of transient, non-Anglo peoples far from the civilizing forces of the hearth and the Western Australian state capital, Perth.[41]

By the time Serventy and his colleagues studied the Abrolhos in the 1940s, the region had undergone an ecological and economic transformation. The rise and fall of the guano industry left the islands thoroughly denuded and its seabird colonies devastated.[42] Meanwhile, the rock lobster fishery had exploded. Once a supplement to "wet lining" or commercial line fishing for species such as pink snapper (*Pagrus auratus*), catching rock lobster was becoming a thriving industry of its own with the fishery annually producing about one thousand tons in the late 1940s. Prior to the Second World War, most were caught around Rottnest Island, near Perth, with much smaller efforts around the Abrolhos, which supplied local canneries in Geraldton.[43] Wartime demand for canned rock lobster to supply the local and US armed forces stationed in Western Australia encouraged more concerted fishing activity along the coast, particularly around the Abrolhos. This area, a fisher later recalled, "was virgin country" when it came to fishing for rock lobster. With demand for the lobsters' frozen uncooked tails skyrocketing, the fishery expanded rapidly along the coast in the 1950s to support over a thousand fishers taking eighty-six hundred tons of lobster each year, making it the country's most lucrative fishery.[44]

This postwar effort to foster Australia's coastal fisheries for economic development benefited from an international endeavor to explore the great expanse of the Indian Ocean. Scientific interest in the Indian Ocean had increased significantly during the 1950s, particularly in the wake of the International Geophysical Year (1957–1958) and the creation of the Scientific Committee on Oceanic Research in 1957. Compared to the Atlantic and Pacific, this ocean was "unknown territory," according to the official history of what became the International Indian Ocean Expedition (IIOE).[45] Such was the paucity of data that American oceanographer Roger Revelle could declare in retrospect, "Nobody ha[d] studied the Indian Ocean."[46] Its exploration, oceanographers believed, would benefit marine science and foster scientific communities and the exploitation of fisheries in the region. The Indian Ocean had become the latest postwar frontier for resource extraction—a marine foil for the exploration of outer space.[47]

Although the humanitarian dimensions of the IIOE encouraged a focus on the coasts of Asia and East Africa, the involvement of Commonwealth Scientific and Industrial Research Organization (CSIRO) scientists and research vessels ensured that Australian waters were not entirely overlooked.[48] For over a year, the HMAS *Diamantina* patrolled a line of stations along the 110th meridian east from Java to south of Australia each month. At each station the crew measured the physical, chemical, and biological properties of the ocean, from which they constructed a time series of the seasonal variations of those properties over a broad swath of the eastern Indian Ocean. Between stations, the vessel could stray. For example, lead scientist Harry Jitts remembers his colleague, the oceanographer David Rochford, "just had to find out where all his water masses were coming from, so he persuaded the RAN [Royal Australian Navy] to let HMAS *Diamantina* wander right across the Indian Ocean looking for them." The data they collected provided further insights into the coastal currents that Serventy and local fishers had earlier observed: the oceanographic survey established the presence of a strong poleward current between the 110th meridian east and the Australian mainland.[49]

In 1964 Gilbert Whitley, who had been a member of Serventy's team on the Abrolhos back in the 1940s, gave his presidential address to the Linnaean Society of New South Wales. In his survey of Australian ichthyology, he declared: "Sea temperatures and water movements, knowledge of which is so important when dealing with migratory fish, are still incompletely understood in Australasia." But along the Western Australian coast at least, the interceding decades had shed some light on the characteristics of the current systems of the eastern Indian Ocean. By the early 1970s, Gentilli could draw on both local research into Western Australia's coastal waters and data gleaned from the surveys of the HMAS *Diamantina*. His synthesis of these findings would see him remembered by oceanographers as the "grandfather" of the Leeuwin Current.[50]

For Australian oceanographers like George Cresswell, access to NASA's meteorological satellite program and the US National Environmental Satellite Service offered further insights into the mysteries of the deep. These technologies tracked the movement of buoys across the world's oceans and provided satellite imagery of ocean currents such as the Gulf Stream.[51] These tools revealed to Cresswell the flows and eddies of what he named the Leeuwin Current in 1980.[52] Although these waters would remain uninhabitable, their growing legibility helped facilitate the emergence of quantitative imaginaries of the Leeuwin Current and its resources.

THE LEEUWIN CURRENT AS FISHERIES INDEX

Fisheries scientists had suspected since the 1960s that coastal currents off Western Australia influenced the life cycle of the rock lobster.[53] It was not until the late 1980s that they could shed further light on this relationship. Parallel developments in fisheries and climate science suggested the possibility of devising a means to estimate the size of the lobster catch from the strength of the Leeuwin Current. By the 1990s collaborative scientific efforts had found a close correlation between the annual recruitment of juvenile rock lobsters and sea levels, which were a proxy of the current's strength. A stronger current was associated with a greater recruitment of juveniles. Establishing this connection provided the basis for a more accurate means of predicting the lobster catch. Earlier predictive models had not taken such environmental variability into account. It meant that the Leeuwin Current could be imagined as an index of the fishery's productivity and magnitude. The current's measurement, therefore, reflected the growing economic significance of the western rock lobster fishery and the demand for more accurate statistics to enable the more efficient exploitation of its resources.

This imagination of the Leeuwin Current as a predictive index was a product of the sustained and growing influence of market-based approaches to fisheries management around the world.[54] Although resource economics played a role in commercial fisheries from the mid-nineteenth century onward, the rise of mathematical fish population modeling after the Second World War reinforced a growing faith that economic principles would allow for the protection of the commons and the achievement of the emergent goal of "maximum sustainable yield." As environmental historian Jennifer Hubbard explains, such an approach relied on the application of mathematical models "to analyse and predict the dynamics of fished populations." For Hubbard, these mathematical representations of fish, markets, and demand resulted in the erasure of both fish and fishers in favor of quantitative abstractions.[55] Although a similar case might be made in the Western Australian context, for the Leeuwin Current at least, the outcome was quite the opposite. Identifying a correlation between the current's strength and the future population of western rock lobster—and quantifying that relationship—instead inadvertently made visible a marine body that had long eluded fishers, fisheries managers, and marine scientists. Now recognized for its material impact on the western rock lobster, the newly mediated Leeuwin Current became a vital input into fisheries modeling for scientific and commercial decision-making.

Efforts to devise a means of predicting the size of the rock lobster catch were driven by the demands of a changing industry. In 1963 the state government introduced new restrictions on the fishery after the average catch per boat began to fall. Western Australian officials were already aware of the potential limits of local fisheries. As the superintendent of fisheries A. J. Fraser had noted a decade earlier: "We certainly have many species of fish, but we have a smallish number of individuals of each species and they could possibly, without proper management, in the long run become depleted. It is essential that we take very good care of what we have."[56] A US market hungry for western rock lobster had encouraged the entry of more boats into the fishery (836 in 1964), and with the advent of more sophisticated technology, stocks were under unprecedented pressure.[57] These were symptoms of a great acceleration in the world's fisheries that had already taken a toll on Southeast Asian waters and would soon manifest in the North Atlantic and the Pacific.[58] In addition to the existing restrictions on the industry, such as minimum size limits and closed seasons, the Western Australian government stopped issuing new boat licenses and limited the number of pots (or traps) per fishing vessel.[59]

These regulations ushered in a new era for the fishery. As fisher George Bass later observed: "making it a limited entry industry put value on your license. . . . [B]ecause the fishing industry started running as a business you have to be more efficient, and you have to be competitive so you have to have the equipment as well to enable you to do that." This new mind-set also had implications for the management of the fishery. Bernard Bowen, who joined the state's fisheries department in the early 1950s, reflected: "It would be certainly much easier for us if, indeed, the economic drive wasn't there. But that's really not the real world either, because the economic drive is there, and we have to take that into account."[60] In addition to securing the livelihoods of the fishers and the future of the fishery, the longevity of the industry was also tied to political ambitions for the development of the states more sparsely populated north.[61]

Commercial, geopolitical, and conservation imperatives therefore encouraged closer scrutiny of the western rock lobster. Biological knowledge of the creature remained limited in the early 1960s, although the species formed Australia's most lucrative fishery. It was only in 1962 that local zoologist Ray George identified the lobster as *Panulirus cygnus*, which is endemic to subtropical Western Australia.[62] Still, questions remained about the lobster's life cycle once the eggs hatched during summer in the saline waters along the coast. What happened to these phyllosoma larvae once they hatched, and how did they reappear as young juveniles (puerulus) in coastal waters nearly a year later?

The IIOE surveys of the HMAS *Diamantina* went some way to answering these questions, which had troubled fisheries scientists since the 1940s. Studying the "topography of the sea surface," CSIRO oceanographer Klaus Wyrtki identified a clockwise eddy west of Perth that he suspected was a permanent feature along the coast.[63] For his fisheries colleagues, this finding shed further light on the larval phase of the rock lobster: where Wyrtki had found a permanent eddy, Ray George and Peter Cawthorn had recently placed the planktonic larval phase.[64] It was this eddy that provided a mechanism by which the larvae would be transported out to the open ocean (up to fifteen hundred kilometers), before carrying them back to the coast as young juveniles where they would mature further. These were indeed "marathon migrations," as Cresswell later observed.[65] This correlation suggested that the biology of the rock lobster was closely tied to the current systems of the eastern Indian Ocean and encouraged further investigation into this relationship.

Questions remained. Chief among them was the reappearance of the lobster larvae on the continental shelf in puerulus or young juvenile form. It was there on the coastal reefs that the lobster juveniles would mature and, in four years, become an adult rock lobster of legal harvest size. In order to count the number of puerulus that returned to settle on the reef, in 1967 CSIRO fisheries scientist Bruce Phillips developed artificial seaweed "collectors," which he moored at intervals along the continental shelf.[66] After monitoring these stations each month for two years, Phillips concluded that puerulus settlement, therefore, provided a means to predict the size of the catch four years later when the lobster would be ready for harvest.

Phillips's "settlement index" proved a valuable resource for the industry almost immediately. When the 1969–1970 season produced low catches and a low settlement of puerulus, Phillips's index showed that catches would again be low in 1972–1973 and 1973–1974. Although this was disappointing, the index suggested that the stocks would eventually improve. Moreover, fishers (and their creditors) could plan financially for the lean years ahead.[67] The demand for such information reflected the increasing professionalism of the industry, the greater investment in technology, and the growing integration of local fishing enterprises into global financial networks. Fisher Greg Roach observed, "we've now got the onset of the GPS satellite navigator, we've got the colour radar, [and] . . . echo sounders have now come in and most boats now carry that sort of equipment. As I say, the boats have become bigger and faster." The average length of fishing vessels increased from 8.7m in 1962–1963 to 9.16m in 1973–1974 and reached 10.93m by 1990, with the number of pots per vessel growing from 90 to over 102 during this thirty-year period.[68]

These bigger and faster boats placed more pressure on the fishery. With this "sophisticated gear," fisher Sonny Healy explained in the late 1980s, "they can fish a lot harder. Rough days we had to stay home. Now it doesn't make much difference whether it's rough or not, [with] the boats and equipment they've got they fish every day, especially through the winter. They don't have days off, like sometimes we'd have a week off if it was too rough to go out or something like that in the winter time. But now they, the boats they've got they can fish every day." The annual lobster catch grew from nearly eight thousand tons in the mid-1960s to over twelve thousand tons in the late 1980s. Under these increasingly competitive conditions, fishery managers had to be more vigilant than ever, and they used the settlement index to allow the lobster stocks to recover.[69]

Yet the 1982–1983 season produced a situation that defied the expectations of fisheries scientists. A study in the late 1970s had concluded that puerulus settlement was independent of breeding stock. This season, however, seemed to suggest otherwise: the highest catch year on record (12,884 tons) had also produced an especially low number of puerulus settling on coastal reefs.[70] Although Phillips retained faith in his settlement index, he joined with CSIRO oceanographer Alan Pearce to investigate what might be learned from the 1982–1983 season. Were the poor settlement statistics a product of overfishing, or were there other influences at play?[71]

Recent oceanographic studies had suggested that events in the Pacific Ocean could affect the Leeuwin Current via a mechanism that would later be named the Indonesian Throughflow.[72] This connection that joins two ocean basins is the only one of its kind.[73] If the Leeuwin, as CSIRO oceanographers Stuart Godfrey and Ken Ridgway concluded, could no longer "be viewed in isolation," then it was entirely possible that Pacific phenomena could influence the life cycle of the

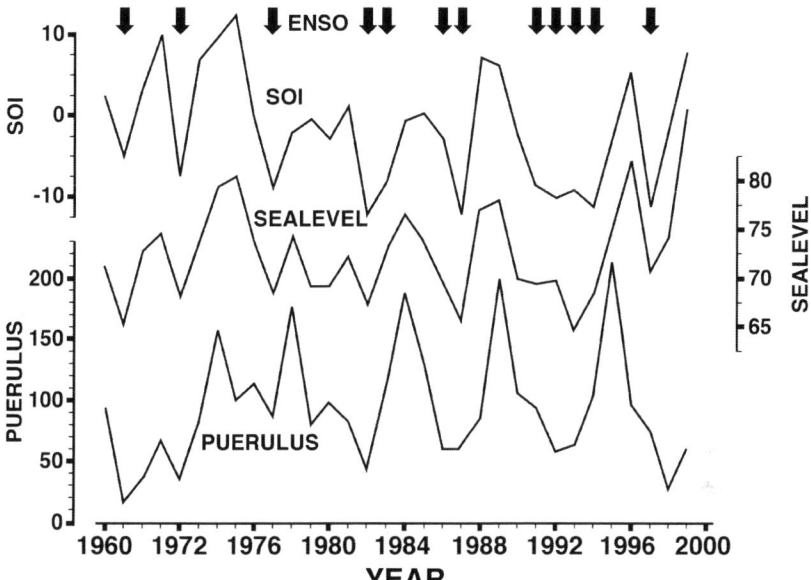

FIGURE 4.3. Annual mean values of the Southern Oscillation Index (SOI), Fremantle mean sea level, and puerulus settlement at Seven Mile Beach, Dongara, Western Australia. ENSO periods are indicated by arrows. Reproduced from Nick Caputi, Chris Chubb, and Alan Pearce, "Environmental Effects on Recruitment of the Western Rock Lobster, *Panulirus Cygnus*," *Marine and Freshwater Research* 52 (2001): 1167–75, with permission from CSIRO Publishing.

western rock lobster.[74] That ENSO could wreak havoc on vital fisheries was not in doubt—the El Niño event of 1972–1973, for example, had contributed to the collapse of the Peruvian anchovy fishery.

Oceanographers and fisheries scientists continued to refine their understanding of the influence of ENSO on the Leeuwin Current and the western rock lobster fishery. As Pearce and Phillips later noted, theirs was the first study to consider the influence of ENSO events on the Western Australian coast as catastrophic drought and bushfires had focused meteorological attention largely on the continent's eastern seaboard. Drawing on the outpouring of international research on the marine and terrestrial impacts of the El Niño–Southern Oscillation (ENSO) during the 1980s, Pearce and Phillips argued in 1988 that the strength of the Leeuwin Current was closely correlated with this global phenomenon and the productivity of the rock lobster fishery.[75] Using coastal sea levels as a proxy of the current's strength, they found that during a "normal" (non-ENSO) year, the current is strong and the settlement of rock lobster larvae—an indicator of the size of the fishery four years later—is relatively high. The strength of the Leeuwin Current, therefore, contributes to the return of the larvae to the coast, where they can later be harvested. Consequently, Pearce and Phillips confirmed that the conundrum of 1982–1983 was the result of ENSO, as the low puerulus settlements of 1972–1973 and 1976–1977 were also (see figure 4.3).

By the turn of the twenty-first century, however, the relationship between the Leeuwin Current's strength and the prospective harvest had become a cause for concern, even though fishers had caught a record catch of more than fourteen thousand tons in 1999–2000. Fisheries scientists wondered whether the greater frequency of ENSO events since the early 1990s (producing a weakening trend in the Leeuwin Current) would result in average-to-low puerulus settlement and therefore, smaller catches in subsequent years. Perhaps, they speculated, rising sea surface temperatures near the coast might offset these conditions by ensuring the warm conditions in which the lobster could thrive.[76] Their concerns were only heightened in 2008–2009, when the puerulus settlement was the lowest in forty years, suggesting there were at least four lean seasons ahead. In fact, puerulus settlement had been well below average since 2006–2007 and this trend continued in 2012–2013. What confounded researchers was that the Leeuwin Current had been strong in 2008 and 2011, which should have produced above average settlement in those years at least. As they investigated the causes of these settlement numbers, fisheries managers acted to prevent overfishing by slashing the allowable catch in half and introducing a quota system.[77] Meanwhile, uncertainty about the effects of future ENSO events raises questions as to whether the Leeuwin Current will continue to be considered a predictive index for the fishery in the future.

THE LEEUWIN CURRENT AS CLIMATE MECHANISM

Over the austral summer of 2010–2011, marine scientists noted unusually warm waters off the Western Australian coast. These temperatures exceeded the average by such an extent that they constituted a marine heatwave in this global biodiversity hotspot, decimating the abalone (*Haliotis roei*) fishery and precipitating widespread coral bleaching in the Ningaloo Reef, a World Heritage area.[78] For the western rock lobster, these temperatures were overwhelming, causing deaths in the fishery near the Abrolhos and "unusual behavior" in the survivors.[79] Meanwhile, the Leeuwin Current was unseasonably strong as a result of one of the strongest La Niña events on record. Further heatwaves in subsequent summers led CSIRO oceanographer Ming Feng to dub the phenomenon the "Ningaloo Niño" in 2013, noting its similarities with events in other eastern boundary currents such as the El Niño of the Pacific and the Benguela Niño in the Atlantic.[80] This particular association of the Leeuwin with these processes drew the current into a global imaginary of a planetary climate. In this configuration, the Leeuwin was conceived as a climate mechanism that localized the impacts of planetary processes in the eastern Indian Ocean.

This reconception of the Leeuwin Current transformed what had been perceived as an isolated stretch of ocean (albeit one with significant implications for Western Australian fisheries) into an important component of a global climate system. Reimagining the Leeuwin as a climate mechanism grafted planetary horizons onto the previously local and regional imaginaries of a fluid frontier and predictive index. Envisioning the current in such a way directed fisheries managers and climate scientists toward a network of global climate knowledge-making. Connecting to this network would provide insights that they could then "re-localize" into "locally relevant information" to guide fisheries management and climate policy.[81] Yet a growing scholarship on the implications of this approach to climate knowledge-making, a symptom of what STS scholar Clark Miller describes as "globalism," suggests that such a shift in the alignment of expertise may not produce the outcomes for which experts hope. Instead, the instrumentalism that globalism encourages can come at the expense of local fishers and their alternative "ways of knowing," while the information that it produces can privilege some fishing groups over others.[82] The imaginary of the Leeuwin as a climate mechanism in the early 2000s, therefore, reflected and reinforced a rift between experts and Western Australian fishers, particularly those working around the Abrolhos.

In the wake of the 2011 and 2013 Ningaloo Niño events, local fisheries scientists were certain that more would follow as a result of anthropogenic climate change. CSIRO climate modeling anticipated that there would be more frequent and more severe La Niña events in the Pacific, which would continue to stimulate Ningaloo Niño conditions in the eastern Indian Ocean.[83] This area off Western Australia was already getting warmer—it had become so warm that Feng described it as one of three "hot spots" in the Indian Ocean where temperatures had risen faster than the basin average since the 1950s.[84] Fisheries scientists already suspected that this trend was affecting the western rock lobster—for instance, the adults were smaller and they were migrating to deeper water earlier than they once had.[85] Having observed the impact of the 2012 heatwave in the Northwest Atlantic on the American lobster (*Homarus americanus*) fishery, Western Australian fisheries managers recognized the importance of reviewing water temperatures to better monitor and prepare the western rock lobster fishery for especially warm conditions in the future.[86]

To examine the effects of climate change on the western rock lobster, fisheries scientists had to look beyond the warming trend in Western Australia's coastal waters. In the early 1990s, local lobster specialist Nick Caputi had identified a positive relationship between winter storms and juvenile or puerulus settlement: strong westerly winds and their attendant rainfall were associated with greater numbers of puerulus on the continental shelf.[87] Over the subsequent decade, however, climate scientists identified a drying trend across southwestern Australia, covering a region that extended from Geraldton in the north to Esperance on the south coast. When the poor settlement trend continued after 2006–2007, Caputi and his colleagues revisited this relationship. They found that the lean seasons of puerulus settlement coincided with 2006 and 2010, two of the driest years on record. Furthermore, less rainfall meant less runoff, which reduced the nutrients flowing into the already nutrient-poor eastern Indian Ocean. With this drying trend expected to continue, combined with a warming ocean, heatwaves, and a forecasted weakening of the Leeuwin Current, the management of the western rock lobster fishery faced significant challenges.[88]

Having overcome threats in the past, fisheries scientists were confident in their ability to guide the fishery through these troubled waters. But many fishers were anxious that their livelihoods might become untenable. Concerns about the future of the fishery were not unprecedented—fisheries scientists voiced their alarm at overfishing in the 1940s, and older fishers feared the impact of improved technology on fish stocks in the 1980s. What was different was the possibility of the distribution of the western rock lobster drifting south, away from the Abrolhos.[89] As one fisher from Geraldton mused in 2013: "If the water gets too warm and the lobsters move further south or that it wipes them out or something well we're all buggered."[90]

Fishing families were already feeling the pinch after fisheries managers introduced further restrictions in response to the poor puerulus settlements of the mid-2000s, which had halved the rock lobster catch and the number of fishers as a result.[91] From the 1980s to the early 2000s, annual catches had been around eleven thousand tons, but from 2008–2009 onward, management limits reduced the annual catch to about six thousand tons.[92] As a fisher explained: "three years on and the sense of community has diminished dramatically with the changes to fishery management policy, the Abrolhos season is longer and fishers come and go at the Abrolhos in time with the 'price.' This has seen many community events cease as there are not enough fishers to support them at any given time."[93] The family oriented lifestyle and local livelihood that had once been associated with the western rock lobster fishery is giving way to a more tightly regulated, profit maximizing industry—a fishery that is increasingly connected in material and imagined ways to global change, commercially, scientifically, and ecologically.[94]

LEEUWIN IMAGINARIES

In December 2015 the Scientific Committee on Ocean Research (SCOR) and the UNESCO Intergovernmental Oceanographic Commission launched the second International Indian Ocean Expedition (IIOE-2) in Goa, India. For oceanographer Raleigh Hood, the chair of the research committee that planned IIOE-2, "The Indian Ocean truly is one of the last great frontiers of oceanographic research [—it] remains one of the most poorly sampled and overlooked regions of the world ocean."[95] The persistence of a frontier imaginary of the Indian Ocean continues a scientific tradition of exploration and discovery with a new emphasis on regional food security and the coastal impacts of anthropogenic climate change.[96] The workings of the Leeuwin Current and its implications for Western Australia fisheries are among the items on the research agenda for the IIOE-2. Along the coast, the western rock lobster industry faces the quandary of record numbers of adult lobsters but a drop in demand from China, now its biggest market.[97] Meanwhile, the Southwest of the continent continues to experience drier conditions, with the reduced runoff causing concerns for regional water supply managers.

What was first a rugged frontier for fisheries exploitation, the Leeuwin Current was reimagined in the 1970s and 1980s as a predictive index in response to anxieties about the viability of the fishery. The scientific findings that fostered this new quantitative imaginary—the relationship of the Leeuwin Current to the ENSO phenomenon—gave rise to another view at the turn of the twenty-first century. This new instrumentalist imaginary envisioned the current as a mechanism in a global climate system that both materially and conceptually localized planetary environmental change off the Western Australian coast, with uncertain consequences for local fishers. Although lived experience no longer sustains the notion of an endless frontier for resource extraction, the imaginary of the *fluid* frontier persists—its very nature inviting efforts to make visible, known, and quantifiable what is otherwise elusive to humans.

How this marine atopia is interpreted and understood is the product of historic geographical imaginaries, which continue to shape particular narratives of productivity, labor, and environmental management. Examining this spatiotemporal triptych of Leeuwin imaginaries shows the fluidity of such representations of spatial forms and processes as they absorb and blend with earlier and alternative outlooks. Furthermore, as study of the Leeuwin's imaginaries shows, these acts of representation impact material practices and geographies, such as fisheries and climate science, and the management of the rock lobster fishery. Studying these imaginaries over time and the material impacts they affect returns humans and the material world to histories of the ocean and climate and reminds us of the significance of local experiences in stories of planetary change.

PART II

HORIZONS OF EXPECTATION

5

IMAGINED GEOGRAPHIES OF CLIMATE AND RACE IN ANGLOPHONE LIFE ASSURANCE, C. 1840-1930

James Kneale and Samuel Randalls

AS A NUMBER OF HISTORIANS HAVE DEMONSTRATED over the last decade, nineteenth- and early twentieth-century life assurance can be seen to have played a surprisingly important role in many aspects of British and American history.[1] This is particularly clear in terms of its contribution to the shaping of medical knowledge and practice.[2] Our own work has begun to explore the extent to which Anglophone, particularly British, life assurance firms similarly engaged with, and possibly helped shape, understandings of climate science.[3] We would like to suggest that actuarial discussions of the risks involved with foreign travel and residence did more than just reflect wider concerns about dangerous locations and diseases, even where the tropics were concerned. To some extent this was because life assurance firms had their own source of data with which to answer these questions and their own ways of establishing the value of other rival sources of information—including climatological and medical science. Actuaries compared the physical laws of the sciences with the "laws of mortality" they detected in their "experience," the recorded mortality of the *lives* insured by their companies.[4]

Their engagement with contemporary medicine was, therefore, partial and unpredictable. As a consequence their judgments on the key climate issues of the day—acclimatization and the significance of "race"—were both highly significant, as these judgments affected the finances, movement, and security of thousands of people, and at the same time rather different to what we might think of as mainstream climatology. In this way life assurance firms possessed their own distinctive imagined geographies of climate. These imagined geographies were developed in and through the practice of defining rates and expectations for lives overseas rather than simply through the adoption of scientific and other discourses about acclimatization, race, and the effects of climate on health. Exploring these geographical imaginations in sites of insurance practice provides fresh perspectives on how climate and health were understood and, more importantly, acted upon in business applications—suggesting, too, how these perspectives might have traveled into policyholders' everyday lives.

Before exploring these geographies it is worth explaining the practice of life assurance as it was practiced in the second half of the nineteenth century and the beginning of the twentieth. An officer in the navy or army or a civil servant, for example, might receive a posting to a colonial station overseas, or perhaps a businessman might need to travel to or reside somewhere beyond Britain. If they already had a life assurance policy—and many British men did at this point, especially those of the middle and upper classes—they might find that they needed the permission of the firm to travel and live overseas. Herbert Thiselton's 1893 review of the seventy-three largest insurers found that only sixteen companies did not mind where their policyholders traveled.[5] The other fifty-seven offices set out zones where the life of the assured was thought to be at risk in some way. Twenty of these firms simply listed these dangerous areas (Japan [five firms] or the Cape Colony [seven firms], for example); the other thirty-seven defined them through lines of latitude, often adding a set of additional danger spots as well. The most common latitudinal definition of risk placed it in the area between 33° north and 30° south, the choice of sixteen firms, which meant that the nontropical world was free of risk—although many offices then made exceptions for other parts of Asia, Africa, and Australasia lying in this "safe" area. If our colonial officer needed to travel somewhere beyond these "free limits," he might find that his office would charge him an additional premium for the duration of his stay, and perhaps after it. The period of the extra premium might cover only the period of acclimatization or might last as long as the policy. This extra premium was meant to cover the additional risks involved in voyaging beyond these "free limits," as life assurance concerns balancing the income from premiums against the cost of paying out when the assured dies or reaches a certain age. Ignoring potential risks beyond those associated with ordinary mortality meant endangering the finances, and perhaps the existence, of the firm. In this way firms mapped—sometimes literally—the risks associated with climate.

Companies were used to dealing with the economic consequences of en-hanced risks, particularly in terms of additional medical risk, and had been using physicians to assess the health of applicants since the middle of the nineteenth century; in fact the medical examination itself has its origins in the insurance doctor's evaluation (physicians were not, at this point, used to examining patients who did not feel there was something wrong with them).[6] The suspicion of a serious illness such as tuberculosis might mean that the applicant was rejected; other ailments might mean an extra premium to cover the additional risk. The size of this premium was calculated after consulting two registers of evidence: existing medical opinion and the "experience" of the firm. The importance of the former can be seen in the many medical examiners' handbooks written by insurance doctors to train medical referees to spot potential hazards of this kind and to explain their significance. The latter refers to the records of mortality kept by the firm, and the records shared between companies in industrywide efforts to improve knowledge. In Britain the Institute of Actuaries became a crucial center of calculation for these efforts, with actuaries presenting papers that were then published in the institute's journal, the *Journal of the Institute of Actuaries* (*JIA*); actuaries also played important roles in the London (and then Royal) Statistical Society, in medical conferences and publications, and in efforts to improve the quality of statistics collected in British colonies.

If this collected evidence suggested that policyholders with a particular illness tended to die earlier than expected, firms might well begin charging extra for new applicants with that problem. The same principle was extended to risky occupations—seamen, miners, publicans—and to travelers or emigrants to foreign parts. Here, again, the company's decisions about whether to reject the proposal, charge an additional rate, or allow the applicant to proceed were based on medical evidence, the firms' experience, or on other data—for example, government reports from censuses to inquiries into dangerous trades, as well as less quantifiable opinions about risk or its absence. These decisions did not merely chart popular understandings or prejudices; they had real effects on the mobility, choices, and practices of applicants. Through life assurance, life itself became a financial object to be carefully managed economically for the benefit of one's family or employer, and risks to lives were mapped and charted in ways that did not always agree with medical or popular constructions of the tropics.[7]

In this chapter we examine the forms of knowledge deployed by Anglophone life assurance firms to evaluate the risk—in terms of climate and disease—of overseas travel and residence, and the practices by which decisions were made about individual applicants. We draw upon actuarial discussions of these risks, on the evidence presented in medical examiners' handbooks, and on the records of individual firms in order to demonstrate that life insurance possessed its own imagined geographies of climate. We make links to histories of climatology and of tropical medicine to demonstrate points of departure and consensus between these fields—particularly concerning questions of "race"—while examining why these imagined geographies did not always fit mainstream scientific understandings of climate and disease. In the end we will see that arguments about the best way to harness scientific understandings were always compromised by disagreements about the best way to conduct life assurance as a profitable business.

MAPPING RISK

How did life assurance firms assess the risk of foreign travel or residence? We argue that these companies constructed geographical imaginaries of climate risks by drawing on existing scientific and imaginative resources as well as their own statistics and experience. Presenting a paper on "Life Insurance in the Tropics" in 1921, the eminent doctor Sir James Cantlie wondered how the extra premiums had been calculated fifty years ago. "Was it from statistics of the percentages of deaths published by the local governments in Crown Colonies, etc., or was it from information received from travelers? If not, one cannot conceive what were the sources from which the knowledge was derived. . . . I have described the tales we hear of these distant tropical lands as ship captains' gossip gathered from the common talk of those lands when the ship goes in harbor." With all this talk of the "white man's grave," Cantlie concluded, "one wondered how any Briton ever went out at all."[8]

Indeed until the 1840s foreign travel and residence were thought to be very risky and it seems likely that most travelers would not have obtained additional overseas travel cover. In 1841 the UK Temperance and General Provident Institution charged one traveler an extra twelve guineas (perhaps five hundred pounds sterling today) for a return trip to Jamaica and forbade him to be out of the United Kingdom for more than twelve months.[9] But as the market for life assurance expanded and more travelers left Britain, firms grew more confident and set lower rates. Between 1840 and 1930 actuaries and medical advisors made decisions on thousands of lives traveling to and residing in climates beyond Britain. While they would never be entirely confident about their decisions, they were able to turn to four kinds of information for guidance, demonstrating the developing dialogue between climate science, medicine, and life assurance.

Their first resource was in their own records of mortality of travelers and emigrants, which would give them concrete evidence of premature deaths in particular areas. However, for most companies, the numbers of policies affected in this way were relatively small. In the London Assurance company, for example, nearly 10 percent of policies had been exposed to extra risk as they "had either visited tropical climates, or had been military or naval officers by profession, or officers of the mercantile marine." However, seamen probably outnumbered tropical travelers within this group, and this figure was smaller than the proportion (14 percent) that had been given additional premiums as "diseased lives."[10] These small numbers would lead to frequent suggestions that firms should pool their experience of tropical lives in order to establish a better sense of risk. Addressing the Actuarial Society of Edinburgh in 1877, Spencer C. Thomson shared the mortality figures of an unnamed company insuring British lives in both the East and West Indies; the following year John Stott presented the mortality of the Scottish Amicable Society in the West Indies over thirty years; and in 1888 the *JIA* published Hardy and Rothery's discussion of the experience of the Barbados Mutual Life Assurance Society.[11] Each of these discussions built upon previous work, establishing the Institute of Actuaries as a center of calculation for this enterprise, as one of the discussants of these papers noted: "New results had been obtained, and those results gravitated naturally to the Institute, whence they were immediately distributed, as it were, to the profession at large."[12] However when Herbert Thiselton presented his review of actuarial practice in 1893 he was still convinced that more could be done to share the experience—and the risk—of foreign travelers and residents among companies. Still, company experience was better than nothing. In 1907 Henry Lutt's own survey of forty large firms pointed out that "To the public spirit of offices like the 'Standard,' [the Standard Life Assurance Company] which from time to time have placed their experience of foreign climate risks at the disposal of the profession, the modern life assurance world owes practically the whole of the information which it possesses upon the subject."[13]

Without extensive records of their own, then, firms turned to a second source: mortality statistics collected by government agencies. In 1877 Thomson had advised, "When there are materials for doing so, it is better and fairer for an assurance office, accepting a large number of risks in any particular climate, to base its premiums on a table of mortality founded on the death rate of the country in question," the raw materials that had allowed insurers to establish the risks of British lives.[14] This was possible in colonies like Victoria, Australia, that possessed census returns and records of mortality.[15] In places where figures were not already available, actuaries turned to existing sources such as the pension funds established in India by the East India Company for its Military, Medical, and Civil Services, or such as the military statistics collected and analysed by men like Henry Marshall, the Deputy Inspector-General of Army Hospitals, and Alexander Murray Tulloch.[16] And of course these figures were compared to a British baseline, to tables based on the mortality of lives in Carlisle or Northampton, and later on to the collected experience of many British companies as a larger, comparative dataset was established. Despite this effort, the medical geographer James Ranald Martin argued that opportunities to improve the collection of colonial statistics had been missed at sites like Calcutta's General Hospital, "through the neglect of the controlling medical authorities."[17]

Beyond these sometimes disappointing records of mortality, firms could turn to a third source: writings that imagined particular geographically specific climates as more risky or dangerous than others. Although the situation had improved a little since the influential Belgian statistician Adolphe Quetelet had complained that "climatology, taking the word in its most extended sense, is a science still too little advanced to engage our attention," actuaries continued to worry about a lack of reliable data.[18] We have examined the use of other climatological works and ideas in insurantial contexts elsewhere, but it is worth considering a number of other discussions of tropical risk here.[19] Thomas Bond Sprague, the most important British actuary of the second half of the nineteenth century, included climatology and the risks of tropical climate among his many interests.[20] Sprague noted that "the Insurance Offices have occasionally to quote extra premiums for residence at various places in Tropical Africa, such as the West Coast, Sierra Leone, Zanzibar, Lake Tanganyika, &c, but they have, I believe, very little information as to the real risk attaching to such residence," which led him to publish records of European mortality along the Congo River from Henry Morton Stanley's *The Congo and the Founding of its Free State* (1886) in the *JIA*.[21] Ten years later Thomas Sprague's son Alfred Ernest Sprague, also an actuary, reviewed R. W. Felkin's *On the Geographical Distribution of Tropical Diseases in Africa* (1895), arguing that "any contribution to our knowledge of the effects of climate on the constitution and prospects of life, should be welcomed by the actuarial world."[22]

Climate, medicine, and insurance were also brought together in a final source: the handbooks written for medical examiners employed by assurance firms. In his early handbook William Brinton acknowledged some difficulties in disentangling the effects of climate from those of endemic disease, but he also set out some "general influences of climate, concerning which we can group our knowledge into something more akin to law."[23] Migrating to cooler climes brought the risk of pulmonary or tubercular disease, this risk increasing as the difference in temperature increased; moving to a warmer climate threatened the digestive organs and risked fatigue and infection by endemic disease. Later textbooks listed the risks—heat, disease, intemperate habits—associated with particular climates, often paying close attention to India, one of the largest markets for tropical life assurance, and to West Africa.[24] In most of these works, climate presents a useful way to distinguish between different risks and potential markets, but it was endemic disease, rather than the climate itself, that received most attention from actuaries and medical examiners. By 1919 life assurance had become such a significant part of tropical medicine that Castellani and Chalmers's *Manual of Tropical Medicine* included a chapter on "Fitness for Tropical Life." Governments and commercial employers also sought to select only the healthiest lives for tropical service, and, the authors stated, "every medical practitioner appointed for the purpose of selecting men for service in the tropics will go faithfully through a routine examination more or less based upon that adopted by the best type of assurance companies."[25] Life assurance set the framework of practice and knowledge used by other actors when assessing tropical risks. These handbooks brought together medical, climatological, and actuarial knowledges and forms of evidence, with mortality rates often acting as proxies for the health of different areas, a metric that traveled easily and that could be translated from one form of knowledge to another.

Paradoxically, it should also be said that, although these four sources of information demonstrate that assurance firms were strongly interested in climate, what they meant by this term was often uncertain and their sense of its importance changed over the period in question. It is clear that they knew climate was not the sole cause of disease or ill health. Martin explained that many influences—temperature and rainfall, the drainage, relief, and geology of a site, the diet and habits of its inhabitants—affected endemic disease, and similar climates might possess very different medical geographies. However, climate, represented by the first two elements, could still be thought of as an active agent in its own right. William Brinton was also aware that "it is convenient to distinguish between the constitutional effects of mere climate, and the risk of diseases more or less endemic to the inhabitants of a given locality."[26]

In the last decades of the century medical topography was increasingly sidelined in favor of arguments for more effective sanitation, which would provide an immediate intervention to prevent disease, regardless of the environment. By 1921 climate was no longer the key focus of discussions of tropical medicine, with Cantlie stating: "Many declare that the tropics are dangerous to the white man only through their diseases, not through their climate. . . . Whilst not, perhaps, going so far as to say that climate has no deleterious effect, I do claim that in the tropics, if malaria can be prevented, the white man can resist the effect of a hot climate for an indefinite time." He added that this must, in time, remove the need for extra premiums on life assurance policies.[27] This was suggested several times over this period, without any agreement being reached across British firms, to some extent because of a protracted argument over the possibility of acclimatization and the best way to cope with it actuarially. This argument also touched on the relationship between "race" and tropical mortality—and thus on key questions about imperial fitness and progress.

Despite the increasing questioning of its importance, though, life assurance writers until the start of the twentieth century continued to talk about "climate" as the main source of risk, as seen in A. E. Sprague's description of West Africa as "the hottest and wettest region in the whole globe, and as is well known, the climate is deadly to Europeans."[28] Although they were aware of the difference between climate and disease, many writers continued to use "climate" as a catch-all term for the character of a region; in doing so—intentionally or not—they granted it causative powers. This may reflect either established usage or ignorance of the latest medical opinion; it might also represent a lack of interest in the exact cause of tropical mortality.

Insurance Maps

Cartography was an important means through which tropicality could be constructed, with maps sometimes decrying the civilizational capacity, or expounding the dangers, of the tropics.[29] Humboldtian cartography often expressed political ideals as well as scientific ones, advancing certain kinds of Eurocentric geographical imaginations of climates and health around the world i.e. the specific superiority of the European climate as intersecting with European society/culture.[30] Life assurance offices drew on these imaginaries to produce "new" geographical imaginations of risk in the form of maps showing where extra premiums should be charged. Although these maps may simply have been used as quick reference guides by agents, they nonetheless inspired a particularly geographical appreciation of risk, with latitude lines refracted in part through imperial territorial extent and spaces of exemption from extra premiums.

Harrison shows how a medical topography emerged from the 1820s that drew from Hippocratic notions of the role of the environment in influencing health and developed a way of more formally differentiating healthy and unhealthy zones. It was army experience with disease that led to calls to identify healthy and unhealthy zones in India. Climate came to be used as an explanation for physical attributes and military success. Rupke and Wonders note three stages in the development of medical cartography: a textual map (words on a conventional map), a contoured map (lines and shadings to show the occurrence of disease), and (the important point for this paper) an isothermal map. The German pathologist Adolf Mühry, for example, used Humboldtian isolines to develop a noso-geographical classification for miasmas and disease, attempting a scientific definition of those disease-prone environments that should be avoided.[31] Isoline maps like Mühry's showed the northern- and southernmost limits of diseases. As Rupke and Wonders note, "the isoline came to symbolize advanced, cutting-edge science" in the mid-nineteenth century. The French anthropologist Arthur Bordier, likewise, used isotherms to color maps for disease plotted not according to actual incidence but by the isotherms. And Heinrich Berghaus created the largest of the Humboldtian maps—a two-volume set with ninety maps of disease published in 1848.[32]

There were criticisms of such mapping. August Hirsch, a German geographical pathologist, critiqued earlier work by Mühry and especially the medical practitioner Caspar Friedrich Fuchs's division of the world into three disease zones (dysentric, enteromesenteric, catarrhic) because there was too little knowledge about medical topological conditions even in Europe to make such claims.[33] Other authors followed suit. Henri-Clermond Lombard produced the largest publication in medical cartography with twenty-five colored maps without any isolines at all.[34] In Australia, challenges in defining a dangerous tropical zone created debates about how to even define such maps; a latitudinal line in Australia would place one-third of the continent in such a climate, although an isothermal line created a much smaller area, as the medical practitioner Frederick Goldsmith pointed out at the 1902 Australasian Medical Congress. Doctors, Goldsmith continued, preferred a 68°F isothermal line, but geographers claimed that this obscured a crucial difference between high tropical temperatures and the dangerous humidity of the tropics. If the tropics are defined as a winter temperature over 75°F, then very little of Australia qualified.[35] Maps thus became caught up in arguments over which were the appropriate lines for mapping disease.

What mattered in many cases was practical experience with disease as Europeans, particularly soldiers, traveled the world. There were various differing views. Perhaps it was about travel from the place where one was born or acclimated that mattered more, and surgeons such as Playfair, Royle, and Wilson worried that changes in the environment—such as cutting down a forest, for example—would lead to ill health. This drew on experience from Indian medical authorities such as Charaka and Susruta who believed that disease could come from unseasonable weather or vapors from swamps.[36] It would be topographically tricky to thereby identify healthy zones for European bodies given the geographical and climatological complexity implied by such a calculation.

If maps were important in medical climatology, life assurance companies also espied the usefulness of such maps to help define areas of risk.[37] For example, a new company, Colonial Standard, established as a subsidiary of Standard Life in 1846, hoped that its cheaper rates would give it an advantage over more conservative offices when it came to travel risks. Risks were divided into four regions, defined in terms of broad geographical areas, latitude, and season. Other companies, like the Eagle, introduced zones based on Britain's major colonies—a more explicitly territorial geographical imagination of risks. Maps allowed risks to be communicated and helped insurance agents and applicants interpret the level of extra premium required for travel to different parts of the world. Sprague liked Felkin's disease maps, stating, "From this map, therefore, a comparative estimate of the healthiness of the various regions can be obtained at a glance, and a rough idea gained of what would be a suitable extra premium for the corresponding climate risk; and it should be of assistance to actuaries when considering the rates to be charged for residence in Africa. The table is constructed on a somewhat similar principle, and will be useful if it is desired to find out at a glance in what parts of the Continent any particular disease is most prevalent."[38]

Thiselton argued for a single rate for all companies that would be easy to mark on a map, perhaps even just one map. "If it were found impossible to agree to limits which could be expressed as above suggested, we might follow the example of those offices which include a map in their prospectuses, and we might then not only mark the free limits by some bold line, but by suitable colouring show the precise extra premiums charged for other parts of the world. Such maps, if accurately engraved, should have a considerable advertising value, for our prospectuses would then be much more frequently carried about and constantly referred to."[39] The geographical imaginations of insurance offices would therefore circulate via maps. Sales staff and applicants could easily discover what extra premiums would be charged by looking at a map, such as the one included in the Prudential Company's pocket-sized booklet of instructions for its agents from 1882.[40]

Firms do seem to have done as Felkin suggested, broadly agreeing on this map of risks and the rates needed to compensate for them. The colored map printed in the third edition of De Havilland Hall's medical handbook showed "the usual rating adopted for foreign residence," with the danger zone between 33° north and 30° south of the Equator and an unrated zone covering Central Africa and the West African Coast where firms might refuse insurance altogether.[41] Between 1880 and 1914 the rates charged by the Sceptre life assurance office to travelers and migrants to China, India, Jamaica, and Central or South America all agreed with the sums noted on De Havilland Hall's map. Sceptre policyholders who wished to move to West Africa, however, were charged much higher rates and might face additional restrictions on their movement.[42] These risky areas of West Africa are precisely those that Felkin identifies as prone to disease—particularly malaria. These imaginaries of risk thus came to shape what policyholders would pay and where they might travel.

ACCLIMATIZATION AND "RACE"

Insurers drew on existing ideas of climate and disease that had been popularized in imaginations of places by geographers, colonialists, and others. But while the insurance maps sought clarity and consistency in the grading of risks, there was a rather more lively debate about risk than these maps might display. Discussions of acclimatization and race haunted insurers' simple pricing calculations and opened up contestations over geographical imagination of overseas travel risk.

Environmental determinists such as Ellsworth Huntington (1876–1947) feature prominently in discussions of race and acclimatization alongside writers like Griffith Taylor (1880–1963), who constructed a white race climograph in 1916.[43] However, European and American science was not universally orientalizing when it came to medicine and there were at times a range of critical and variegated ideas and practices. Harrison critiques Arnold for offering too simple a story of European views of Indian degeneracy and a pathogenic Indian environment, whereas for him the differences were of degrees rather than kinds.[44] Indeed, if the Indian environment were so dangerous, there was little place for progress toward development, and in practice this would have implied a real limitation on the possibility of imperial power. In 1865 the *Report of the Commissioners appointed to inquire into the Sanitary State of the Army in India* had doubted "whether, with the mortality rate of the last 40 years, it would be possible to keep up an army of 70,000 men in India."[45] However, the Indian environment could not be completely pathogenic if it could also be shaped and exploited by imperialists who could actually live in the country.[46]

These pathologizing geographical imaginations made Africa particularly dangerous, an idea bolstered by medical statistics suggesting that between one-quarter and three-quarters of Europeans arriving in West Africa died within a year. Central southern Africa, however, generated a reputation for being much safer as a result of reduced mortality experience, and missionary and medical records provided a more variegated and complex understanding of the geographies of disease through the African interior. It was therefore in practice, in lived experience, that a singular geographical imagination of "the tropics" or "Africa" broke down. This was reflected in insurance practice.[47]

Hardy and Rothery's paper on the Barbados Mutual Life Assurance Society provides an opportunity to explore a long-running disagreement about the possibility of European acclimatization. Responding to the paper, Benjamin Newbatt—actuary for Clerical Medical and soon to be president of the Institute of Actuaries—noted that Hardy and Rothery had argued "the effect of tropical climate upon the mortality of Europeans increases with the period of exposure," while other authorities such as Stott and Spencer Thomson thought migrants stood a good chance if they survived long enough to become acclimatized. "Those two opinions were diametrically opposed," Newbatt concluded, "and the question was a much more open one than would appear from this paper."[48] Several other members of the audience were disappointed that nothing was said about acclimatization, given the common practice of charging an extra premium for the first few years of assurance. One questioner wanted to know if "race" was the explanation and whether the Barbados office insured "lives born in Europe, or native-born descendants of Europeans, or from a mixed stock, and if there were many negroes pure and simple?" The answers given by Hardy and by Mr. Da Costa, one of the firm's directors, made it clear that almost all of the lives in question were "acclimatized lives," but "natives" rather than "British-born."[49] By the end of this discussion the definition of "acclimatization" was no clearer, but for decades the idea continued to hold a fascination over actuarial imaginations of climate.

In fact James Ranald Martin had since the 1850s been sceptical of the idea that Europeans might adapt to new climatic conditions, or that "intertropical natives" prospered in "our own latitudes."[50] "Man" could not physically adapt to new conditions, but "different habits of body" made all the difference between "the seasoned and unseasoned."[51] In fact, while the first few years of residence could be very dangerous, "all statistical evidence goes to prove the increase both of sickness and mortality in relation to length of residence." The mortality of retiring (white) officers of the Indian Army was about one-third higher than that of men of the same age in England.[52] Martin's voice was an important one in the worlds of tropical medicine and life assurance. However, this argument had to be repeated over and over again for another fifty years—perhaps because it appeared to contradict both the experience of assurance firms, which showed high mortality rates during the first few years of residence, and the myth of the "white man's grave." Over a decade later, Arthur Bailey had to remind his audience at the Institute of Actuaries that "the interesting question of acclimatization was still unsettled. Sir Ranald Martin, the medical examiner at the India Office, has no belief in acclimatization, and thinks that a new comer is better able to resist the ravages of the climate than an old resident of the same age."[53] Despite what Newbatt said in his rejection of Hardy and Rothery's argument, Spencer Thomson *did* think that while the initial risks of living in the East Indies could be offset if the assured "learned the ordinary rules of living . . . after a time he begins to succumb to the climate, and the longer he continues to reside in the East the less easily does he appear to be able to resist it, and eventually he very frequently breaks down under an enfeebled constitution, which becomes unable to battle longer against its injurious powers."[54]

Similarly, Sprague's review of Felkin's medical geography noted "the author points out that the subject of acclimatization is one that is very largely misunderstood, and exposes the fallacy of the popular idea that a European migrating to a foreign country becomes 'acclimatized' in the course of a few years."[55] Brockbank, writing in 1908, echoed Martin's idea of the "seasoned" body or Thomson's "ordinary rules of living" when he wrote that "a man runs most risk of getting ill in the tropics in the first five years of his life there. After this he has learned wisdom; he appreciates the dangers which surround him, recognizes the advantages of good health and takes more care of himself."[56] Different colonies presented different risks, though; West Africa was dangerous for Europeans because "the whites have not as distinct a residential quarter, separate from that of the natives, as they have in India, and sanitary measures are hard to carry out where natives live."[57] Brockbank said nothing about the subsequent effect of climate on mortality. However, Castellani and Chalmers sided with Martin in summarizing the opinion of Caddy, who "in 1912 came to the conclusion that the European does not acclimatize well in the tropics, meaning by this statement that he is unable to rear healthy strong children in India, and that he becomes debilitated by residence there." They concluded that, although the risks were greater in the first few years of residence, they did not disappear altogether after this.[58]

Martin's ideas—supported by the experience of some firms—continued to struggle against older ideas of acclimatization. It is interesting that Sprague's paper, published in the *JIA* in 1895, had to remind actuaries that this sense of acclimatization was a myth. It may be, of course, that life assurers were *themselves* the source of this myth, through the insistence of many of them on charging a higher premium to cover the initial risk of residence in a new climate. Policyholders may have assumed that there was no additional risk associated with residence once this period was up. So it is possible to see several definitions of "acclimatized" here: the meaning that emerged in discussion of Hardy and Rothery's paper, of "native lives" brought up in this climate; the sense that after an initial period of risk, with heavy mortality, survivors would be acclimatized in the sense of "seasoned" or perhaps "wise" in terms of expected behavior, and therefore as good as British lives at home; and the idea established by Martin and repeated throughout this period suggesting that, although emigrants might be over the worst risks, the new climate (standing in, perhaps, for an assemblage of risks of disease, warfare, etc.) would continue to take a toll on life that exceeded "normal" (British) mortality.

By the 1920s geographical perspectives appeared outmoded in medical literatures with a greater focus on the assemblage of risks and the specific transmission of individual diseases, making foreign residence a much less frightening prospect for settlers.[59] Imaginations of risk in part became less about climate and place and more about civilization, though this still left no easy conclusions about life assurance prospects. For instance, Dr. J. S. C. Elkington, writing in 1905, argued that it was the insanitary practices of local residents that would need to be monitored, and interventions initiated where necessary, to prevent risks to settlers.[60] By 1925 Sir James Barratt (1862–1945), a doctor and a past president of the British Medical Association, could declare that there was nothing to stop white people from settling in tropical Australia as medical science would make the space safe for the white settler body.[61] On the other hand, Robert DeCourcy Ward, in his final paper (1929), implied that the tropics would need absentee administration to succeed as the tropical people could not progress in civilizational terms by themselves and yet, at the same time, the tropical environment was not conducive to the settler body even after many decades and generations of settlement.[62] White settlement was needed to civilize the land, but it could still be insufficient in terms of reducing the risks to the policyholder.

Of course "race" still figures in this discourse, but it is in the sense of white lives at risk—in the tropics, at least. As we might expect from histories of the links between environmental determinism and scientific racism, early twentieth-century life assurance and tropical medicine were shaped by ideas of "race." Ideas about natural immunity pervaded many accounts, suggesting, for example, that Filipino natives should automatically resist tropical diseases, Spaniards would have some resistance as a result of their native southern climes, and Americans would be particularly vulnerable.[63] For some writers "race" became simply another source of risk or at least of identification. Brockbank, for example, considered the assessment of Jewish and Armenian applicants, pronouncing them generally good lives.[64] The Minnesotan medical officer Charles Lyman Greene thought "the American Indian, the negroes, particularly those of mixed blood, mulattos and the like" were particularly susceptible to tuberculosis, as were recent immigrants of Celtic and Scandinavian origin, but he did not consider questions of "race" beyond the United States.[65] Lister divided non-British Europeans into Scandinavians, whose lives were as good as the British, and less healthy southern Europeans, which included Germans.[66] And, of course, within the United States life assurance played a particularly important role in supporting, as well as opposing, eugenicist arguments about the "natural" fitness of African Americans, other Americans, and newly arrived migrants.[67]

"Race" also figured in discussions by British firms of the insurance of non-British traveling lives. As Lister pointed out, "Every race in the world is insured by English offices. Every racial peculiarity is, therefore, a matter of importance to the insurance examiner who may be employed in Europe, or in the colonies, or in the dependencies, to examine either native lives, or those who were born in the old country, and for some time of their life have had to live abroad."[68]

However, for some writers, "native lives" seemed to be *inherently* inferior lives, in at least three different ways. First, as many authorities had noted, residents of tropical climates often fared badly in more temperate locations, while some experts believed that Europeans *might* be able to acclimatize. As Lister explained, "a negro or an Indian life resident in England [is] not as good as the average English life. This arises partly from the fact that they are not living in the temperature adapted for their race, and that there is what some actuaries consider to be an 'attenuated vitality' resulting from the previous generations of their ancestors having lived in a hot climate."[69] Second, the life expectancy of Barbadians was inferior to white British migrants, and in the early twentieth century, at least, the "racial death-rates" of "Natives" in the Bombay Presidency were twice or three times those of (white) Europeans.[70] Third, Cantlie noted a reverse acclimatization in Chinese inhabitants of Hong Kong; "it is the natives who have become 'Anglicized' to a greater or less extent that usually come forward for insurances." This became a problem when these applicants had adopted "British ways of living" when it came to eating and drinking alcohol. "The ignorant and simple living native might be as good a life as any European—as a matter of fact he seldom is—but when he departs from his simple habits, and takes to 'foreign' ways of eating and drinking, his life is not a good one."[71] Cantlie had been expressing this opinion for some time, and it would be repeated, with some changes, by Brockbank, Lister, and others. Despite being identified as a "native," then, it seems that the non-Europeans could not be healthily at home in either their country of origin or abroad.

This does not entirely capture the consensus on this topic, however, though it may fit the early twentieth century better than the late nineteenth. Although James Ranald Martin certainly borrowed the language of ethnology to describe "types" of Indians, the figures he presented suggested that "racial" mortality rates in India unraveled as they were cut across by class, caste, diet, and habit. He noted that the recent cholera epidemic had been no respecter of racial types, and in Bengal the annual mortality of European soldiers from cholera had been more than five times that of the Sepoy Army; the "boasted strength" of the former had been more than offset by intemperance and overindulgence.[72] P. M. Tait, Director of Indian Business for the Albert Life Assurance Company in Calcutta, discovered some similarly "remarkable" results in his analysis of a pension fund open to both "Eurasians" (his term for those of mixed Indian and European heritage) and white Europeans, usually considered to be "the most select class of lives in India"—the mortality of the former was superior to that of the latter.[73]

Race mattered in these geographical imaginations, because it shaped the specific geographies of risk for a particular racial body.

Insurers were therefore interested in these debates about disease, climate, and bodies, and the debates came to shape their geographical imaginations of risks, but never with a single, uniform conclusion. Indeed by the twentieth century some insurers were suggesting that it might be possible to ameliorate the effects of climate on disease and life expectancy. What actuaries sometimes called the "universal law of mortality"—where firms' experience suggested that age, and age alone, determined the proportion of a population who died at a particular age—was, at times, thought to transcend "race" and climate, a point noted in many of the life assurance accounts. Similarly, the emphasis on sanitation, hygiene, and lifestyle meant that some forms of acclimatization might be available to anyone. This being said, insurers retained a considerable interest in climate and race, particularly for those traveling overseas. This suggests that there were multiple imaginaries of risky lives operating within actuaries' office spaces and neither a universal law of mortality nor a geographical climatic imagination won out. The tension between specificity and universality played out time and again in practice, more than in theory, in making decisions on individual lives.

CONCLUSION

In many ways life assurance companies echoed debates about acclimatization and race in the medical geography literatures and reflected some of the changing debates through this period: from a concern with the geographical climates of risk to the concern with the specific bodies inhabiting the environments; from a concern with the possibility of disease in white bodies in the tropics to a concern with any body transposed from its native climate; and how to price the risks of mixed-race bodies. But as with maps of geographical risk, acclimatization and race represented both business opportunities and risks. Firms competed to win business but also looked at their own experiences of mortality in their records to determine the pricing and acceptability of particular risks. Actuarial practice and experience again trumped pure scientific evidence. Life assurance companies then operated with and created particular geographical imaginations of climate and disease. These were not strictly scientific imaginaries, though they drew on medical evidence to a great extent; they were also imaginaries in practice, which connected changing debates about acclimatization and race with business experience.

These geographical imaginations came to shape the extra premiums charged to policyholders and, on occasions, restricted movement within the affected zones. Life assurance firms may have helped further substantiate the imagined geographies of risk through this period, but these risks were also acknowledged to be more complex than an insurer could fully take into account. The case of acclimatization demonstrates this, with varied responses to the question of length of residence and the quality of life on return home. Life assurance companies followed mortality experience more than any naked political ambition or imperialism, but clearly their pricing reflected imperial contours through the additional experience in those areas. It is important to examine both how geographical imaginations are produced and the effects they had, not all of which can be simply assumed from the visual representation of maps or data.

It is therefore important not just to look at scientific and popular accounts of climatic determinism, race, and the tropics but to see how these imaginations are translated in the context of a business that had to make practical decisions about the risks to lives in particular places. Some ideas stuck, though they were not always supported by medical evidence; others failed to find any purchase. In other words, the tropics were not a discursively homogenous orientalized space with singular effects; rather, business experience and pricing strategies shaped and reshaped the ways that individuals traveling overseas experienced the "risks" of the tropics. While a simple system of grading climate risks would have made the administration of residence risks much easier, "insurance, after all, was always a local problem" and tropical risks constantly shifted into matters more closely understood as imperial fitness, sanitation, and development, or local epidemiological conditions.[74] Explorations of geographical imaginations importantly need to be connected to practice to explore how these are translated in the everyday spaces of business (and other) decision-making.

6

THE BRITISH WOMEN'S EMIGRATION
ASSOCIATION AND THE CLIMATE(S)
OF SOUTH AFRICA

Georgina Endfield

OVER RECENT DECADES, there has been a "remapping" of the colonial land-scape to incorporate women.[1] Marked shifts in the work on the histories of empire have emerged as the fields of imperial and women's histories have inter-sected, informing a wave of scholarship focusing on subaltern women's voices, and the contribution that white women made to the British imperial endeavor, particularly through the production of popular geographies of "other" places.[2] Feminist scholars in particular have addressed the involvements of white women and the nature of their location and agency in the process of British colonial domination, offering insight into settler women's attitudes, experiences, and social roles.[3] Other work has highlighted the role of British missionary wives and women as "actors" within the metropolitan gender order and in the sphere of empire more generally, addressing their role in the so-called white woman's burden to bring civilization to her "fallen" sisters in otherwise dark imperial spaces.[4] Such work has broken down traditional dichotomies between home and empire as private and public spaces respectively; it has also helped to demolish the coupled assumptions that the empire was a predominantly male space and that the colonies were no place for white women.[5]

The activities of women as imperial propagandists and agents has also received attention.[6] Scholarship has focused on the many—often overlapping—women's emigration societies of the late nineteenth century and the way in which these societies sought to at once address a prevailing perception of there being surplus women "at home" while also promoting the development of Britain overseas through "womanly imperialism."[7] During the late nineteenth and early twentieth centuries, thousands of British women emigrated to a wide range of destinations, but many to British imperial fields in Africa, New Zealand, Australia and Canada. Indeed, between 1884 and 1914, around twenty thousand women emigrated under the auspices of British female emigration societies.[8] Both women emigrants and emigrators in late nineteenth century, therefore, were well placed to be active participants in the imperial endeavor, earning recognition for their labor at a time when women's work in the arenas of domestication (cultivation, governance, and control) and domestic spaces (national landscapes, gardens, household interiors and management) was thought to be critical to the processes of imperialism.[9]

While there has been work on the domestic aspects of these female emigration efforts, as Chilton has argued, detailed "transnational histories of British female migration still remain to be written."[10] Moreover, there has been limited investigation of the way in which these societies, their activities, their publicity materials, and the emigrant women themselves were influenced by—and helped to frame and shape—imaginative geographies and climatologies of colonial spaces, although scholarship has addressed the relationship between empire and propaganda.[11] These themes will be explored through the example of the British Women's Emigration Association (BWEA), founded in 1901, and its offshoot organizations, the South African Expansion Committee and the South African Colonization Committee.

Between 1901 and 1910, assisted through various emigration schemes, in excess of four thousand single women left Britain for South Africa, and many of them were settled in the Transvaal.[12] The collection of the correspondence, minutes, administrative papers, and published documents produced by and about the BWEA and now held in the Derbyshire Records Office (DRO) provides a unique insight into the association's role in promoting an imperialist ideology that constructed South Africa as a place that was at once heroic and healthful for Europeans; as somewhere ripe for—and indeed in need of—targeted "women's work," which offered opportunities for female emancipation but which also afforded opportunities for spiritual, moral, and bodily improvement and "climatic therapy." Collectively, these sources represent media through which to investigate the role of space and place in the production, circulation, reception, and application of knowledges about weather and climate around the end of the nineteenth and beginning of the twentieth century. They also allow us to explore how the climates of the overseas colonial fields were imagined and articulated and how these articulations were informed by and helped to inform discourses surrounding climate and corporeality. The "healthiness" of place was very much centered on the effects of climate, including positive effects, on the bodies of those women who put themselves forward for the emigration schemes. Two interconnected discourses underpinned and justified female emigration initiatives in the nineteenth and early twentieth century.

FEMALE EMIGRATION IN THE NINETEENTH CENTURY

As Kathrin Levitan has noted, William Rathbone Greg's article "Why are women Redundant?" published in the *National Review* set in motion a pervasive debate over what became referred to as the "surplus woman problem."[13] Basing his analysis on the 1851 census, Greg had noted that there were some five hundred thousand more women than men in Great Britain at the time, which raised concerns about those women who remained unmarried and, more specifically, about the decline of the family unit as the moral centerpiece of society at a time when large populations were seen as essential for the maintenance of British imperial and military supremacy.[14] Marriage was perceived to be the highest moral and social condition that a woman could attain, while the family remained the basic institution of society.[15] Single women, in contrast, were held to be "dangerously unproductive" and as such, Greg argued, placed a burden on the national economy. This situation was, as Levitan has highlighted, pitched as a distinctly middle-class problem because it was women of this particular demographic who faced challenges in supporting themselves, unlike women form poorer backgrounds who had always worked. As Levitan states, "the options available to middle class women who did not marry were limited."[16]

Around the same time, there were concerns about the recognized shortage of white European women in the colonies. Their presence and their labor was regarded as crucial to the survival of settler societies across Britain's dominions.[17] These coupled dilemmas fostered the possibility of emigration of "surplus" females from Britain in order to bolster the imperial endeavor, and attention began to focus on how surplus women "at home" could be made more "usefully" reproductive (in a moral and physical sense) overseas. While feminists advocated fundamental changes to cultural values that included more educational and career opportunities for women, possible solutions for those women who were labeled "surplus" were thought to lie in organized emigration to the colonies.[18]

The Colonial Office had sponsored specifically female emigration since the 1830s, and in the 1860s Maria Rye founded the Female Middle Class Emigration Society, as a predominantly philanthropic endeavor to assist impoverished gentlewomen in finding opportunities of "women's work" overseas. The Girls Friendly Society (GFS) was established in 1874 under the leadership of a group of middle- to upper-class women and intended to draw together predominantly working-class women, many of whom were domestic servants, "of unblemished moral character for the purposes of mutual help," support, and moral "improvement."[19] During the 1880s the GFS developed offshoots in the colonial spheres of Canada, New Zealand, Australia, and South Africa (establishing also an emigration department in 1885). Shortly afterward, the Primrose League was established by followers of the former prime minister Disraeli to promote Conservative Party principles and imperialism, taking as its emblem the primrose—Disraeli's favorite flower.

The United Englishwoman's Emigration Association was founded in February 1884, as a direct follow on organization from these earlier initiatives. Its aims were to emigrate women of good character, to ensure their safety during and after their travel, and to maintain contact with them for some time after their arrival (with Ellen Joyce, Adelaide Ross, and Grace Lefroy at the head of the organization).[20] By 1888 the group began to work in cooperation with the Scottish Girl's Friendly Association and the Scottish Young Women's Christian Association (YWCA), and the following year the United British Women's Emigration Association changed the original constitution, centralizing what had been a loose grouping of independent workers and outlining their responsibilities, roles, and relationships. The majority of emigrants who passed through this emigration association in the 1890s were, at this stage, destined for Canada, New Zealand, or Australia, but toward the end of the century the flow of emigrants to South Africa increased so much that it became necessary to set up a South African Expansion Scheme Committee. In 1901 the parent organization dropped the "united" element of its name, and the British Women's Emigration Association (BWEA) came into being, drawing many of its members from the association's previous incarnations, including Lady Louisa Knightley of Fawsley who was an important figure in GFS's imperial activities and a founding member of the Primrose League. She would become editor of the BWEA flagship monthly journal, the *Imperial Colonist*.[21] Another key founder was Edith Lyttleton Gell—wife of Philip Lyttleton Gell of Derbyshire, the director of the British South Africa Company (1899–1917, 1923–1925), its chairman (1917–1920), and president (1920–1923). Under this leadership the BWEA despatched seventeen thousand young women overseas in its first three decades of service as a means primarily of consolidating the British Empire.[22]

The question of women's emigration for those societies and organizations of the second half of the century was one of quality as well as quantity. Prevailing eugenic propaganda aiming mainly at the middle and upper classes focused on the need to propagate superior stock to populate the British Empire There was an expectation that women would act as agents of "British civilization" on the colonial frontier.[23] Indeed one of the primary roles of white women in settler societies was to promote an ideology of imperial motherhood. Social and biological reproduction of Britain overseas was a potent force. The BWEA placed emphasis on educated middle-class British women who sought work as shopkeepers, dressmakers, teachers for farm schools, elementary schools, and high schools, as governesses, and as "mother helps," nurses, typists, and accountants. Attempts to recruit young middle-class women to the ranks of domestic help were challenging but assisted in two ways. First through the notion of preparation and training for domestic work, and for life in the colonies generally; and second through subtle adjustments in terminology, recrafting positions in domestic service as "Companion Helps."[24]

The BWEA sent many of these women to Canada and New Zealand, but the emigration of single women to southern Africa around the turn of the twentieth century was regarded as especially important. There was something of an imperial pessimism in the wake of the Boer War tragedy. As Chilton notes, "the association of imperialism and female migration was heightened after the Boer War when it became obvious to many observers that South Africa could be won by war only to be lost through failure to assert cultural imperialism."[25]

The war had in fact already provided hundreds of British women with "unprecedented opportunities to participate in imperial conflict" and "had a powerful politicising effect on elite women in Britain and the white dominions."[26] It turned some women, particularly those with South African interests such as Edith Lyttleton Gell, into active imperialists and gave rise to networks of women's organizations devoted to imperial propaganda and patriotic charity. Moreover, there was support in the form of powerful figures with interests in South Africa who were prepared to invest in emigration schemes and exert political pressure to promote female emigration.[27] These included Field Marshall Viscount Wolseley, Susan Countess of Malmesbury, the Prince of Wales, and Lord Baden Powell, all of whom wrote letters of support for the BWEA, published in various issues of the *Imperial Colonist* in the early years of the BWEA's existence.[28] With their support, at the conclusion of the Boer War, the South African Colonisation Society (SACS) was founded as an offshoot of the BWEA, with Lady Knightley as chair, with a mission to encourage and facilitate the emigration of single British women to South Africa.[29]

FROM CLIMATE THREAT TO CLIMATE THERAPY

Such female emigration schemes operated in a context of significant change with respect to ideas about climate, health, and empire. Geographical discussions of climate matters in the nineteenth and early twentieth centuries focused on centering patterns of variation in levels of human civilization within a regional climatic framework. Ethnic constitution was riveted to regional climatic characteristics. Temperate climates apparently encouraged the development of "superior" races, and pernicious climates such as those of the tropics (or "Torrid Zone") were thought to propagate "inferior" stock.[30] These moral discourses were interwoven with theories of European acclimatization or the ability of European civilization to adapt to "alien" environments.[31] Certain aspects of the climate were thought to have a deleterious effect on the health of Europeans. The heat, the sun, and the supposed humidity then associated with the tropics were also considered to induce a "low sense of morality."[32] There were fears that prolonged residences in such environments would result not only in physical deterioration, particularly among women and children, but also moral degradation among Europeans.[33]

The climates of Africa were considered to be among the most threatening for human development and among the most hazardous for European constitutions.[34] Some parts of the continent were constructed particularly hazardous. As Curtin mentioned, West Africa, for example, was imaged and represented as the "White Man's Grave" not least because of the prevalence of malaria and yellow fever.[35]

Many early nineteenth-century British observations about Africa, however, suggested that the continent generally, as well as its inhabitants, would benefit inestimably from active intervention and European colonization. Anticipative pathological geographies of the first half of the nineteenth century had positioned African climates as among the most dangerous for human development and among the most hazardous for European constitutions. The intense heat in areas south of the Sahara was thought to preclude the existence of civilized and physically and morally healthy society.[36] The "pathological potency" of these regions rendered them particularly hazardous for European constitutions and, in as much, represented an obstacle to European imperial investment of any kind.[37] As Crozier has pointed out, a clear divide was established between African and European health practices, so that "diseases such as malaria were identified, through constant references to the natives, as classic African diseases, even though they had long been prevalent in the temperate world."[38] Women and children, who were considered to be weaker and more vulnerable than men, were thought to have much greater proclivity to deteriorating physical health, mental stress, and moral breakdown in such areas. As Bell has highlighted, medical authority "pointed to the higher mortality of European women and children than that of European men," the explanation lying in biological difference—women were physiologically the weak sex "and could be more easily upset by tropical heat."[39] The same was true of children, whose immature constitutions were thought to be vulnerable to physical and moral deterioration if they remained in the tropics beyond five years.

These perceptions informed and framed many of the imaginative geographies of European travelers and colonists headed for Africa. Yet throughout the second half of the nineteenth century, a growing appreciation among the public of the spatial variability in the climatic conditions and salubrity of different regions was informed by improving the knowledge and experience of living and working in different parts of the continent. While central Africa presented very real threats to the well-being of European settlers, as Jennings has illustrated, some parts of South Africa, and particularly the Cape coast, and upland areas more generally, because of the temperate qualities of all such locations, were thought to offer more temperate climate conditions and hence were imagined to be healthful, and even therapeutic, for Europeans. As Savage has highlighted, actual engagement with other parts of the world resulted in the identification of favorable and unfavorable, healthy and unhealthy locations comprehended on the basis of similarities or differences to the temperate world.[40]

Long-held preconceptions of the homogenous climate threat that Africa posed to European health were further eroded with major shifts in climatic-medical science. The dissemination of germ theory in the second half of the nineteenth century led scientists to question whether climate was the predominant explanatory variable in mortality and morbidity.[41] Susceptibility to disease became something that could be controlled through effective management of the environment and adaptations in personal behavior, through protective clothing, technology such as mosquito nets, medicinal preventatives such as quinine, abstinence from alcohol and an exercise regime.[42] Moreover, by the later nineteenth century, medical science had begun to recognize that emigration overseas could provide some relief from the chronic and degenerative diseases then being recognized, particularly among urban societies, in the northern temperate latitudes. Emerging fears of overcivilization and degeneracy consequent upon "diseases of decadence" in Britain enhanced the attraction of some overseas destinations, and a "change of air" to certain overseas destinations began to be seen as potentially therapeutic.[43] The association of some parts of Africa with pathological benefits including some coastal "resorts" that benefited from sea breezes and opportunities for bathing, and upland locations where temperate restorative climatic conditions could be found, coupled with the apparent need to spread civilization, both fostered and justified imperial investment and intervention in such areas.[44] The work of emigration societies was central to this endeavor and had a powerful role to play in making the case for female emigration to places that had hitherto been constructed as uniformly dangerous for European settlement.

Changing understandings of climate, disease, and the healthiness of place relating to South Africa were harnessed and represented in the administrative documentation and correspondence of the BWEA and its associated organizations. The organization's publicity material and particularly its popular magazine—the *Imperial Colonist*, published monthly between 1902 and 1924—both constructed and promoted a "new" imagined geography and climate of Africa. These materials acted as vehicles for mobilizing and popularizing South Africa as a both heroic and emancipatory place for single female emigrants and, importantly, as an attractive and therapeutic place climatically. Letters of application and correspondence from women wishing to emigrate under the auspices of the association show how conditions in South Africa were anticipated and how this helped to inform the screening of applicants for emigration. Though they are relatively scarce, notes and letters written by female emigrants who did relocate, and also correspondence from their employees in South Africa, give insight into actual female emigrants' experiences. This material reveals three themes—an emerging appreciation for the differential climates, a varied medical topography characterizing different parts of South Africa, and a prevailing narrative that continued to position climate as the main determinant of health, despite the emergence and acceptance of germ theory.

ATTRACTING AND RECRUITING "THE RIGHT SORT OF WOMAN"

Marketing strategies for South Africa around the turn of the century focused on a combination of both exciting and familiar images and representations in order to attract women as well as skilled men, and positioned South Africa as a place suitable for "overflow" female populations.[45] The *Imperial Colonist* carried features written by eminent colonial male officials as well as women emigrators, promoting a sense of responsibility but also adventure, inducing hundreds of single women to emigrate for a new life. Although the journal was run for and by women, stereotyped gender roles were very much reinforced within its content. As Bush has highlighted, ideas were "borrowed" from male platform speakers and "saturated with gendered language and imagery."[46] A message printed on the inside cover of the first issue of the *Imperial Colonist*, from George, Prince of Wales, reinforced the point: "I would appeal to my fellow countrymen at home to prove the strength of the attachment of the Mother-land to her children by sending out only the best."[47]

From the outset of the BWEA, the intention was "to emigrate only such women and girls as are of good character and capacity." It was made clear that "No single woman who has had an illegitimate child or whose conduct has been unsteady, intemperate or dishonest can be accepted."[48] The colonies, it was argued, needed "all round, sensible, active individuals who can turn their hands and heads to all kinds of practical work, therefore genuine wage-earning women alone need apply." Moreover, the BWEA suggested, "No women is suited for this life unless she has done household work at home, or has been through some sort of training in cooking, dressmaking, cutting out material and other household matters. If she is musical it will be a great gain as it helps to make the home life cheerful." It is clear then that the association encouraged "the right sort of women," those that could act as role models and who could adapt to a working life, and firmly discouraged "those women who were considered wholly unsuitable."[49] This is also expressed in a letter from Lord Milner to Lady Malmesbury and that appeared in the first issue of the *Imperial Colonist*: "I am as you know a great believer in increasing the British population of South Africa, which is miserably under-peopled. To do this we need not only men settlers but women settlers. . . . In helping respectable and competent women to come out you will be doing South Africa a great service, besides opening a fresh and brighter career to many who are blocked here."[50]

There were issues associated with recruitment and the management of applications from the "wrong sort of woman." There were clearly some applicants who also seem to have played the system, securing passage out and then simply disappearing. Louisa Brown, for example, seems to have taken advantage of the scheme by accepting a place, emigrating but not participating in the BWEA initiatives in any way. The case was discussed at length, but a note to Edith Lyttleton Gell, dated April 27, 1904, from Louisa Rhodes, noted, "I think Louisa Brown has behaved very badly, but unless 'as an example' I do not see we shall gain much by proceeding against her now. But if we ought to take proceedings to deter others from making use of our association for their own ends I am willing to agree to whatever the committee decides should be done."[51] Such moral challenges of this genre were presumably germane to many emigration schemes.

Yet "suitable women" were also those considered to be capable of bearing the apparent strains of living, and of course working, in an African climate. The BWEA placed emphasis on the health of their applicants, requiring responses to a set of probing health questions, relating to physical (and moral) health, disease, vaccination histories, and proclivity to common ailments. "Satisfactory certificates as to health" were required from every woman with ambitions to settle in South Africa.[52] Notes from a SACS agreement with the Scottish branch, for example, indicate that "only single women of unblemished moral character . . . , and who can show a doctor's certificate for good health and freedom from infirmity and who are capable in their trade and occupation, are eligible for South Africa."[53] As Blakeley has also highlighted, girls also had to provide a medical history of their entire family to ensure that "every danger of hereditary or incipient disease" was known to the medic assessing the cases.[54]

Applicants who were considered to have potential for teaching work had to undergo medical examinations to provide further medical proof that they were capable of adjusting to life in an African climate. In the case of the women going under the government scheme to the Transvaal, for example, it was agreed that a special certificate "has to be obtained from a doctor appointed by the Colonial office." One of the functions of the education subcommittee of the South African Expansion Committee, which was founded to "introduce teachers to meet the growing needs of the South African Colonies," was not only to "maintain the very highest standards, as regards professional efficiency, character," but also "physique in the teachers recommended for South Africa."[55] In fact, women applying for all manner of roles in South Africa were excluded immediately if they were considered by the SACS committee to be "morally and physically unsuitable."[56]

The detail on candidates' applications and medical certificates, including those rejected, reveal long-held preconceptions of female ability to acclimatize in a South African climate. For example, Miss Ella Heanley, who applied to work in Rhodesia in the spring of 1903, was found to have lung problems. A note dated May 28, 1903, from the acting secretary of Rhodesia Railways Ltd to Edith Lytteton Gell suggests, "As regards Miss Ella Heanley, if the pulmonary trouble . . . is of a character likely to be cured by the dry air of Rhodesia, the Directors would I feel sure be wishful that the free pass granted to her should remain in force, but if not I fear it must be cancelled and she would not in such cases constitute a desirable colonist." A note dated May 25, 1903, from the Colonial Office medical assessor Dr. Neville Wood, however, highlights additional concerns: "I think she should serve as a serious warning to all of us who hold responsible positions. . . . Her appearance is robust, but on examination I found a spot of lung disease and I further found in her expectoration abundant consumption germs, so that she might be a source of infection on board ship and in the hostel."[57] For those who, unlike poor Miss Heanley, did pass the stringent medical hurdles, the next stage was to take advice about what these robust and predominantly young women should prepare for in terms of travel to—and ultimately life in—South Africa.

IMAGINING AND REPRESENTING THE PATHOLOGICAL GEOGRAPHY OF SOUTH AFRICA

Given the long-held views about the pathological dangers facing Europeans in Africa, it is not surprising that advice for those intending to travel to such areas was myriad and plentiful. There were many medical guides and handbooks, and,

Crozier has argued, "nearly every health account offers guidance on what to wear, what to eat, how to act and how to stay well."[58] Advice on how the European might avoid constitutional impairment also considered "the virtues of bathing, the benefits and evils of alcohol."[59] The BWEA and SACS also offered such counsel and were clear in their goal to only send out women "under carefully guarded conditions" and with the benefit of sound knowledge on climate and maintenance of health during travel and for life and work once settled in a new destination.[60]

Emigrants were encouraged to attend residences at BWEA hostels at embarkation (Liverpool, London) and disembarkation points (Bullawayo, Salisbury) where BWEA matrons and lady superintendants would welcome and assist the emigrants with the logistics of departure and arrival and to prepare the women for a life overseas. Advice was also offered to all new "leavers" through other mechanisms. The Offices of the South African Expansion committee, for example, in 1903 published a pamphlet entitled "Advice to women and girls settling in South Africa," priced one penny and authored by Mrs Adelaide Ross. This offered "a few simple statements about climate, health, dress and employment" designed to help women judge for themselves "what life in South Africa means, how travelling can be managed and what the gain of going there will be in the end, and what the discomforts you might possibly encounter." While those likely to purchase Ross's pamphlet were those generally declared fit for travel, Ross advised on a suite of "preparations," to avoid climate related health impairment, from "showing your teeth to a dentist" as "change of temperature and of food often causes toothache and in the new country you may be far from anyone to relive the suffering," through to the specifics of day-to-day clothing and what to pack.[61]

As has been discussed by Johnson, to combat the apparently "deadly but desirable landscapes, the British built a considerable commodity culture around the preservation of white European health," and clothing was, in fact, "one of the most important and essential items in their kits."[62] Ross suggested that women "should travel in a stiff skirt, taking either a bodice to match or a warm blouse and some cool ones. A tight fitting jacket and a hat are best and also a shady straw hat. A wrap for the neck and shawl or rug for the knees and a small hand bag are needed," while "it is never wise to be without warm coverings." Once a woman reached her destination, it was argued, "it is also much better to wear flannel or knitted vests next the skin all year round, for wool not only keeps out the cold in winter, but absorbs the excessive action of the skin in hot weather and will make it safe for you to wear the cool loose washing blouses which are so pleasant when it is sultry. You must prepare not only for the heat but for the cold, and in winter nights you will find flannelette night dresses useful." A list detailing suitable "outfit" that women should take was supplemented with the comment, "as you read the description of different parts of the country you will see why such and such articles are especially recommended."[63]

As this last point reveals, by the time female emigration was being encouraged, there was a growing awareness of the climatic heterogeneity of South Africa. There was also an associated appreciation of the variable of healthiness of place in different parts of the colony, subverting a previously widespread geographical imagination of uniformity in the African climate. "If you look at a map of South Africa," Ross argued, "you will see that the British possessions occupy the southern and south eastern portion. In that large area, there is a great variety of climate." The Cape, in particular, had long been associated with more temperate conditions and more health and recuperation for traveling Europeans. As the nineteenth century progressed, and as the geography of the interior began to be better understood by Europeans, the perceived "locus of healthiness" shifted inland and to other parts of South Africa. Thus Ross considered "the climate of the Orange River Colony" to be "very dry and healthy," and it was suggested that "People go there who are threatened with lung disease and though they would have died had they remained in England, they live there in health for many years." Precautions were recommended "as the nights are cool even in summer and *very cold* in winter." Nevertheless, this provided a "peculiarly exhilarating effect so that daily life becomes a pleasure."[64]

Betraying an interest in social engineering and countering prevailing concerns that had centered on the degenerating effects of African climates on children's development, Ross added that, in the Orange River Colony, "children grow up robust and strong and of a goodly stature." In the Transvaal, in the northeast of South Africa, she suggested that "the land lies so many feet above the sea that the nights, though refreshingly cool in the summer, are bitterly cold in winter. Do not therefore be misled by the accounts of the burning sunshine in day-time, as if that represented the whole temperature." Drawing on personal links with colleagues already based in other parts of South Africa, Ross also added, as extra evidence, how contrary to what was often anticipated "in Africa," "a friend of mine, an English lady, who lived many years in Johannesburg, told me she was delighted to have three blankets on her bed and to wear a warm cloak in the early hours of the day-time. She found her blue serge dresses invaluable."[65]

The pages of the *Imperial Colonist* reveal that influential BWEA supporters, powerful military and political figures, and philanthropists also shared their representations and accounts of regions, towns, and cities that were suitable for the woman emigrant seeking work. In an article on "Natal before and after the War" by Captain MFM Meikeljohn VC, for example, Durban was described as "delicately built, seems to be a city of perpetual sunshine—sunshine which is never oppressive, even in summer for the sea breezes temper it. In its midst, symbolical of the blessings a noble reign bestows on travel and colonisation, rests 'clear cut in perfect air' a statue of the great White Queen." This place of "perfect air," temperate sea breezes, and perpetual (but crucially, nonoppressive) sunshine was also held to be "a country of immense fertility . . . the imagination is quickened as much by the look of quiet contentment on the faces of those who cultivate the great sugar fields." In continuing, Meikeljohn described a scene of transformation, a place of health, wealth, and production, but, importantly, he argued that it could only be rendered so through colonial intervention:

> What was the surface of the land before it was resolved into the orderly fields and cultivated terraces that stretch out on either side the railway? Take a walk to the sea from the well-ordered house of one of the great sugar planters of Natal and one passes through a two mile belt of wonderful vegetation, where plants, ferns and flowers, beautiful and rare catch the eye at every turn. Chattering monkeys swing on the branches that overhang the murmuring brooks; squirrels scold from the tree trunks; and here and there a black bursts from the covert. Through the bush and down across the sugar fields, and behold, we are on the west coast of Ireland. The never ending surge and thunder of the ocean are prophetic and then suddenly blue waves with fleecy tops are breaking on black and glistening rocks, to lose themselves on a beach of dazzling whiteness.

Even further inland, and "up through hill and dale and swelling valley where the grass seldom loses its fresh green," conditions were no less welcoming. Even the oppressive heat of Pietermaritzburg could be relieved by a trip for climate therapy—"the hills about Mooi River are not far off," where "rosy-cheeked children bear testimony to what the fresh breezes can do for them."[66] Similarly, in an article entitled "Openings for Women in South Africa," published in the *Imperial Colonist* in July 1904, it was noted that "it must always be remembered that the climate of Griqualand West [in central South Africa] is particularly beneficial to anyone suffering from consumption, especially in its earlier stages." Cape Town, in contrast to earlier reports, was not considered to offer "a suitable climate for consumptive persons, as the damp in winter and dust in summer are injurious to delicate lungs and throats."[67] Somewhat dismissively, the risks, however, were apparently quite limited as other articles reveal: "Five hundred women sent by SAX have passed through the corresponding committee in Cape Town. Of these only five have proved unsatisfactory; one died, another fell ill, and three left their situations. Of these latter, two returned and did well, the third is living with her brother in Durban."[68]

It was not unusual to encounter similarly positive accounts of other locations across South Africa in the pages of the *Imperial Colonist*, though many are written in such a way as to highlight the transformative potential of a British—and particularly British female—presence. Rhodesia, for example, was referred to as "an Infant Prodigy" by the Honorable Mrs. Lyttleton Gell, in a 1904 issue of the journal, on the basis of the fact that it was the "latest babe of the mother country, youngest member of the British establishment." Here could be found environmental paradoxes galore. This was a place "vast in extent, though comparatively thinly populated; at once savage and sophisticated, fertile and arid, backward and progressive; like a precocious infant, she surprises at every turn."[69] The ideology of imperial motherhood infuses such accounts. As Sandy Krogulski notes, imperial discourses like this "drew heavily on familial language that upheld Britain as the Mother Country to her colonial children."[70] Here was a place not yet mature, volatile but capable of being shaped and nurtured through a civilizing maternal hand.

The climates of South Africa were generally considered to be far more "healthy" for Europeans than those prevailing in other British interests. India was often considered to be among the most pathologically dangerous for Europeans and was held as a negative yardstick against which the comparative overall healthiness of South Africa was highlighted. Ross notes, for example, how "it is the coolness of the night temperature that invigorates the body, and it is for this reason that Africa though so hot by day is healthy and entirely different from India and its enervating temperature."[71] Yet through such comparison, and notwithstanding the clear differentiation that much of the BWEA reports sought to provide on the climate and healthiness of South Africa, one might argue that there appears to have been something of a rehomogenization of the African climate in support of the BWEA's and SACS overall mission.

EMIGRANT WOMEN'S EXPERIENCES

Livingstone has highlighted how the imagined geographies of imperial spaces were frequently at odds with actual experience.[72] In many cases, letters and articles published in the *Imperial Colonist* would indicate that experience exceeded expectation. This is not surprising given the role the journal was intended to play in enticing women to emigrate. Excerpts of emigrant letters were almost certainly selected and accounts tailored to be as positive as possible. Most accounts of the journeys taken by emigrants highlight the beneficial and restorative effects of travel, common to many travel writings of the eighteenth and nineteenth centuries.[73] Writing from the SS *Ingeli* en route to Natal on May 13, 1902, for example, Gertrude Golding wrote back to the BWEA describing how "we had a very smooth passage in the Bay—not at all rough. . . . We are enjoy-ing the voyage immensely—the weather is perfect, the sea is so blue—a lovely blue—the evenings are so warm—it's so strange and interesting altogether. I feel so grateful to the B.W E. Association for sending me out—and shall do my utmost to please in every way, when I get settled."[74] Similarly, a letter published as correspondence from "a refugee passenger" and simply signed E.M.B, notes: "we had a most enjoyable passage over and all feel much benefited by the sea trip. Everything was so pleasant on board that we were quite sorry to reach our destination."[75] The act of travel was, in this case, considered to be therapeutic and restorative.

In common with the accounts designed to attract female emigrants, the let-ters written by British women based in South Africa that made it through to the pages of various issues of the *Imperial Colonist* extolled its climatic and healthful virtues. One letter from a Mrs Shepstone, for example, speaks quite generally of the South African environment, noting how emigrant women "will gain much great freedom of life—in many cases probably greatly improved health, a fine climate, an easier time generally, and certainly in most cases."[76] Other letters offer comparisons with England. Having found a base and employment in Durban, for example, one correspondent highlights how she was "most agreeably surprised with Africa as far as I have seen it, and one scarcely imagines that Durban is a foreign town it is so English and like home."[77] News from emigrants to Basutoland (now Lesotho) was no less favorable. A letter from a Sister Henrietta published in the *Imperial Colonist* in July 1904 notes that "I am in Basutoland where everyone is clamouring for mother's helps, governesses, family friends." She described "a kind of rough plenty, not the skimping that people with small incomes in England have to practice" and added that in Basutoland could be found "a glorious climate, not hot and dusty like Kimberley, but bright and clear and cool, with glorious sunshine, rocks and hills and great pastures." Another emigrant, Florence Wall, writing to Edith Lyttleton Gell at the BWEA from Bulawayo on February 13, 1903, highlighted how "The sun here is so lovely, although it is sometimes terribly hot."[78]

Betraying a stereotyped understanding of the apparently unhealthy climates associated with parts of Africa, some women emigrants expected to fall ill having been exposed to the elements and were in fact pleasantly surprised when they didn't. One Irene Shene, for example, who had been posted to Luona in northeast Rhodesia in the 1905, noted in a letter to Gell dated April 19, 1905 that "I have had no fever whatever either on the way out or since I have been here." Irene made every effort to re-create England abroad. "Ever since I have been here," she continues, "I have been busy making the house look pretty. I brought out plenty of paint, cretons and muslins etc. and have changed it really into a pretty English house. My husband said the other day when he came into the drawing room 'To look at this room you would never think we were in the wilds of Africa.'"[79]

It is clear from the correspondence between the BWEA, SACS, and female emigrants, however, that the experiences of women who applied for and were successful in being recruited were not always as positive as indicated in the published letters. For some women emigrants, the journey to South Africa represented the first stumbling block. A letter written by one emigrant, Florence Dawson, to Miss Tarleton at the BWEA dated November 27, 1902 highlighted several inconsistencies between expectation and the reality of life and work in South Africa. Florence was writing from the Hostel at Rosebank, Cape Town, and was supposed to be traveling onto Bulawayo. She noted, however, how she had to instead stay at Cape Town because "the insanitary conditions and bad food on the Delphic (which proved to be only an emigrant ship) have left me so out of health that a few days rest is a necessity. A journey under such conditions would have been a strain had one been in the best of health. Coming after a month of constant hard work and excitement, the strain is exceptional." She makes it clear also, however, that her experiences were counter to what she had been led to expect. "Besides this," she continues, "so much that was told me by Mrs Gell and other members of the Society that sounded very satisfactory has since proved to be very inaccurate. . . . I feel that it would not be wise to risk a long journey until my mind is set . . . on certain points."[80] She was clearly not alone in her disillusion. Securing suitable work and appropriate places to stay, with time-restricted residencies in BWEA hostels in South Africa, and securing alternative billeting posed very real obstacles for many new emigrant women, forcing some back home after very limited time actually in Africa. Detail from unpublished correspondence between a Mrs. Smith—an employer in South Africa—and Lyttleton Gell, regarding the state of health of one emigrant, Florence Brown, reveals that the assessment of the relative robustness of emigrants for African living was sometimes questioned. Smith, for example, notes that Florence "seems so anxious to please me in every way [but] . . . she does not appear, however, to be very strong." Lyttletton Gell's response is defensive: "As regards her health, all our travellers have to pass a very severe medical examination made by a medical man who is appointed by the colonial office to examine their candidates. He thoroughly understands the S. African climate, and is most particular in rejecting anyone not suitable for it, so I hope you will find it was merely the fatigue of the journey and the new conditions."[81]

It is clear that failing health was more of a problem for other women who took up posts for a new life overseas, some it seems having already suffered ill health in other previous postings. A letter of May 24, 1904, written to the BWEA from one of the BWEA assessors, Emily Fox, details the illness of a Miss Dopping Hepenstall who appears to have suffered a heart attack while in Rhodesia. According to Fox: "It was after fever in China she first had a heart attack, and since she returned she has seemed quite strong, could bicycle last summer in Ireland thirty miles, without fatigue and as you know she worked very hard at Clapham [training college] without being knocked up. It was because of her having fever in China that we begged her not to go back there and I fully believe if it had not been for this most unfortunate attack of fever her heart would not have been bad at all and she would have got on quite well."[82]

More generally, it seems, there was a level of disillusionment among some emigrant women. One particular example is a poem, written by Margaret Grogan based in Johannesburg, entitled "The Emigrants Return." This fairly vitriolic attack on the emigration project as a whole appears to have been intended as a submission for a journal referred to as the "Xmas Owl." Selected verses reveal the way in which the lifestyle and also the climate and environment she encountered stood in stark contrast to the expectations set by the emigration societies:

"The Emigrants Return" (for the Xmas Owl)

You said you told me everything
You thought there was to know
When we parted on the platform
Now just eleven months ago

My head was very weighty
With the things I thought I knew
When I shook your hand in friendship
From the train at Waterloo

I shudder now and tremble
Everytime that I compare
All the pictures that you painted
And the daubs they really were

You showed me but in fancy
Such a calm and peaceful shore
With the palm trees in the background
And the hammocks in the fore

You told me that the climate
Left no more to be desired

One never felt too hot or cold
And never overtired

But you quite forgot to mention
Such a thing as violent rains
And the Monday heat which almost
Rivals India on the plains

And you never even told me
There was such a thing as dust
That of all the many subject
There was only one discussed

I am sure you mentioned nothing
Of a never ceasing wind
You did your duty nobly
But it really was not kind

The author's verse continues to explain how she was assured she would find wealth and fortune through emigration, though actual experience proved otherwise and also highlights the difficulties of finding suitable accommodation. Her closing verses reveal a growing disillusionment that finally fueled her resolve to return home, out of pocket and disappointed.

My courage failed daily
I was more and more at sea
Till I suddenly discovered
It was not the land for me

So I broke the resolutions
I had only lately made
And the breakage brought my temper
Down to normal in the shade

So I bade farewell to Afric,
And the charms I should not miss
To imaginary palm trees
And imaginary bliss

When you bade me make that journey
Was it you or I who scored?
Take the remnants of my temper
As your plentiful reward

Margaret Grogan, Johannesburg[83]

CONCLUSION

The BWEA and its ancillary organizations, including the SACS, constructed an overwhelmingly positive image of South Africa, but it was in their interests to do so. The imaginary was underpinned by the need to secure a British presence and race in overseas colonial territories and was supported by concerns over so-called redundant gentlewomen at home. There was an apparently "inexhaustible supply" of high-character candidates available "at home," and it seemed that South Africa was a perfect destination for women wanting to contribute to empire through "women's work." British women were a "repository of British cultural values upon which the racial future depended."[84] The emigration project afforded a means of promoting an ideology of imperial motherhood. Moreover, the imperial dominions were "less formal than British society" and were constructed and then imagined "as spaces of social and physical freedom, with the very attractive combination of an exciting but familiar environment, but also as places where [female emigrants] could contribute a civilising influence."[85]

This construction of South Africa is most obviously evident in the pages of the BWEA's flagship journal, the *Imperial Colonist*, but can also be seen in some of the more public correspondence between the BWEA committees and the scheme's often influential supporters. Through these discourses, colonies were positioned and imagined as heroic places, in the "field," where women could actively engage as partners in imperial enterprise so often dominated by males. Emigration was presented as an opportunity, and those opportunities in turn afforded mechanisms for the cultural assimilation of single British women emigrants into this corner of empire at a time when there was growing competition with other imperial powers. The women's presence ensured a racial solidarity and a means of re-creating Britain overseas.[86] Southern Africa was constructed not only as a place of hope and opportunity but also as a place of salubrity, where there were opportunities for moral, spiritual, and physical well-being and improvement. Organizations like the BWEA and its offshoots appropriated and made use of climate as a "tool" of empire—a form of propaganda to support the female emigration project. The timing of the BWEA initiatives to South Africa was, however, important in this respect, coming at a point of transition in climatic understanding. For much of the nineteenth century Africa, with the exception of specific locations such as the Cape Coast, was presented in colonial discourses as being uniformly hazardous for European settlement, on the basis of an imaginative geography of homogenously hostile climatic conditions. By the turn of the twentieth century, this homogeneity was being dismantled and new imaginative geographies started informing discussions and debates among societies and organizations having responsibility for promoting Britain and Britons overseas. Moreover, germ theories had begun to challenge long-held climatic understandings of health and disease. The discourses highlighted how prevailing climatic understanding was harnessed at a range of spatial scales, from the microclimate immediately affecting the emigrants' bodies through to the claims about the climatic "character" of southern Africa more generally. The publicity material of the organizations trace the changing imaginations and integrated representations of climate, disease, and the healthiness of place, and reveal how scientific knowledge, cultural politics, bodily experience, and spatial imaginaries were combined to construct an image of South Africa as climatically healthful and yet ripe for the emigree's intervention. A more informed understanding of the spatial variability of African climates certainly seems to have shaped the accounts circulated through the propaganda machinery of the BWEA. Accounts from the turn of the century, however, reveal that long-held theories linking climate, health, and disease proved more persistent, representing widely circulated and enduring frames of reference, informing popular knowledge and also expectation. Climate and health remained intertwined in the imperial discourses the BWEA drew on and compiled. Indeed, concerns over the generic health implications of the African climate framed recruitment and selection procedures, though there was a clear awareness of the differential healthiness of different locations. South Africa in general was represented as overwhelmingly healthful for British women emigrants, and recognition of the heterogeneity of the climates across the region as a whole allowed the BWEA to promote some destinations as places of climate healthfulness, with climate being presented not so much as threat but, rather, as therapy.

Relative to the availability of publicity material and propaganda extolling the many benefits and indeed virtues of emigration for single females, the correspondence from emigrants themselves is somewhat more limited. As Van Helten and Williams have argued, it may be that long-term contact was not that well maintained between the association and emigrant women.[87] It may also be the case that women, once in their new destination, failed to keep to their "predestined path." There were certainly examples of women who ended up "lost" to the organizations. The degree to which women emigrants' actual experiences in South Africa met the horizons of expectation engineered by organizations such as the BWEA therefore remains an open question and remains a theme worthy of further investigation. It seems likely that the imagined climatic experiences, the opportunity for climate therapy and well-being, which represented key "tools" of empire, may have been at odds with the reality of on-the-ground experience.

7

RACE AND RAINMAKING IN TWENTIETH-CENTURY SOUTHERN AFRICA

Meredith McKittrick

AS CAPE TOWN GRAPPLES WITH THE SPECTER of running out of water, history reminds us that this fear has long been with people living in southern Africa's semiarid environments. Diverse African communities engaged in rain-making activities for centuries, if not millennia. From the seventeenth century onward, Dutch and, later, British settlers prayed for rain but also speculated about how their land-use practices affected rainfall. In the early twentieth century, some white southern Africans were drawn to the idea that human intervention in the water cycle might increase the amount of rain that fell or make its arrival more predictable. The viability and desirability of creating "artificial" rain were debated in white communities across southern Africa, from the arid margins of the Namib Desert in the west to the "Highveld" around Johannesburg to the east.

Scholars have increasingly been drawn to the historical development of ideas about both the drivers of climate and the potential for humans to change it.[1] A robust literature on imperial knowledge-production, with its focus on how scientific ideas circulated within colonial networks—and between former and current British colonies—has informed the field.[2] But although climate historians have also thought about the transnational circulation of knowledge, they have reached beyond the experts to explore realms of popular knowledge—as with the weather "makers and fakers" that are the subject of James Rodger

Fleming's history of climate engineering.[3] European-descended farmers and citizen-scientists had their own ocean-spanning conversations, reflecting what Saul Dubow has called "a spirit of catholic scientific interest" in the colonies.[4] In the recent volume *Climate, Science and Colonization*, Stephen Legg argues that debates within the popular press over the role of forests in climate "annihilated distance" by presenting global ideas to the Australian public. Kirsty Douglass demonstrates the parallel development of settler ideas about climate in Australia and the American West—a theme also taken up in the same volume by James Beattie, who notes that rainmaking experimenters in early twentieth-century New Zealand drew inspiration from the western United States.[5] American influence spread to many places: Charles Hatfield, the self-proclaimed California "moisture accelerator," was well known to white South Africans, who repeatedly investigated the possibility of hiring him.

The factors that shaped popular climate knowledge in the nineteenth and early twentieth centuries were diverse. Settlers were influenced by ideas they brought with them from Europe. Local environmental factors also affected the reception of scientific knowledge. The temporary increase in rainfall on the US Great Plains in the 1850s and 1860s helped shore up the idea that "rain follows the plow."[6] Geomorphological features of semiarid environments—signs of water-based erosion and large dry riverbeds and lake beds—reinforced settler beliefs that these places had experienced a prior pluvial golden age.[7] And people drew on their own experiences in accepting or rejecting expert knowledge. In South Africa, government scientists' insistence that rainfall was not declining in the early twentieth century was widely rejected by farmers, who insisted that droughts had become more frequent, that water tables were plummeting, and that their own rainfall records contradicted official knowledge.[8]

Strikingly, most of the new histories of imperial climate knowledge deal with settler colonies—Australia, New Zealand, North America, and North Africa—where the presence of white settlers complicated the production and circulation of imperial knowledge. Aside from creating the conditions for the circulation of climate knowledge among imperial spaces, how did the fact of white colonization—and the dynamics of settler-indigenous relations—matter?

Meteorological knowledge claims and geographical imaginations changed markedly as they moved to settler colonies beyond metropolitan centers in the global North. A particularly striking example of these claims and imaginations can be found in climate engineering schemes during the first half of the twentieth century. While they were often popular in semiarid and arid environments, these schemes were also notably popular in a distinctive sociopolitical context: settler colonial societies. Notions of race and European superiority were central to these colonizing projects, and they also shaped the development of meteorological knowledge and associated climate engineering schemes.

Southern Africa was one such context and provides a compelling case study of why place matters in histories of weather and climate. Early twentieth-century debates over weather modification cannot be separated from questions of race and power. In this context, "artificial rainmaking" became a means not just to fight drought but also to assert a white identity grounded in modernity, scientific rationality, and mastery of technology. The distinction between magic and science was important to people who strongly identified with the latter. In his 1926 study of attempts to modify climate, William Jackson Humphreys of the US weather bureau stressed the difference between supernatural and scientific rainmaking.[9] The racial politics of early twentieth-century southern African settler societies increased the importance of this distinction and energized attempts to enforce it. In a place where "magical" rainmaking rooted in African cosmologies continued to flourish alongside supposedly "scientific" rainmaking, attempts to control the climate were simultaneously about seeking security from drought and asserting white dominance over a black majority.

The border between rainmaking rooted in "superstition" and rainmaking rooted in "science" proved to be as unstable as other supposedly clear categories of racial difference. By the twentieth century, both white and black South Africans prayed for rain in their churches, and the prime minister declared days of "prayer and humiliation" during droughts and days of thanksgiving during abundant rains. Meanwhile, many white—and uniformly male—citizen-scientists claimed to have discovered methods for making artificial rain. The existence of African rainmaking, combined with the dubious science employed by famous rainmakers such as Charles Hatfield and the less famous white South Africans, threatened to erode distinctions between Africans and white settlers. In the first decades of the twentieth century, government scientists did what they could to reinforce those distinctions, by inviting into their conversations white men who proposed artificial rainmaking techniques grounded in science, no matter how apparently outlandish their ideas. Missionaries might lecture Africans about the falsity of their rainmaking techniques, but government officials politely thanked white men who proposed everything from dragging metal mesh behind airplanes to spraying saltwater in the desert.[10]

Two letters in the popular agricultural publication *Farmer's Weekly* from the years just after the creation of the Union of South Africa directly address the importance of white desires in shaping government policy. A correspondent writing under the name "Unity" wrote in 1912: "The Government is ourselves, the members being elected by us, to act for us, and paid by us to further our (the public) interest. They are our servants, and must do as we wish. . . . Be white men. There can only be two classes, white and black, and white must dominate and cooperate and work for one end, and that is the public benefit."[11] The next year, a farmer living on "Progress Farm" complained about an unrelated government policy and asked: "When will the Government realize that it is no earthly use to pass laws which are not in accordance with the wishes of the majority of the people who are going to be affected thereby?"[12] The unspoken rule—that the ideas and desires of all white men deserved to be treated with respect and taken seriously—shaped politics and public opinions about science in tandem.[13]

Rainmaking produced distinctive conversations, knowledge claims, and imaginaries in different spaces. In the context of South Africa, concerns about sustaining white dominance and about reinforcing unity between Afrikaners and British South Africans shaped rainmaking and its reception. Charles Stewart, the government meteorologist from 1912 to 1931, justified his involvement in rainmaking experiments on the basis of white popular opinion: "As this idea of causing rain by such means seems to have taken hold of the imagination of quite a large number of farmers, it seems desirable that we should put ourselves in a position to give a definite answer to its supporters."[14]

White rainmaking reveals the dreams of white settlers and the hardening of racial hierarchies in South Africa. Popular enthusiasm for rainmaking was rooted in an imagined future in which white dominance of the rural areas would be secured. But it also helped produce those hierarchies by opening up a treacherous but potentially productive terrain for government officials seeking to consolidate a unified white identity in the decades after Union. White rainmaking could help reinforce a rationalistic, modern white identity: it was "artificial" and distinct from nature, relying on airplanes, electricity, and chemicals, and it purported to harness natural processes that had been demonstrated in laboratories. But white rainmaking carried within it a danger as well: "artificial" rainmaking experiments had to be subjected to scientific methods and standards of evidence. It turned out to be very difficult to prove the efficacy of "scientific" rainmaking, which meant that attempts to create artificial rain always carried within them some of the same hints of magic and deceit that whites attributed to African rainmaking.

EARLY RAINMAKING

Southern Africa's twentieth-century rainmakers inhabited a land where humans had sought to coax water from reluctant skies for millennia. The western half of the subcontinent receives less than twenty inches of rainfall a year.[15] But even in better-watered areas whose climate is shaped by the Indian Ocean, rainfall can be highly variable, and rainmaking has long been practiced in communities across the region. Rainmaking ceremonies are a prominent theme in rock art attributed to hunter-gatherer communities who were the region's first human inhabitants. The agropastoral societies that stretched from the Indian Ocean coast to Angola and Zambia also engaged in ritual activity to bring rain. While household heads and clan leaders might perform rainmaking ceremonies, in many societies rain was ultimately the responsibility of rulers whose legitimacy rested on their ability to ensure the health and fertility of land and people. Some royal lineages claimed to possess secret rainmaking medicines, while others used the services of specialized ritualists. In times of severe drought, when local ceremonies did not yield results, famous rainmakers might be summoned across long distances. In the Eastern Cape, which became a center of white rainmaking in the twentieth century, Xhosa rulers routinely called rainmakers from afar when the rains failed.[16]

Indigenous rainmakers used a variety of techniques and materials in their work. They frequently sought to mimic the physical processes and sensory experiences of precipitation—what has been called "homeopathic" rainmaking.[17] Ritualists re-created "clouds" by burning green branches; they sacrificed black animals, whose color evoked storm clouds; and they sprinkled water or human sweat in sacred places to reproduce the phenomenon of raindrops. Climate and politics were intertwined. Dead chiefs were associated with rain (and their graves often served as rain shrines), but so too were autochthonous populations supposedly conquered by those chiefs.[18] Certain kinds of animals were associated with rain, especially birds and some types of antelope. Some of these associations are linked to hunter-gatherer cosmologies that are thousands of years old, while others seem to have emerged later, in agricultural communities.[19]

The efficacy of particular rainmaking agents or activities was undoubtedly debated, particularly when human efforts failed to break a drought. And oral histories refer to the discovery and incorporation of new rainmaking techniques. But what Todd Sanders has referred to as the a priori "world-as-it-is"—"the frame of knowledge and its conditions of possibility"—reflected a shared belief that humans could influence rainfall through their actions, and that some had more power over the rain than others.[20]

Missionaries were among the first outsiders to observe and write about southern African ceremonies to call rain. They understood immediately that the practice rested on religious foundations—what they termed "superstition." They derided rainmakers as charlatans who preyed on the desperate. In the 1840s the missionary Robert Moffat described a rainmaker in a Tswana community as "the demon of mendacity" who "kept the chiefs and nobles gazing on him with silent amazement."[21] Moffat claimed that the rainmaker admitted privately that he only "pretended" to have power over the rain because he was poor and needed the payment proffered, while another confided that only wise men could be rainmakers because it required great wisdom to deceive the public.[22] The Wesleyan missionary William Shaw put "rainmaker" in quotes and called one who was summoned to a Xhosa community an impostor.[23]

Missionary accounts are replete with reconstructed dialogues between themselves and rainmakers.[24] Many of these accounts end with a scene in which the missionary supposedly persuades an assembled audience that the rainmaker is a fraud. Shaw recounts one such encounter, in which he used biblical passages as evidence that only God has control over the rain. The rainmaker replied, "The Book speaks truly. I also say that God gives the rain." Shaw claims that the assembled chiefs chuckled at this "admission," but in fact this was the fundamental belief system of all communities that engaged in rain ceremonies and partly explains why rainmakers were not routinely blamed when drought continued, despite missionaries' attempts to highlight their failures. As one man told Moffat when the efforts of a powerful foreign rainmaker did not produce results, "there is a cause for the hard-heartedness of the clouds if the rain-maker could only find it out."[25]

The rainmaker that Shaw met faced a similar dilemma, as the drought raged on despite his efforts: "There is something that turns or hinders the rain," he told Shaw, who repeatedly asked what that "something" was. The rainmaker finally responded, "You are the hindrance. . . . When the clouds come up from the sea, and spread over all the land, and the rain is ready to fall, *that thing* which you have brought into the country and set up on a pole on the hill . . . goes *tinkle—tinkle—tinkle*; and immediately the clouds begin to scatter, they disappear, and no rain can fall."[26]

In this morality tale, told from the perspective of the missionary, Shaw boldly and publicly accuses the rainmaker of lying, a special day of fasting and prayer is held at the mission, and it pours rain for three days.[27] But missionaries were not always so lucky. Across southern Africa, in societies from the Namibian-Angolan border to the Indian Ocean coast, missionaries—and European hunters and traders as well—were accused of driving away the rain not just by ringing church bells but also through their prayers and singing, the movement of their wagons, the firing of their guns, the killing of certain animals, the importation of strange products into the country, and more. Sometimes, missionaries were killed on charges that they were the cause of drought and starvation—that, in effect, they were practicing witchcraft.[28]

Missionaries were suspect in part because they openly and vocally rejected African rainmaking activities and insisted that their own prayers for rain were fundamentally different from indigenous rainmaking. Maintaining this distinction was vital to reinforcing a worldview in which indigenous religion was false and

Christianity was true. This binary extended to encompass debates over rainmaking in ways the missionaries had not intended. In the twentieth century, the distinction between African rainmaking rooted in "superstition" and white rainmaking rooted in "science" became part of a larger package of purportedly fundamental racial differences. But the legacy of missionaries also troubled those racial distinctions, because both white and black Christians prayed for rain in churches in ways that looked remarkably similar and that rested on nonscientific epistemologies.

ARTIFICIALITY

In 1905 a Los Angeles newspaper reported that the California rainmaker Charles Hatfield had "entered into contracts with nine owners of diamond mines and sheep ranches in the Transvaal, which are supplemented by a contract with the British government." Charles Stewart, the chief meteorologist for South Africa, wrote that the Bathurst Farmer's Association had also sought to recruit Hatfield in 1905. In 1913 a man living in the Orange Free State proposed taking up a subscription to bring Hatfield to South Africa. When a leading conservationist criticized this idea, a man living in the semiarid Karoo defended Hatfield's record. Both Stewart and the editor of *Farmer's Weekly* publicly attempted to refute Hatfield's claims of success, but in 1915 the farmers of Victoria West tried to recruit him.[29]

In the pages of *Farmer's Weekly* in 1921 white South African farmers debated whether Hatfield's experiments were successes or failures. The heated exchanges—the *Farmer's Weekly*'s editor eventually refused to print additional correspondence on the subject—reveal the extent to which white farmers were collecting information on meteorology and rainmaking from North America. One man wrote to the US Weather Bureau asking for information about Hatfield (and was informed that he was a fraud). Another insisted he possessed data that showed "there is a certain amount of truth about Hatfield's claims." *Farmer's Weekly* declined to print the lengthy account of Hatfield successes that the correspondent included.[30] Charles Hall, a farmer in the Karoo with a keen interest in artificial rain production, corresponded with a Canadian farmer in Medicine Hat, Alberta, where Hatfield had been contracted to make rain.[31] A representative for a farmers' association in the Karoo wrote South Africa's chief meteorologist to ask about Hatfield.[32] A group of citrus farmers, living in the Sundays River community of the Eastern Cape, invited Hatfield to come to perform rainmaking experiments—but also wrote an agriculture professor at the University of California, Berkeley, to ask about Hatfield's credentials.[33]

Hatfield never did come to South Africa. But favorable reports from farmers in Canada encouraged Charles Hall to continue his own investigations into the production of rain. Hall had been trained as a civil engineer but had returned home to run his family's farm. He seems to have been inspired by Hatfield but also by the experiments of the British physicist Oliver Lodge, who had demonstrated that electricity could aid in forming water droplets. Hall also drew ideas from the Australian rainmaker John Graeme Balsillie and the rainmaking experiments of the Cornell chemistry professor Wilder Bancroft in Ohio.[34]

In 1919 *Farmer's Weekly* published a series of articles by Hall that sought to explain the drivers of southern Africa's climate and suggested that artificial rain could be produced by using electricity. Hall's initial article drew the attention of Francis Kanthack, the former director of irrigation, who told Hall in a personal letter, "I am perfectly willing to stand sponsor to you."[35] Hall thanked Kanthack for his support and noted, "A man who comes along with a rain making idea is bound to be looked upon as a humbug or a madman, so I expect to have to write a good deal more before anyone would listen to me."[36]

Hall's disarming acknowledgment of rainmaking's dubious reputation, combined with his obvious familiarity with basic scientific concepts, seems to account for the willingness of South Africa's government scientists to engage him. Charles Stewart, South Africa's chief meteorologist, wrote a ten-page private response to Hall's first article, exploring in detail the current state of knowledge of atmospheric electricity and its relationship to precipitation and reviewing experiments performed around the world. The memo is both didactic and collegial; Stewart corrected Hall on many points but assumed he could follow a scientific discussion. And Stewart added an encouraging letter, thanking Hall "most heartily for the excellent exposition of the effect of the coastal topography on the moisture in South-east winds; your discussion of this subject ought to remove a good deal of the misunderstanding and misconception which has arisen in connection with this particular subject."[37]

This letter marked the beginning of a decade-long correspondence between the chief meteorologist of South Africa and an eccentric tinkerer who erected "rainmaking boilers" on his farm and later purchased an airplane for cloud-seeding experiments. Hall sent Stewart newspaper cuttings about rainmaking experiments around the world and expressed his optimism about the prospects for artificial rain: "I predict that before ten years are past, rain making stations will be as plentiful as railway stations are at present, but will be infinitely more profitable."[38] Hall also predicted catastrophe for South Africa's drought-stressed livestock farmers.[39] And in 1920, he persuaded Stewart's department to support his experiments.

Hall developed a reputation in the Eastern Cape, where he was only one of several would-be climate engineers.[40] In 1927, the *Rand Daily Mail* reported that desperate farmers in the Sundays River valley had turned their attention from Hatfield to Hall and were asking the minister of agriculture to support Hall's experiments over their farms.[41] The Sundays River farmers also lobbied the governor general of South Africa. The much publicized irrigation scheme that was supposed to bring them prosperity had not amounted to much, because the reservoir that was meant to supply the water had not filled since its construction five years before. "Owing to scarcity of water we are always in danger of ruin," one farmer wrote. "If there is anything in artificial rainmaking, Mr. Hall may be able to make our fortunes for us or anyway save our fruit trees!"[42] The writer emphasized that the region's entire irrigation board had signed the appeal, "including our two Dutch members"—an allusion to the supposed antiscience religiosity of Afrikaners and an assurance that scientific rainmaking and religious belief were not in conflict.[43]

In spite of these assurances, the governor general's aides were wary. The problem was not rainmaking's opposition to the world of belief but its apparent kinship with that world. One aide wrote that artificial rainmaking "is a rather problematical question and may be held to savor somewhat of witch-doctor methods!"[44] The response reveals the importance of a white identity grounded in rationality and modernity—characteristics that supposedly distinguished the "European" community from a backward and superstitious African majority. Did "artificial rainmaking" fall into the category of science or the category of "witch-doctor methods"?

The existence of African rainmaking meant that the question of *how* rain was produced was crucial. Artificiality became a necessary feature of rainmaking practice in South Africa's white community. Articles on white rainmaking experiments all used the word "artificial." The term graces the archival files that contain correspondence on rainmaking. Stewart, the chief meteorologist, admitted his fascination with "artificial" rainmaking.

The distinction between natural and artificial processes of precipitation was not unique to South Africa. US meteorologist James Espy, in his book *The Philosophy of Storms* published in 1841, wrote that under the right conditions "any cause which produces an up-moving column of air, whether that cause be natural or artificial, will produce rain."[45] The British physicist Oliver Lodge had asked in the 1880s: "Why should not natural precipitation be assisted artificially?"[46] Hatfield, too, referenced artificiality: "There is always a certain amount of rain material in the air, and all we do is bring it together by artificial means, with the result that it falls on account of its weight. . . . It is a physical phenomenon, as natural as the telegraph or the telephone."[47] In the context of southern Africa, this distinction took on great importance. The "artificiality" of rainmaking rested in the crucial feature of human intervention into "natural" physical processes that were demonstrably real. For whites, this distinguished what they were doing from what African rainmakers were doing—although the distinction itself was somewhat artificial, as many African rainmakers also claimed they were harnessing and augmenting natural processes. But nothing in the archives suggests that Africans were seen as producing "artificial" rain; this designation was reserved solely for the activities of whites.

For the governor general's aide, however, the boundary between artificial rainmaking and "witch-doctor methods" was disturbingly unstable. And this representative of the British Empire was not alone. One reader of *Farmer's Weekly* scoffed at Hall's proposals to generate rain: "Personally, I have as much faith in a Kaffir rain doctor's powers, as in the dust theory."[48] The proposals of other whites also were placed in the framework of African rainmaking by their critics. One skeptic of a scheme to stimulate precipitation through the creation of large bodies of water titled his self-published rebuttal of the scheme "The Fallacy of Schwarz's Kalahari Rain-Making Magic."[49] A columnist who regularly wrote on farming matters for *Farmer's Weekly* drew the same parallel in a 1933 article: "Apart from the incantations and ceremonies of the professional rainmakers to be found in all savage tribes, we are told of the pseudo-scientists of more civilized countries whose methods, while apparently based on ascertained facts and accepted theories of rain production, are as vain and futile as those of the 'inyanga' or medicine-man who claims to have supernatural powers to cause the rain to fall."[50] And *Farmer's Weekly* in a headline called Hatfield "the California *Inyanga*."[51]

Other South African whites insisted that there was a clear difference between "artificial rainmaking" and "witch-doctor methods." The Sundays River farmers reported that they had been told Hall's ideas were scientifically valid. And the meteorologist Stewart was only one of several men trained in science who were skeptical of whether artificial rain could be generated economically, but who argued that the idea was grounded in scientific principles.

The notion that theories of "artificial" rainmaking could be subjected to experimental tests offered a justification for supporting them. As R. T. A. Innes, the Union astronomer, noted when discussing Hall's proposals in *Farmer's Weekly*: "It must be conceded that the causes which lead to rainfall are so slightly known that there is much room for experiment and investigation. The chances that the puny efforts of man can control the mighty movements of our atmosphere are remote; nevertheless when such simple and inexpensive experiments as are required by Mr. C. K. Hall's processes can be made they should be."[52] His words brought Hall firmly out of the category of "a medicine-man who claims to have supernatural powers" and into the category of a fellow scientist whose theories could be tested using the scientific method. But could this distinction survive actual practice?

THE BURDEN OF PROOF

In 1920 Stewart heeded Hall's entreaties for assistance in testing his rainmaking technique. On 5 October 1920, the Defense Department sent an airplane and pilot to the Irrigation Department, and the chief meteorologist oversaw the release of dust into a cumulus cloud near Pretoria. Hall argued before the experiment began that the quantity of dust, limited by the size of the plane, was too small to have any effect. And indeed, little happened. But this did not settle the debate over artificial rainmaking. As Stewart presciently noted, it was going to be difficult to prove Hall wrong. "Although this experiment failed to shew any visible effect," he wrote in his report, "the only conclusion that can be safely drawn from it is, not that this negative result disproves the theory; but that the amount of dust employed was insufficient to bring about the effect anticipated."[53]

Hall noted that he had predicted the experiment would fail. "Real tests," he said, would have to await better equipment. Hall opined that the Irrigation Department should not be blamed, as it was "trying to deal with all the irrigation schemes and cranks of half a continent, with few hands and less money"—a statement that allowed him to appear generous in defeat while also distinguishing himself from "cranks" with irrational ideas and demands.[54] But the fact remained that he, like African rainmakers, could produce an explanation for virtually any experimental failure. Ultimately, assessments of the efficacy of artificial rainmaking would come down to faith. Hall himself acknowledged this:

> When a man like Hatfield or Balsillie comes along and gives a demonstra-
> tion of rain-making, he has either got to be crucified by the high priests
> of erroneous belief, or the meteorologists have got to admit that they have
> subscribed all their lives to an absurd theory; and I leave it to the reader to
> determine which is going to happen, remembering that there is nothing
> definitely to connect the resultant rain with the apparatus of the rain-mak-
> er. The man who observes and understands is convinced; but the man who
> does not understand, and the man who does not choose to, can take refuge
> behind the theory of coincidence.[55]

Both supporters and opponents of artificial rainmaking struggled with the question of how to prove or disprove efficacy. Innes, the government astronomer, recognized this difficulty months before Hall's idea was tested: "Unfortunately, it is easy enough to make an experiment, whilst it is very difficult to draw correct deductions from it."[56]

Hall wanted to conduct his experiments in conditions already conducive to rainfall, and this did not seem unreasonable. The goal of artificial rainmaking was to intervene strategically in natural processes. Almost no one argued that rain could be generated in the dry season or in a cloudless sky. Yet African rainmakers had long been charged with engaging in their craft only when conditions were already ripe for rain. How was modern rainmaking different? And if it happened to rain after one of Hall's experiments, what conclusions should be drawn? Innes noted that, for such results to have credibility, multiple experiments would have to be run, and neighboring districts used as control groups, in order to prove that artificial rainmaking was effective.[57]

The difficulty of establishing whether natural forces or human intervention had caused rain was used by critics to question the legitimacy of artificial rainmaking. These debates raged everywhere, but they had a distinctly racialized edge in southern Africa. A man writing under the pen name "Desiccated Didymus" offered his tongue-in-cheek thoughts on the news that Hall had been offered a government airplane for his experiments. "I trust that Mr. Hall will not risk failure by allowing himself to be unduly hurried. If he will take the advice of one who has had some small experience of drought, he will make no attempt to produce rain until March or April [the height of the rainy season]; while, if the prospects do not appear sufficiently favorable then, it would be far better to wait . . . than to run any 'silly fool' risks while the drought continues."[58] Like African rainmakers, Hall was accused of awaiting ideal conditions before commencing rainmaking activities.

This dilemma became more pronounced in the late 1920s, when the Sundays River farmers sought assistance in hiring Hall to make rain for them. In 1928 the farmers managed to obtain a government plane and pilot. Over the course of three weeks, the pilot flew through small clouds as conditions permitted, while Hall and George Cox, a meteorological assistant dispatched by Stewart, observed. But it proved difficult to know what they were looking for. While Hall and the representative of the Sundays River farmers bickered over details of technique and equipment, rain was falling all around, and the sky changed so rapidly that the targeted clouds would often disappear.[59]

The experiments, Cox wrote, were largely "confined to periods when rain was imminent, and this was causing me some anxiety. Any claim, or even suggestion, that rains and experiments were connected would be accepted as proof of connection by a large number of people, despite all evidence to the contrary." When Hall telegraphed a newspaper that it was raining, Cox demanded that Hall send another telegram, clarifying that the rain was unconnected to his experiments. Cox concluded that the experiments were "an utter failure" by any measure, but, he added, "At the same time I do not consider this any proof that electrical treatment of a cloud would produce no evident result, since both the apparatus used and the methods employed left much to be desired."[60]

The problem of proof dogged rainmaking experiments around the world over the next few decades. An Australian government memo noted that large-scale experiments carried out there between 1955 and 1963 were inconclusive. Natural variability in precipitation over short distances meant that even in controlled experiments it was difficult to distinguish artificially produced rainfall from that occurring naturally.[61] A South African official wrote in 1952 that an American firm had approached the Southern Rhodesian government in 1951 offering rainmaking services. "As you know the rainfall in Rhodesia this summer has been particularly high and had arrangements for artificially inducing rain been concluded this would have been hailed as an outstanding success for the firm concerned." The official noted, "In weather science the truth is often very hard to establish and the need to find it and use it are often so great that the public will grasp at weak analogies, engaging in half-truths and delusory conceptions and the problems of weather forecasting and weather control are peculiarly open to wishful thinking."[62] In short, the line between artificial rainmaking and African practice was scarcely discernable. Both rested on belief and the near impossibility of an objective interpretation of results.

In the absence of any definitive proof that the artificial production of rain did not work, dreams of inexpensive climate control retained their seductive power.[63] In the north-central farming area of South West Africa (Namibia), a territory governed by South Africa until 1990, white farmers banded together in 1953 to raise money for the rainmaking activities of a man named Herbert Bartaune. "Every farmer in that district has paid a considerable amount of money into the fund," an official wrote as he relayed their request for government support.[64] The director of works noted that Bartaune planned to release silver iodide from an airplane into the clouds. Meteorologists in South Africa, however, stated "that although there have already been large sums of money spent in America to artificially increase rainfall, the results are such that scientific analysis cannot prove any increase."[65]

The proposal was not rejected on this basis, however. Instead, the director suggested that, since the farmers were "prepared to risk their money in the venture," the government should consider offering financial support. In exchange, the rainmaker should submit a report at the end of the rainy season on his methods, his expectations, and his actual achievements.[66]

Bartaune, a German citizen born in South West Africa, was a former member of the Luftwaffe—a skilled pilot whose scientific qualifications were apparently limited to a brief stint researching the effects of silver iodide on rainmaking in Australia. He eventually achieved fame not for rainmaking but for ferrying Nelson Mandela and other African National Congress exiles from Botswana to Tanzania.[67] Yet colonial administrators were prepared to accept him as someone who had the potential to add to a body of scientific knowledge about artificial rainmaking, however unsuccessful these techniques had been to date.[68]

CONCLUSION

Try to imagine, for a moment, an African rainmaker receiving this sort of response—or any response at all, save chastisement for engaging in deception—from a government body, and the role of race in ideas and practices of climate control becomes clear. The aridity of the region and the severity of the droughts that plagued it from the 1910s through the early 1950s do partly account for the public enthusiasm around white rainmaking. But political imperatives mattered as well. The intelligence, rationality, and authority of white men simply were not questioned. Suggestions that men such as Hall offered for changing the climate received serious and measured responses. Correspondence between citizens and government officials was one way of performing the stated equality of all white men before the state and helped reinforce the supposedly natural separation of white and black.

But the fertility of white imaginations—and, one suspects, the existence of intellectual traditions that do not make an appearance in government reports or in the letters of educated men—made it difficult for government officials to manage the parameters of respectable rainmaking. In 1958 a man named F. W. London, also based in the Eastern Cape, sent the secretary for agriculture a rambling four-page epistle in which he claimed to have had power over the rain since he was a child. He requested that the South African government hire him on a "no rain no reward system." "I break the drought without floods not an easy job [sic] and my system cannot spoil the usual rainfall sources, as I use the Natural System." London wrote that he had been injured and unemployed since 1944, and that wealthy farmers had spurned his services during the last drought. Poor farmers, however, had pleaded with him for help, and he had relented out of sympathy for them—a sympathy born of his own ancestors' struggles against drought. He also hinted at a list of enemies, including "parties who claim to represent the Church aspect," although he insisted, "I use no unholy principle." He ended his letter with: "I will of course be in danger from scientists, etc., as in the past."[69]

Officials debated how to best respond to the series of letters London sent to the Department of Agriculture over the course of several years. And London was not alone: A. J. H. Stander, a man writing from another Eastern Cape Town near the arid Karoo, also wrote to ask if the government had funds to pay "rain people" (*reën mense*). Stander stated that he had been a professional rainmaker for many years and felt he should be compensated, since he had made rain voluntarily and without payment.[70]

One imagines that such colorful characters amused and exasperated bureaucrats. But it is also telling that someone scrawled "Communism" across the top of one of London's letters and added a note on Stander's stating that he was the "same class" as London. These self-proclaimed rainmakers who offered justifications for their work in rambling, often incoherent, terms but who alluded to the unmet needs of poor white farmers were not categorized as eccentric, possibly insane, local characters but as *Communists*—an identity that was effectively illegal after 1950. In the 1950s all political dissent in South Africa was repressed under the banner of fighting Communism: Nelson Mandela and 155 others were placed on trial for treason for organizing nonviolent protests and boycotts. In this context London's mild disparagement of rich farmers and the church and Standers's observation that the rich benefited from the uncompensated services of others justified calling them Communists. And yet these suspected Communists received a polite, bureaucratic response, not a legal summons. Stander was told the Agricultural Department did not have funds to pay rainmakers. London was eventually shunted off to the Department of Water Affairs and told that this was the proper agency to consider his request.[71]

London claimed power over the rain in terms similar to indigenous African rainmakers, a combination of natural ability and acquired knowledge. In a 1962 letter he stated that he was a "born rainmaker."[72] And yet he also invoked the language of Western science, calling himself a "rain technician" who had a knowledge of meteorology. "I do not claim to be a Scientist," he wrote, but "I do practice Weather Science."[73] The blurring of these categories, of African *inyanga* and Western scientist, seems to have left officials uncomfortable and perplexed. It took them three years to respond to the letters London wrote them—and when they had to place a label on him, they settled for Communist.

London's theories were only slightly more bizarre than those of another man who professed to be able to make rain. G. J. van de Waal, a water engineer from Holland, wrote Prime Minister Hendrik Verwoerd in 1959, arguing that artificial rainmaking was the logical extension of increasing human mastery of the earth: "Rivers are being diverted as if it is nothing, and it will not be long before people will also have complete control of the climate."[74] The engineer proposed spraying seawater into the air and building enormous evaporation basins filled with saltwater that could float on the ocean's surface—both aimed at disrupting the cold Benguela Current that prevented precipitation along South Africa's west coast.

Verwoerd, popularly known as the architect of apartheid, was enthusiastic and forwarded the letter along with a personal note to the director of water affairs: "Although one may tend to cast such proposals as fantasies, it would be worth investigating it carefully. To the layman it does not look so shabby."[75] A Weather Bureau employee responded that the ideas underpinning van de Waal's proposal were "pure nonsense."[76] The engineer received a polite letter explaining the flaws in his idea, and there the matter rested. And other rainmakers continued to write with their ideas, into the 1960s.

Even as its government scientists fended off rainmaking requests and schemes, South Africa's neighbors assumed that those scientists had superior knowledge of artificial rainmaking. Like the famed rainmakers called from great distances in times of drought, South African expertise was sought by faraway communities in need of rain. In 1957 the agricultural service in Réunion wrote the South African secretary for agriculture to inquire about research into artificial rainmaking.[77] In 1960 an association of sheep farmers in the Eastern Cape asked the Agriculture Department whether the government had run trials in artificial rainmaking and, if so, what kind of rain had fallen—violent storms or "soft rain"?[78] In 1962 a man at the Kakamas irrigation scheme and another living in Calvinia, in the arid Northern Cape, both inquired about technologies for artificial rainmaking.[79] In 1963 a farmer in Southern Rhodesia (Zimbabwe) wrote the South African agriculture minister asking where he could obtain "rain rockets"; the following year, the Ministry of Agriculture in Southern Rhodesia wrote that the white farming community was interested in using such rockets for cloud seeding. Both were informed of a company in Cape Town that manufactured them.[80] There were also letters from Australia and Holland, including one from a man in Sydney asking about reports that Hall's rainmaking theories had been tested successfully.

The stature of South Africa in a larger quest to make artificial rain speaks to the importance of technology in distinguishing "legitimate" rainmaking from that which took place all across southern Africa at the start of every rainy season. African rainmakers used "incantations" and "ceremonies," but modern-day white rainmakers relied on chemicals and electricity, rain-rockets and airplanes, and "boilers" built on their farms to coax precipitation from the sky. They ran experiments with observers who wrote reports. Optimism prevailed: optimism that technology was the answer to human travails, and optimism that the segregation that white South Africans sought to perfect could be applied to the border between white and black rainmaking activities. And yet a definitive answer to whether human intervention could cause rain to fall remained elusive, and white farmers continued to discuss the possibilities of "artificial rain."

8

WEATHER, CLIMATE, AND
THE COLONIAL IMAGINATION

Meteorology and the End of Empire

Martin Mahony

IN THE IMMEDIATE AFTERMATH of the Second World War, Great Britain was suffering the effects of a worldwide shortage of edible fats. The country's larders were low on margarine, and this lacuna in the British diet risked undermining both the health and the morale of a population straining to recover from the trials of wartime. The war also threw new light on the practices and rationales of British colonialism. With the moral justification of imperial rule being questioned with renewed vigor in Britain and beyond, the postwar Labour government advocated a new, more proactive stance toward colonial development. These two political-economic trajectories came together in the East African groundnut scheme, a program to produce groundnuts on an unprecedented scale in Britain's East African territories using the latest in scientific knowledge and agricultural technology. Under subtropical skies, Britain's food economy was to be revived and its imperialism rebooted.

The groundnut scheme was, however, a notable failure. So notable in fact that it became a totemic example of governmental hubris, recalled to this day by opponents of socialist interventions in economic development as an example of the state blindly overreaching, trying to plan, direct, and innovate in settings where only the market can provide optimum solutions.[1] In even more prosaic terms, the groundnut scheme became a catchall referent for governmental blundering, with Spike Milligan famously labeling the Concorde—the Anglo-French project to develop a supersonic passenger jet—a "flying ground-nut scheme" in a cartoon for the satirical weekly *Private Eye*. Various explanations for the scheme's failure have been put forward, from simple administrative incompetence to the unsuitability of the technologies employed to clear, sow, and harvest the land. But one issue that dogged the scheme's planners and practitioners throughout was that of climate. Like a microcosm of the European imperial experience more broadly, the groundnut scheme is a story of competing climatic expectations, troubling extremes, and a dialectical confrontation between the necessity to adapt and the will to control.

CLIMATE AND COLONIALISM

Ideas about climate have been shown to have influenced the ideologies and practices of empire in profound ways.[2] Recent historical scholarship has forcefully shown how ideas about climatic *change* entered into the scientific, political, and moral discourses of imperialism and colonialism, often with the intention of emphasizing the relevance of historical scholarship to contemporary debates about anthropogenic global climate change.[3] However, although this historicization is clearly important and necessary, the emphasis on the long history of worrying about climatic change may obscure the equally significant history of how ideas and knowledges about climate were woven into the imaginaries and practices of imperial rule and colonial government. The groundnut scheme offers an opportunity to examine closely how thinking about climate intersected with late imperial British stances toward colonial development, through close scrutiny of how the constitution of this particular site of colonial practice was constituted by a number of broader intersecting trends of colonial knowledge and power.

I conceive of the groundnut scheme as a particular "knowledge space," the product of a dialectical shaping of knowledge and social space pursued through the assemblage of heterogeneous knowledges, imaginaries, forms of expertise, technical practices, and social labor.[4] Knowledge spaces emerge, cohere, and stabilize through practices of making things equivalent, and we can see how the rendering of equivalence and difference is a dialectic at the heart of colonial worries about climate. The "tropical" was always a radical Other, at once both threatening and alluring in its alterity, and empire boosters promoted outward migration from the metropole with promises of climates that would offer the sensory and moral security of the weather back home. Imperial expansion and colonial management saw the production of maps of climates awaiting domestication, cartographies that both delineated the uninhabitable and offered zones of geographical equivalence where crops, people, lifestyles, and moral orders could be transplanted.[5] Yet these imaginative climatologies were always hampered by a scarcity of data and by the related challenge of reliably extending inferences from point observations to wider planes of climatic reality. This move from points to planes, this creation of spatial coherence, was a perennial tension in scientific debates surrounding the groundnut scheme, and the question of how many point observations one needs to make a reliable claim over a wide space is still a matter of some statistical controversy in the atmospheric sciences.[6]

Such controversies are not solely statistical. Historians of meteorology and climatology—and of the geographical sciences more generally—have shown how epistemic indeterminacy is a product of social disagreement over different forms of testimony and the assumed trustworthiness of its bearers.[7] Investing trust in an empirical or experimental claim means investing trust in an expert witness, a profoundly social act shaped by considerations of class, race, gender, expertise, and moral standing. In the atmospheric sciences, "a composite of many different forms of observing and recording, each with its distinctive criteria of what constitutes good or bad work," this sociology of testimony has played out in a field where different epistemic practices have jostled for position as the most reliable means of sensing the atmosphere.[8] From glancing at the sky to diligently recording the weather's instrumental traces, and from scanning a synoptic map to modeling atmospheric change far out into the future, different ways of knowing the weather have always existed in tensile relationships with each other, with the primacy of certain ways of knowing achieved not as a straightforward triumph of technique or theory but as a social process, shaped by the cultural politics of particular times and places. I suggest that different ways of knowing the climates involved in the groundnut scheme were shaped not only by disciplinary differences among groups of experts but also by broader commitments to certain ways of knowing and imagining colonial space. I focus in particular on the conflict between Albert Walter, a government meteorologist in British East Africa, and groundnut scheme officials, who clashed over the environmental wisdom of the scheme and over the epistemology of agricultural climates. I suggest that these clashes were rooted in competing imaginaries of colonial development—one that placed great faith in the power of science and technology to rationalize, accelerate, and homogenize economic development and another that put climate, and its careful monitoring, at the heart of attempts to mold developmental pathways to environmental limits. I suggest that the notion of "sociotechnical imaginaries" can offer analytical purchase on these competing visions of colonial order, with implications for how we reconstruct the imaginative geographies of empire.

IMAGINING AND PRACTICING COLONIAL DEVELOPMENT

The postwar reorientation of British agricultural policy marked the emergence and spread of a distinctive "socio-technical imaginary" or a "collectively held, institutionally stabilized, and publicly performed" vision of a desirable future, "animated by shared understandings of forms of social life and social order attainable through, and supportive of, advances in science and technology."[9] The concept of sociotechnical imaginaries marries the imaginative and the material and offers a way of comprehending the embedding of broader idealizations of order into the spatial arrangements of economic production, social life, and knowledge-making.[10] The origins of the distinctive sociotechnical imaginary of postwar British colonial developmentalism can be traced to the efforts of Joseph Chamberlain, as secretary of state in the Colonial Office between 1895 and 1903, to supplant the laissez-faire approach of earlier imperial policy with a much more interventionist attitude to colonial development. The "new imperialism" of the late nineteenth century saw science and technology move to center stage as tools not simply for opening up conquered spaces to free trade commerce but for advancing the moral and material conditions of colonized populations. Chamberlain's developmentalism was given greater urgency years later by the Great Depression and the worldwide collapse in commodity prices. Faith in free trade economics waned, and the state took on new roles in engineering the health and wealth of its citizens. The Colonial Office embarked on a period of sweeping policy reform, which culminated in the Colonial Development and Welfare Act of 1940. As Joseph Hodge points out, the most striking feature of these developments is the emphasis placed by the Colonial Office and the wider colonial service on scientific and technical expertise, with nearly half of all new colonial service recruits being appointed to technical roles in agriculture, chemistry, forestry, geology, meteorology, and other fields where the survey and the field experiment were becoming key techniques of colonial development practice. "In many ways," Hodge argues, "late British colonial imperialism *was* an imperialism of science and knowledge, under which academic and scientific experts rose to positions of unparalleled triumph and authority."[11]

This imperialism of science and knowledge was far from monolithic. Helen Tilley has argued that technical experts in the colonial service developed detailed knowledge of—and intimate relationships with—colonized places and peoples and were often the harshest critics of the social and environmental consequences of imperial ventures.[12] Colonial power was characterized by factionalism and disagreement, by competing forms of knowledge, and by experts battling for authority and influence both in the colony and in the metropole. Nonetheless, the Second World War created a new urgency in the intervention of the British state in colonial development projects, buttressed by the bold statism of the post-war Labour government. New projects, hastily proposed and arranged, promised to annex large tracts of marginal colonial land and render them productive through the application of capital and technoscientific expertise. Circumventing the primacy of private capital, the state would become the primary motor of production, exercising control over the colonial commodity trade like never before. The immediate postwar period saw the assertion of a rebooted sociotechnical imaginary of colonial developmentalism, which deepened the connections that had earlier been made between science, technology, and the engineering of new social relationships in colonial societies.

THE TURN TO GROUNDNUTS

The seed of the groundnut idea appears to lie in a 1946 meeting between Frank Samuel, of the United Africa Company (UAC), part of Unilever, and R. W. R. Miller, director of agricultural production in Tanganyika. Miller tried to convince Samuel that "development" could not be achieved through the proliferation of trading companies—which the UAC then was—but through cooperation with companies prepared to sink capital into the soils of Britain's overseas territories. Miller put it to Samuel that a few thousand acres of groundnuts could be cultivated by the UAC, with success guaranteed by wholesale mechanization. Samuel ran with the idea, taking heart from the success of mechanized groundnut production in the United States and encouraged by an overflight of lush-looking East African vegetation on route from Johannesburg to London.[13] He managed to convince the new minister of food of the wisdom of the idea. The Cabinet agreed and despatched John Wakefield to East Africa to begin detailed planning.[14]

Wakefield was an experienced colonial agricultural officer who'd spent a number of years serving in Tanganyika before going to head up agricultural operations in the West Indies. The report of the "Wakefield Mission" proposed a massive program of mechanized agriculture on a series of sites around East Africa, totaling over three million acres. With the guidance of relevant experts and "scientific farming methods," the authors believed, an annual yield of eight hundred thousand tons was achievable by the early 1950s, significantly boosting the supply of fats to the British market.[15] But the significance of the scheme was not only to be measured in margarine. It would also stave off what was increasingly being seen as a Malthusian crisis of mismatched population growth and food production in colonial Africa, by innovating a broader set of agricultural techniques that could be rolled out across Britain's tropical colonies. Founded on the latest scientific knowledge and driven by cutting-edge technologies, British colonial agriculture would be radically modernized in a whirlwind of developmental transformation; the groundnut scheme in this view was but a modest test bed for a new suite of practices with global applications.

The groundnut scheme can be situated within the public unfolding—and, later, unraveling—of this imaginary, as an instance of its public performance and experimental application. The scheme was to be an "ocular demonstration" of both the social and technical elements of this imaginary, demonstrating very publicly the power of mechanized agriculture and representing, as Frank Samuel put it, "an opportunity to teach the African to be an economically productive unit."[16] It was the product of an imaginary that placed great faith in the economics of scale. During the 1940s it was widely held that scaling up agricultural activities through the application of capital and technology would automatically produce greater yields. The experiences of American agriculture appeared to bear this out, not least on the large peanut farms of the South. The Wakefield Mission produced yield estimates by obtaining sample yield data from areas of indigenous cultivation and a handful of experimental sites in northern Rhodesia and Uganda and then "multiplying by the number of acres to be cleared without regard for possible soil and rainfall differences over the hundreds of square miles involved."[17] The simple assumption was that the sociotechnical organization of mechanized plantation agriculture would create such efficiency and homogeneity so as to override the geographical variety of the proposed sites and to render insignificant any differences with the sites from which yield estimates were drawn. Capital-intensive mechanized agriculture was to be an assemblage of knowledge and practice that could roam freely, a homogenization machine to be let loose wherever "unproductive" native agriculture was holding back colonial development.[18] The groundnut scheme was thus to be, as Hodge put it, "a showcase of what science, technology, and the state could achieve," overcoming what Wakefield described as "the despair of hand and hoe" with as the "sure hope of mechanisation." This was a hope held out not only for the development of indigenous populations but for the future of British imperialism itself. As Wakefield put it to Samuel, "The present opportunity to effect such a revolutionary change must not be missed—if it is I foresee the complete failure of the Colonial system."[19]

Environmental conditions were known to be trying, but the scheme was framed publicly as a war with nature and "an economic Battle of Alamein," turning the leftover rhetorical and material resources of the British war machine against the bush, soil, and climate of East Africa.[20] Machinery and supplies were bought up from military surplus stores at points around the empire, most notably in the form of tractors that were repurposed for bulldozing bush vegetation and ploughing the stubborn soil. When these later proved unreliable, the scheme turned to surplus American Sherman tanks, refitted by the Vickers Company into custom-made bush destroyers, christened "Shervicks." But while the task of clearing the bush and ploughing the soil was figured as one of technological conquest, dealing with the climate was a question of minimizing risk. The Wakefield plan described the "erratic rainfall" upon which East African agriculture was largely dependent and expressed the idea that, "in order to lessen the risks of crop failure," it was desirable to "spread production over more than one set of climatic conditions."[21] Occasional droughts seemingly occurred only on local scales, and continental droughts were rare, so the groundnut crop could be made robust to the risks of rainfall shortages and the uncertainty of the rains by spreading activities far and wide. This "rain insurance" by spatial dispersal had been impressed upon members of the Wakefield Mission during their tour of potential sites in Tanganyika in July 1946.[22] David Martin, a UAC plantations manager, wrote that "Everywhere ... we have been cautioned about the erratic behaviour of the rainfall and the calamitous effect it has had on the local crops, more especially in the past few years." While the scheme's other challenges—boring for water, organizing labor, awkward topography—could be overcome by "human ingenuity," it wouldn't be possible to "coax rains," and a "proper geographical spread of risk" was required to insure against this most intractable of variables.[23]

The team was determined not be overawed by the naysayers. "Whilst not ignoring this evidence, we are agreed not to be unduly influenced by it," reported Martin. He explained how native cultivation produced respectable yields despite "due care" not being given "to selection and treatment of seed, correct time of planting, and, perhaps most important of all, to the building up of the moisture retention capacity of the soil."[24] But with careful planning of the topography of the planting zones to minimize runoff and the addition of organic matter to maximize soil moisture retention, the groundnut scheme could be made robust enough to counter a fickle climate.

Despite this confidence in the inevitable triumph of the mechanized and the scientized, the mission members heeded the advice of Sir Charles Lockhart, former economic adviser to the East African governor's conference, to "consider two or more widely scattered areas rather than one large composite block" so as to minimize the risk of failure from drought.[25] Potential sites were therefore identified across the territory. In the south and west of Tanganyika, areas in receipt of around thirty or forty inches of annual rainfall were earmarked for development, along with Kongwa further north, in the Dodoma region of Central Province, which could boast only around twenty inches of rain. The Marenga

Makali—"harsh lands"—around Mpwapwa were described as "marginal" in the Wakefield Plan, but local groundnut production had apparently yielded from twelve thousand to twenty thousand pounds per acre, even with supposedly suboptimal varieties and cultivation techniques.[26] However, detailed information about the "physical conditions and climatic hazards" to be faced by the project were acknowledged to be "scanty," and mission members hoped that experience and direct observation would provide more reliable knowledge.[27]

The initial recommendation of the Wakefield Plan was to start in the south first, where more reliable and bounteous rainfall could be expected, and then initiate production further north the following year.[28] Wakefield commissioned the East African Meteorological Service to analyse the rainfall data for the proposed sites, with the director of the service, Albert Walter, taking a keen interest in the work. Wakefield's team was confident that the required twenty-two or twenty-four inches of required rainfall would be met at all sites including Kongwa in the north, despite a 1918 publication warning of the "shortness and uncertainty of the rainy season" and an official rainfall map from 1942 placing the Kongwa area "well within the yellow band marking the territory's lowest rainfall."[29] Sure enough, Kongwa came bottom of the rainfall league in the Wakefield plan, but the apparent success of native agriculture in the wider area, along with the area's convenient proximity to the central railway line, meant that the possibility of the scheme commencing there, rather than in the south, started to grow. Sensing this shift in momentum, Walter wrote to Wakefield: "The analysis of the rainfall in the drier patch north of Mpwapwa on our rainfall chart has been pushed forward. Three or four stations are completed and I am not very happy about the conditions. Do not push too far north as a plan until the analysis is complete. I have thought it advisable to warn you of this in case planning is proceeding too rapidly for this area."[30]

Others, however, were not convinced by Walter's admonitions. Hugh Bunting, a South African agricultural botanist and the new chief scientific officer on the scheme, conducted more detailed ground surveys of the area around Mpwapwa, while Martin took to an aeroplane to survey conditions from above, and Wakefield "saw it from the top of a ridge of hills."[31] Commenting on a copy of Walter's terse letter to Wakefield, Bunting argued that "There is no reason, for a start, to call the area 'drier.'" The vegetation to the north of Mpwapwa, around the hamlet of Kongwa where groundnut units were being planned, was of "a wetter type than to the South, no matter what Walter's analysis shows." The analyses of vegetation patterns made by Bunting, Martin, and Wakefield not only threw the empirical reliability of the extant rainfall records into doubt but also caused Bunting to wonder whether "there are any rainfall stations relevant to our area; it looks entirely uninhabited."[32] The few records that Walter had to hand, spanning only a few years, were from stations a few miles away, such as at Dodoma Reservoir and the Kongwa Mission Station, the latter about seven miles south of the proposed growing sites and with only seven years' worth of records.

As Walter later recounted in a letter to the Oxford economist S. Herbert Frankel, the "agricultural experts" agreed, "after flying over the area," that the vegetation suggested much wetter conditions than the rainfall statistics, and that these statistics were unreliable anyway, given the geography of the underlying data. However, Walter countered that "it is almost an axiom in climatology that given the main wind currents and topography over a given region interpolation of the isohyets can be performed with only slight geographical error if rainfall stations surround the region under review. This was the case at Kongwa. The vegetation could not in consequence be a guide in replacing a rainfall chart, in judging the suitability of the area for groundnuts or any other economic crop; but the opinion of the agricultural experts prevailed over my own."[33]

The production of the groundnut scheme as a knowledge space was thus a contested process from the start. Different forms of reasoning—by statistics in Walter's case, by proxy for Bunting and Martin (with vegetation as proxy for rainfall), and by spatial analogy for Wakefield who drew comparisons with a similarly "marginal" yet productive area in South Africa—competed for recognition as the most authoritative means of making reliable claims about a climate that, until then, had been largely outside the sphere of European interest. Being able to bring the landscape and climate of a wide area into a single field of vision, through statistical interpolation or by observing from the air, was naturally considered a prerequisite of rational planning; a colonizing act of making-known and making equivalence, reducing complex landscapes and ecologies into tables and charts ripe for comparison and ranking of their economic potential. However, the exchanges between Walter and groundnut officials make plain that the place of meteorology and climatology as sciences of agricultural colonization were far from secure. The ability of the meteorologist to make claims about the climate beyond the immediate environment of the rain gauge or the thermometer was not taken for granted by a group of experts who were more comfortable with reasoning through a combination of experiential knowledge and direct observation of the climate's apparent effects. Backed up by a faith in the power of mechanized agriculture to overcome the effects of any rainfall deficiencies, these forms of testimony won out over Walter's.

The agricultural experts' views thus smoothed the way to a decision to start at Kongwa rather than in the south, with Kongwa having better links to the railway network that would bring in the heavy machinery and, it was hoped, take out the bounteous nuts. But Wakefield was seemingly more disposed to listen to Walter, and later that summer engineered his appointment as meteorological advisor to the scheme. Despite, or perhaps because of, his early warnings going unheeded Walter enthusiastically took up the position, successfully applying for early retirement from the meteorological service in order to devote his full attention to the new position. He confided to Wakefield that this could be his swan song: "It gives me . . . an opportunity of getting a few of my long cherished ideas into operation."[34] Walter was not just possessed of different epistemic commitments but by a sharply contrasting sociotechnical imaginary of colonial rule.

THE CLIMATIC THOUGHT OF ALBERT WALTER

Walter had been interested in the relationship between crops, climate, and colonial political economy since he arrived as a young observatory assistant in Mauritius in 1897. He had cut his teeth at the Royal Greenwich Observatory, making and computing astronomical and meteorological observations, before his appointment to the Royal Alfred Observatory at Pamplemousses.[35] In among his day-to-day activities of making observations and monitoring and forecasting cyclones, Walter embarked on an ambitious statistical study of sugarcane yields and their shaping by climatic conditions. His most significant conclusions concerned the effects of cyclones. He claimed that with observations of cyclone intensity at opposite ends of the island he could infer, through statistical interpolation, a cyclone's intensity at points in between. This enabled the construction of what he rather clunkily called "isoptherms," lines of equal devastation, with which the effects of a passing, imperfectly known cyclone could be scaled down to the level of the individual farm or plantation. Walter offered this technique of retrospective reconstruction as a basis for a possible insurance scheme. He argued in his 1910 publication, *The Sugar Industry of Mauritius: A Study in Correlation*, that his statistical reasoning would provide a surer basis for insurance claims than the unreliable visual testimony of damage inspectors. But an insurance scheme would not simply redistribute the risk associated with agricultural monocropping in a capricious climate; it would also, Walter supposed, ensure the smooth circulation of capital through the island economy, giving the industry a "new lease of life" as capital that would ordinarily lie dormant between December and April would be put into circulation.[36] His proposals won the backing of the colonial government, and his statistics were hailed in London as a pioneering effort in agricultural meteorology, but the underwriters were initially unconvinced and no insurance scheme with Walter's statistical underpinnings appeared until 1946.[37]

Following his move to East Africa, Walter's focus shifted from defining correlations to defining the spatial envelopes within which conditions were right for certain forms of agricultural development, thus refining his techniques of statistically transforming sparse data into geographically extended inferences. He lamented the loss of his island laboratory, where, free from "the effect of extraneous outside influences . . . the relation of cause and effect" could be readily traced.[38] Now, his isolines would have to stretch further, connecting together an expanding but still spatially diffuse observation network. He worked with agricultural scientists at the Amani Research Institute on the microclimatology of crops and developed new techniques for mapping climates that were favorable for agricultural development. But the groundnut scheme represented something of a return to the island—an opportunity to get closer to the total surveillance of climate, to achieve a panopticism (seeing-all) through the achievement of a dense synopticism (a regulated seeing-at-the-same-time through coordinated observation). Walter proposed a dense, "grandiose" network of first- and second-order climatological stations along with more than one hundred rain gauges to be arrayed across the Kongwa units.[39] He asked for voluminous instructional books and reams of data tables for use by the "one supervising European meteorologist . . . [employed] for inspection and control, 3 Asians for tabulating and computation and 6 African clerks for the routine tabulating work."[40] This close monitoring could show where best to plant groundnuts and could also form the basis of tailored weather forecasts that would dictate the timing of planting.

Having seen how dismissive decision-makers could be of the value of detailed meteorological work, Walter sold his scheme by appealing to Ellsworth Huntington, the doyen of twentieth-century climatic determinism. Early in 1948 Walter shared with the scheme directors and with the Colonial Office copies of his inaugural presidential address to the Nairobi Scientific and Philosophical Society, which he had cofounded. Walter quoted extensively from Huntington's *Civilization and Climate*, drawing on passages that emphasized the ultimate sovereignty of climate in shaping human progress, particularly as that "climatic influence" was mediated by agricultural development. Walter made some familiar—if, by this point, increasingly unfashionable—arguments about climate and racial difference, but he also used Huntington to buttress his call for comprehensive surveys of climatic conditions as the first step on the road to development. He emphasized the fundamental role "which climatology must play in the counsels of those who are to be entrusted with the carrying out of the many development plans proposed" for colonial Africa, with "intensive climatic surveys" to be prerequisites, not afterthoughts, in development planning.[41]

Walter was the product of a scientific culture on Mauritius that was shaped by the intellectual upheavals of prerevolutionary France. In the eighteenth century Mauritius had become a locus of protophysiocratic thought, with colonial officials, botanists, and nascent conservationists promulgating new visions of a social order where agricultural labor was the source of all value. The physiocrats developed critiques of both traditional and capitalist forms of landownership in France and tended toward the "idealisation of peasant and resident landholders."[42] On Mauritius these kinds of intellectual currents were prefigured by the likes of Pierre Poivre, intendant during the 1760s, who argued against the plantation system and in favor of careful monitoring and conservation of natural processes, particularly those—such as plant transpiration—that seemingly encouraged rain. For Poivre and his physiocratic successors, agriculture could be made more productive and sustainable not through intensification but through careful empirical elucidation and maintenance of the "natural order."[43]

Walter was well-read in the intellectual and environmental history of Mauritius and participated actively in the ongoing debates about forests and rainfall that began with the theories and practices of Poivre, Bernadin de Saint-Pierre, and others. His Huntingtonian warnings about climate being the chief arbiter of human fortunes found an interested audience in the Colonial Office, where the idea of careful expert planning was by then well established.[44] In Walter's climatic thought we can detect a very different sociotechnical imaginary to that motivating and animating the groundnut scheme. Rather than a confidence in the ability of science and technology to reshape environments and peoples, Walter was concerned with the more modest tasks of understanding and monitoring natural "limits" to development, with scientific expertise positioned not as the vanguard of environmental transformation but as the protector of a natural order. But it was the Colonial Office's affinity with this more precautionary attitude that led to its being bypassed in the planning of the groundnut scheme, where pace was to be prized over precaution and ultimate control being handed to the Ministry of Food at the outset. By bringing Huntington into the conversation with both the sidelined Colonial Office and the scheme's managers, Walter was responding to what he saw as an increasingly unstoppable momentum. By the middle of 1947 bush-clearing was well underway, with the scheme's directors eager not to miss the first rainy season when it arrived toward the end of the year. Walter gleaned "from newspaper reports that the whole question of organization has been shifted to a higher level. The politicians will not thank the technical officers if they interfere with their pet schemes. I can only hope that some of you are independent enough and strong enough not to allow public money to be squandered to please the politicians if the conditions are definitely unfavourable scientifically."[45]

He had begun to order meteorological equipment from Great Britain and to send spare kit from Nairobi, and by October Hugh Bunting was distributing the first few rain gauges around the Kongwa site as the rainy season approached. As the Kongwa planting areas began to take shape as sites of intensive agroscientific experimentation and observation, the social constitution of the space mitigated against the smooth rolling out of the planned observational infrastructure. Despite his own observation that most of the early rains had come in the form of highly localized thunderstorms, which would require a dense network of rain gauges to capture, Bunting was reluctant to place gauges in areas beyond the overseeing eye of European staff, who could undertake or directly supervise the taking of measurements around the "offices, stores and unit headquarters" where "Europeans work daily."[46] Assembling the groundnut scheme as a knowledge space was social as much as it was technical—a process infused with racial hierarchies as to who could be considered a trustworthy observer of the weather.

CONFRONTING THE CLIMATE

The first harvest was far below expectations. Although this could largely be attributed to only one-tenth of the expected amount of land being cleared and planted, the preponderance of short, sharp rain showers meant the ground dried quickly and baked hard, posing new challenges for the scheme's machinery, and yields per acre were less than half Wakefield's projections. As clearing operations commenced again following the harvest Walter sought to accelerate the rollout of his observational infrastructure, but equipment persistently arrived late or broken, he struggled to get to the site to set up the instruments himself, and he rankled officials by bending procurement rules to make use of his existing favored suppliers in East Africa who could offer custom-built thermometers and rain gauges adapted to tropical conditions. He lobbied for the recruitment of a trained meteorological assistant, recommending one of his African clerks for the job of setting up and monitoring instruments. But Bunting again resisted, arguing that the key roles of such an appointment would be less the managing of equipment and data and more "persuading other departments to do things for us. . . . For this work, I am sure, we must have a competent European."[47]

So with little new data to analyse or staff to manage, Walter's position as a full-time advisor was looking increasingly untenable. In September 1948 the decision was taken in London that "no meteorologist at all is required" on the scheme, and Walter's contract was terminated with immediate effect.[48] This occurred amid a wider drive to centralize the scientific parts of the program in London, much to the displeasure of those in the field, many of whom appeared to support Walter's work.[49] And as if the climate was seeking recrimination for the dismissal of its slighted spokesperson, the following growing season was disastrous. Only twelve inches of rain fell, and yields per acre were less than half those of the previous year. Things improved slightly the next year, but yields were still far below expectations and costs were escalating. By 1949 there were rumblings about abandoning the whole scheme altogether, of cutting the government's losses and avoiding the fruitless investment of further capital.

But in a report submitted to the Minister for Food, it was argued that things could still improve, and that the stakes were too high to quit. Officials agonized that "Our standing as an Imperial power in Africa is to a substantial extent bound up with the future of this scheme. To abandon it would be a humiliating blow to our prestige everywhere—at home, in the Colonies, in the international field." If the scheme was failing to be the "ocular demonstration" of the imagined might of scientific, mechanized agriculture, its abandonment would be an all-too-ocular demonstration of Britain's declining legitimacy as an imperial power. So ministers agreed to let the scheme limp on, but continuing drought conditions at Kongwa meant that its directors, desperate for rain, soon started turning to what seemed to be their last and only hope, to turn the hand of purposeful environmental manipulation onto what had initially been deemed beyond the scope of the scheme's "human ingenuity"—the direct and artificial production of rain.[50]

By this time the management of the scheme had been taken over the by Overseas Food Corporation (OFC), which was charged with implementing a number of similar agricultural schemes around the empire. For OFC joint general manager G. W. Raby, rainmaking experiments held out the hope of saving the scheme and its reputation. "In consideration of the 10 or 12 million pounds already expended in getting to Kongwa and creating our capital assets there, and in view of the somewhat desperate situation in which we find ourselves, due in the main to a meagre rainfall, it is felt that the results accruing from successful experiments would be of great benefit and very far reaching in their implications."[51] The OFC sounded out the Air Ministry and the Meteorological Office for help, along with the physicist and chemist Irving Langmuir, a controversial figure in US atmospheric science who had conducted, through General Electric, a series of rainmaking experiments—of disputed success—with silver iodide across North America.[52] The steady prevailing winds and moisture-laden clouds of the atmosphere above Kongwa were thought to be suitable for application of the American techniques, but Langmuir ignored the OFC's repeated overtures until he was cornered by the British Embassy. His reaction was "apoplectic," insisting "it was not his business to assist groups in the British Empire," but he eventually agreed to share some of his reports, subject to State Department approval.[53]

The OFC tried to persuade the Air Ministry to lend a small plane of the Mosquito type for delivering dry ice directly to the clouds. But the Air Ministry queried the OFC's estimates of the amount of payload involved, insisting that at least two, much bigger, planes would be required. "We only want to seed clouds around Kongwa," protested a telegram from the field, "not [the] whole African continent." But the Air Ministry refused to commit aircraft to a project lacking in military significance.[54] The OFC decided it couldn't afford to charter a suitable plane locally. Little encouragement came from the UK Meteorological Office either, who poured cold water on the idea of seeding clouds for any purpose other than learning about cloud physics.

More favorable meteorological inclinations were found closer at hand. In 1949 David Davies, a future head of the World Meteorological Organization, had been sent from the Meteorological Office to take over the East African service following Walter's retirement. Quickly embracing the experimental possibilities of the troubled Groundnut scheme, Davies responded favorably to OFC requests for help with their cloud-seeding ideas and efforts to impregnate the clouds above Kongwa with smoke-borne silver iodide quickly got under way, with Bunting's Scientific Department helping out with the chemistry.[55] Silver iodide was a cheaper alternative to spraying dry ice from an aeroplane, and charcoal burners were constructed and arrayed on a track seven miles upwind of the groundnut units, where the dense concentration of rain gauges that had eventually been achieved made for an ideal laboratory of atmospheric experimentation: "There can be few areas in the world where such a network of rain gauges exists and in this respect Kongwa is a particularly suitable site for experiments of this kind."[56]

Thus was the "no-man's land" of Kongwa, a space constructed as essentially empty of human endeavor and hence ripe for the sociotechnical experimentation of the groundnut scheme, transformed into an outdoor laboratory for climatic modification.[57] But this transformation in the imaginative and material geographies of Kongwa was itself enabled by the clash between the guiding sociotechnical imaginary of the groundnut scheme and the more physiocratic imaginary of Albert Walter. After all, it was his insistence on closely monitoring the weather, against the homogenization machine of the new developmentalism, that populated the space with densely packed rain gauges and meteorological observers. In so doing he created the laboratory within which the guiding imaginary of the groundnut scheme could be taken to its logical conclusion, of bringing the atmosphere itself within the realm of technical manipulation (see figure 8.1). Smoke from the iodide-impregnated charcoal would, it was hoped, deliver to the clouds a stimulus for ice crystal formation and growth, with the heavy crystals then falling and melting to produce rain drops. But as in all experiments of this nature (see McKittrick, this volume), the challenges were as much epistemological as technical: "Owing to the huge area affected by the seeding, interpretation of the

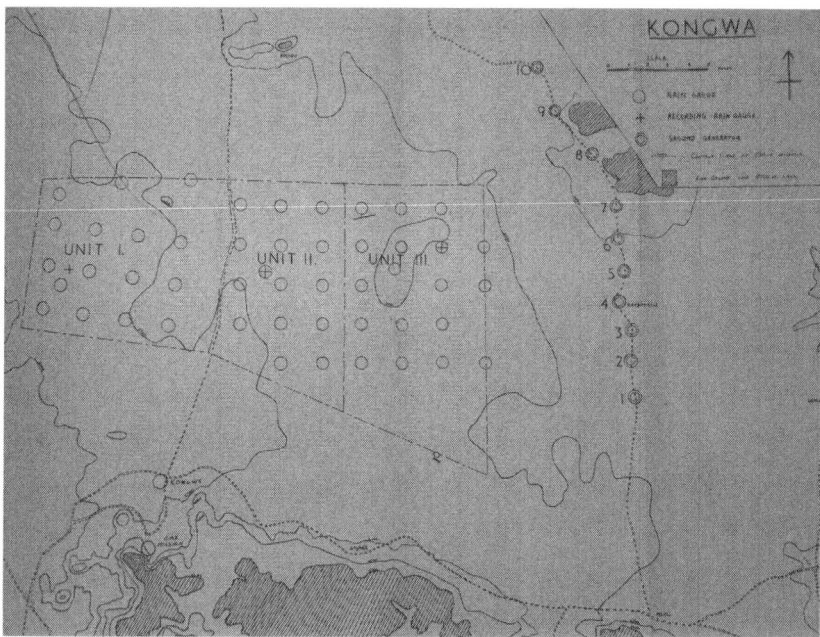

FIGURE 8.1. Walter's field observatory (*center, left*) and the advancing front of climate-altering charcoal burners (*right*). Extract from Davies, Hepburn, and Samson, "Report of Experiments, 1951," Appendix 1.

results is not easy, though this can be overcome to some extent by a statistical approach," Davies reported. Again, statistics, it was hoped, could help extend the reach of observation and inference across space. But the robustness of any statistical analysis was hampered when a plan to seed on alternate days, to enact a form of experimental control, fell afoul of growers' wishes for rain at particular moments, and dryness at others, which "were naturally given precedence over scientific experiment."[58]

With the wind failing to reliably blow the charcoal smoke in the right direction, "balloon-bombs" were developed by tethering aluminium photographic film canisters, filled with a mixture of gunpowder and silver iodide, to meteorological balloons. The bombs, it was hoped, would offer greater accuracy through the ability to target individual clouds as they passed over the site. However, four months of experimentation, closely watched by two dedicated observers stationed amid the rain gauges, seemed only to yield up the suggestion that the cloud-bombing might actually be reducing rainfall by dispersing the clouds, although the smoke generators seemed to be successfully coaxing some extra rain.

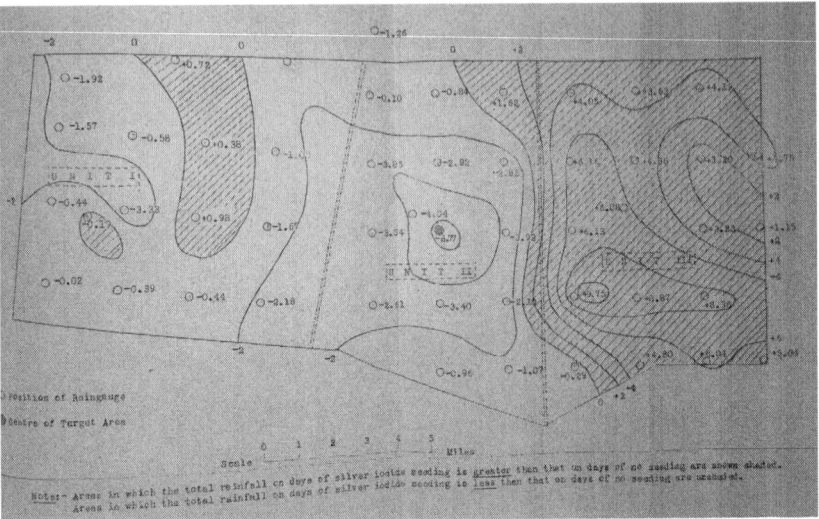

FIGURE 8.2. "Isopleths Showing Difference between Total Rainfall on Days of Silver Iodide Seeding and Days of No Seeding," from Davies, Hepburn, and Samson, "Report of Experiments, 1952," 20.

During the following rainy season hygroscopic salt particles were tried as a seeding agent, having been found to be more successful in inducing rain from cumulus clouds that had not reached freezing level.[59] By this point, however, the British government had at last lost patience and pulled the plug on the groundnut scheme in its original form, and at Kongwa activities were radically downsized and continued on a notionally "experimental" basis. For the rainmakers, this meant that a stricter experimental design could be followed as agricultural activities were wound down. A specially adapted Royal Navy rocket-flare launcher was sourced for more precise cloud bombing, and methods were more thoroughly described for a community of interested virtual witnesses that had grown with the publication of a series of reports in *Nature*, as well in scientific journals in South Africa and India (see figure 8.2).[60]

Whereas the earlier results were presented largely in narrative form, with isohyet maps of rainfall on individual days, the new experimental regularity enabled maps to be made of the average rainfall on days of no seeding, seeding with silver iodide, and seeding with salt, allowing more synthetic conclusions to be drawn. The titles of the respective reports reflect this new experimental confidence, with referencing changing from the "artificial stimulation" to the "artificial control" of rainfall. Hygroscopic days appeared to exhibit rainfall between six and twelve miles downwind of the seeding area, but less rainfall over the seeding area itself. Whether the net effect was more rain or less rain, Davies and company were confident that "the release of seeding agents did have some effect."[61] Later trials elsewhere in East Africa appeared to corroborate the Kongwa findings, albeit without the advantage of the "entirely objective" network of rain gauges at Kongwa, and the subsequent reliance on trained "personal observations" had to be carefully justified in a later report for *Nature*.[62] By this point the rainmaking trials were less a last resort for the troubled groundnuts and more an experimental proving ground for a strategy that still held out a glimmer of hope for British colonial developmentalism as a means of overcoming the apparent climatic limits to its ambitious schemes. This new tentative experimentalism, as opposed to the hubris of the heady days of the Groundnut scheme, was part of a growing realization during the 1950s that the British Empire was a far less significant global power than it had once been. Katharine Anderson has written of how the atmosphere of nineteenth-century India provided a model of disciplined order, extended over a vast imperial space.[63] In Davies's account of the rainmaking trials, the capacity to shape and discipline a natural order on any such scale is diminished, but the hope of a wider replication, a modest practical and moral contribution, still lingers: "Participation in experiments of this kind inevitably leaves the experimenter with a feeling of the enormity of the task he has undertaken in trying to interfere on a large scale with the natural processes of the atmosphere. The enormous benefits that all countries would derive from some ability to control the weather are obvious and this is particularly true of East Africa where rainfall is of fundamental importance. These conclusions, modest as they may be, are believed to be significant."[64]

METEOROLOGY AND THE END(S) OF EMPIRE

For a brief period toward the end of the troubled life of the Groundnut scheme, on which rested so many hopes for Britain's continuing legitimacy as a colonial power, everything came down to a small group of weathermen standing in a field, aiming balloon bombs and a flare gun at any passing cloud, desperately trying to resuscitate not only a few flagging groundnut plants but also the flagging authority of Britain's postwar colonial developmentalism. Albert Walter was by this point watching events from afar and fighting a rearguard action to defend the meteorological reputation of the scheme. S. Herbert Frankel, professor of colonial economic affairs at Oxford, had lobbied for the abandonment of the scheme and wrote a pair of excoriating essays in the *Times* criticizing the marginalization of expert advice: "Before work was begun at Kongwa no time could be spared . . . for adequate investigation of meteorological information bearing on rainfall."[65] Walter wrote a rejoinder, arguing that adequate investigation had been conducted, by himself, which gave a clear steer away from Kongwa, but the Ministry of Food, by this point in acute face-saving mode, successfully persuaded him not to publish his account of the whole messy episode in the *Times*.[66]

Walter's interest in climate and agriculture continued well into his retirement and was shaped by his groundnut experience. He continued to publish on statistical techniques for teasing agriculturally relevant information out of observations that were not necessarily made with that purpose in mind.[67] And he continued to try to speak climatological truth to developmental power even as the locus of that power shifted away from the British Empire. He offered to the US Department of Agriculture and to UNESCO a *Proposal for a research programme designed to lead to the provision of assistance to the under developed countries of the world*. With climate front and center the scheme championed both the scientific delineation of spaces fit for agricultural development and the value of peasant cultivation, arguing that the groundnut areas of Kongwa should have been "divided up so as to establish peasant proprietors on the land who would have been able, under the expert guidance of the eminent men employed to take advantage of the climatic conditions and adopt successful cultivation methods that a centralized mass production could not do." Peasant cultivators could have responded nimbly to the irregular rains, harvesting at the unpredictable end of the rainy season before the ground hardened in the sun. Walter did not deny the power of capital to create profitable agricultural industries but argued that beneficent flows of capital had to be shaped—like in Mauritius—by available climatological information.

Here was a very different sociotechnical imaginary from that which shaped the Groundnut scheme. Rather than faith in the power of capital and mechanization to overcome environmental limits, existing forms of cultivation could be nurtured and rendered more "resilient" (to use a current term) to environmental fluctuations through detailed technical knowledge in combination with the wisdom of farmers. In a Rousseauian flourish, Walter argued that the peasant cultivator "is surely much nearer to nature than any of the parasite industries which attach themselves to agricultural countries."[68] This was an imaginary that resonated with the agrarian developmentalism of the Colonial Office, but that also arguably had roots in the long debates over climate, agriculture, and political economy in Mauritius. The physiocratic roots of these debates emphasized the primacy of the cultivator to the economic health of a country and prized the rational maintenance of a natural order.

Walter can also be positioned within a conservative, paternalist tradition of British imperial thought, convinced of the global significance of the civilizing mission yet deeply sceptical of grand schemes to "force ... African peoples into channels for which they may be unfitted." And it was climate, above all, that determined the contours of developmental possibility and that required the most concerted European intellectual engagement. The local inhabitants of areas such as Kongwa might "look to the higher civilisation for guidance and inspiration," but that "higher civilisation" also needed a more nuanced appreciation of the "cultural and natural environment" of colonial territory and the shaping of both by climate.[69]

Different ways of dealing with climate were woven into the practices and imaginaries of British late imperial colonialism. The groundnut scheme was a contested knowledge space whose geographies were shaped in consequential ways by particular and competing imaginaries of the future, whether in the form of a dominant view of colonial development as the heterogeneous engineering of matter and people through new forms of mechanized agriculture and putative climate control or as Walter's neophysiocratic determinism, with its elevating of climate as chief determinant of human fortunes, its faith in peasant cultivation, and its emphasis on close monitoring of weather and climate across spatial scales. These two imaginaries clashed through disagreements not only about what Kongwa's climate was like but also about how that climate could most reliably be known—through direct or indirect observation, with statistics, or by analogy. The case offers further illustration of how, even in its most confidently homogenizing moments, British colonial power was still marked by profound disagreements among its expert agents not only about the nature of colonial environments and peoples but also about the nature and direction of the imperial project writ large. The notion of sociotechnical imaginaries allows us to capture how these disagreements were built upon very different idealizations of a future contract among science, expertise, and social order and to trace the power of such normative visions in shaping action in historical presents. We can consider sociotechnical imaginaries as constitutive features of broader imaginative geographies, encompassing claims upon "the condition of both the known world and the horizons of possible worlds."[70] These imaginative geographies shaped the material and social geographies of climatic knowledge-making at Kongwa, the spatial practices of making credible claims about colonizable space, and the fate of a particularly ocular demonstration of the (faltering) power of colonial ambitions.

PART III

ATMOSPHERIC ENTANGLEMENTS

9

DARWINIAN HIPPOCRATICS, EUGENIC ENTICEMENTS, AND THE BIOMETEOROLOGICAL BODY

David N. Livingstone

IN RECENT YEARS THE EFFECTS OF CLIMATE CHANGE on human health have increasingly gripped the public imagination. In report after report, bodies such as the World Health Organization (WHO) and the United States Centers for Disease Control and Prevention (CDC) have issued warnings about the impact that elevated temperatures, extreme weather events, El Niño, ozone depletion, and the like will have on the incidence of respiratory allergies, cardiovascular disease, neurological disorders, vector-borne illnesses, and many more.[1] In some ways, if Charles Rosenberg is correct, these concerns represent something of a return to the localism of the ancient Hippocratic tradition with its emphasis on the significance of winds, weather, and seasons for health, and thus on what he calls "the epidemiology of place."[2] By the mid-twentieth century Hippocratic ways of thinking had been dramatically decentered in Western medicine in favor of laboratory-driven diagnostics. But since the 1990s Rosenberg discerns an increasing dissatisfaction with the "abstracted bodies" of medical modernity and a recovery of "bodies situated in specific places" owing to the growing awareness of the health impacts of climate change.[3] This renewed interest in the relationship between local climate and human bodies is thereby contributing in major ways to the remaking of global geographical imaginations in our own day.

One of the clearest manifestations of this localist recovery is what I call "postcode prognostication," an online means of checking health risks by address.

The website of the US Natural Resources Defense Council, for example, provides a facility for keying in your zip code and receiving a report itemizing the health threats arising from climate change for your particular locality.[4] Information is provided for the occurrence of ozone and ragweed, the average number of days per year of extreme heat, the incidence of positive reporting of dengue fever and mosquito vector species, average number of days of low watershed flow, and so on. When viewed in composite the site provides a cartography of climate health risk from area code to continental scale.

The links back to the ancient Hippocratic *Airs, Waters, Places*, moreover, are not simply the diagnosis of Rosenberg the professional historian of medicine. Rather, the Hippocratic sense of place, particularity, and pathology is advertised by many prominent writers on the health threats of climate change. The WHO's *Climate Change and Human Health*, for example, sets current medical concerns in the context of Hippocrates's writings on epidemics and seasonal variation. And Sir Andy Haines, in his Harben Lecture for the Royal Institute of Public Health in 2005 prominently advertised Hippocrates's insistence that "climate has wide-ranging impacts on health."[5]

A lengthy period of time has elapsed between the writings of Hippocrates and the WHO publications—too long to survey even in outline here. What I want to focus on is a critical moment in the first half of the twentieth century when a neo-Hippocratic resurgence crystallized in the development of biometeorology as a species of medical-geographical science.[6] By inspecting the ways in which practitioners and their associates merged Hippocratic medicine with Darwinian evolution and how they reimagined the body as a biometeorological entity—a "cosmic resonator," as one spokesman put it—we can begin to see something of how a blending of climate science and medical research could succumb to the enticements of eugenics. Exactly what lessons are to be drawn from this historical inquiry for today's concerns about climate change and health I leave for others to judge.[7] But I hope to show the ease with which inquiries into the health effects of different climatic regimes could glide toward the pathologization of particular places and people and allow space for the expression of sentiments shaped by the desire for political biopower. Dividing the world's climates into the sickly and the salubrious provided a rich resource for those intent on fashioning a geographical imaginary that could readily serve the aspirations of eugenic enthusiasts who sought to control human populations through managing national immigration and human breeding.

THE BIOMETEOROLOGICAL BODY

The modern portrayal of the human body as a biometeorological object owes a good deal to the writings of William Ferdinand Petersen (1887–1950), a Chicago physician and a professor of pathology and bacteriology at the University of Illinois College of Medicine.[8] Indeed while the advocates of human biometeorology routinely insist that, when viewed in the longue durée, its primary coordinate can be found in Hippocratic philosophy, its modern revitalization is frequently traced to Petersen.[9] Petersen's early work on the aetiology of disease was dominated by a bacteriological perspective.[10] But he increasingly moved toward a more holistic conception of illness and produced a number of works on the health effects of climate. The more than four thousand pages that constituted the seven-volume *The Patient and the Weather* (coauthored with Margaret E. Milliken) represent the most comprehensive statement of Petersen's medical meteorological thinking. Here he reimagined the human body in the language of bioclimatology by reconstruing disease as "air hunger" and insisting that "dysfunction and inadequacy of the mechanism that has to do with oxygen supply is probably the fundamental cause of all disease."[11] In support he elaborated in painstaking, not to say mind-numbing, detail correlations between meteorological and medical conditions. A sampling will suffice to give a flavor of the whole. Attacks of coronary thrombosis peaked during the winter; asthma rose to a climax during late fall and early winter; gastric and duodenal ulcer flare-up was associated with the passage of cold fronts; appendicitis struck one to two days after a cold front has passed. The list goes on. Arthritis, thyroid disease, gallbladder disorder, tuberculosis, coronary thrombosis, cerebral haemorrhage: all these and many more are examined as disorders that are meteorologically conditioned.

Because of the causal connections Petersen believed he had identified between weather and well-being, between seasons and sickness, his understanding of pathology bore a strongly spatial imprint. What might be called the geography of death frequently surfaced in his analyses. Mortalities from angina pectoris, for example, clustered in the northeast United States where there was the greatest meteorological instability; in New England death rates from eclampsia spiked in the spring, whereas in Illinois this occurred during winter and fall; fatalities from appendicitis, closely associated with cyclonic events, were maximal in the Rockies and on the Great Plains.

Petersen marshaled his formidable array of data in support of the conviction that the human body, as he put it on various occasions, was a "cosmic resonator" (see figure 9.1).[12] The "individual is weather conditioned," he declared; "we respond to every whim of the air mass in which we exist." Self-consciously following in the wake of Hippocrates and his successors, he could assert: "The cosmos (as Bodin suggested) lives in the organism as truly as the organism lives in the cosmos."[13] Elsewhere he further elaborated: "I have used the term 'Cosmic Resonator' to indicate the very close integration of the autonomic apparatus of

FIGURE 9.1. "Man as a Cosmic Resonator." Petersen, *Lincoln–Douglas*, 133. Illustration by Jean McConnell.

the body and the influences of the cosmic—and particularly meteorological—environment in which we exist." The organic rhythms that governed nearly every aspect of human health, he insisted, were "largely meteorologically conditioned and reflected even minor change to a surprising degree."[14] Since, for Petersen, medical climatology connected the intimate spaces of the human body with the far-flung spaces of the cosmos, he was engaged in an enterprise to renovate the geographical imagination of the ancient Hippocratics. In pursuing this vision, Petersen liberally deployed meteorograms incorporating medical records to highlight precisely the sorts of association he divined between atmospheric and clinical conditions. In Petersen's hands these medical-meteorological charts became rhetorical devices of persuasion enabling him to give visual expression to how "the organic mass" was supposedly conditioned "by the inorganic environment."[15] It was a comprehensive enterprise. For it was not just the body that came within the arc of climate's imperial reign; mind and emotion were no less governed by its imperatives.

By the time Petersen's *Man Weather Climate* appeared in 1947 he was already operating at quite some distance from clinical medicine. In the preface he made it clear that the ways in which the human body registered "organic reactions" to atmospheric circumstances were "reflected in cultural cycles, in the waxing and waning of religious and ethical concepts, in social revolution, in mass migrations and in wars."[16] In the final chapters he turned his attention to the writings of such figures as Ellsworth Huntington and Cesare Lombroso. He drew, too, on the work of Flinders Petrie, the English Egyptologist who used meteorological data in archaeology to explain the rise and fall of civilizations, and on the work of the German climatologist Wladimir Köppen who, Petersen observed, "predicted the Russian Revolution in 1917 on the basis of solar disturbance."[17]

But this was not a new departure. A few years earlier, in 1943, he had published a study entitled *Lincoln–Douglas: The Weather as Destiny*. Although the work was ostensibly dealing with the lives of Abraham Lincoln and his political opponent Stephen Douglas, its "true subject," as one reviewer put it, was "everyman."[18] Right from the start two things were crystal clear. First, Petersen's purpose was to connect, in causally compelling ways, climatic forces with the life paths of Lincoln and Douglas. His tactic was to focus on critical moments in the experiences of both figures to uncover their susceptibility to atmospheric circumstances. Thus the prevailing climatic conditions when each was conceived (in May 1808 and July 1812, respectively), the pattern of storm tracks in central Illinois and the effects of seasonality were called upon to explain such things as the "complexity" of Lincoln's personality, the "simplicity" of Douglas's, Lincoln's moodiness and depressive tendencies, Douglas's easygoing temperament and indifference to philosophy. Second, the particulars of Lincoln and Douglas were recruited to serve as proxies for far greater generalizations. For example, these two figures were staged as instantiations of more general body types. Lincoln was a case of what he called the "Slender Type," Douglas of the "Broad Type." The former, "tough and wiry" but more likely to be "deficient in muscle and connective tissue," was "distinctly more sensitive to the environment" than the "heavy-boned" and "well-muscled" Broad Type (see figure 9.2). They thus reacted in different ways to the weather, especially so for those inhabiting the northern latitudes—"those regions of the world that are atmospherically unstable."[19]

Petersen's meteorobiography of Lincoln and Douglas was self-consciously tethered to Hippocratic moorings. Chapter after chapter was inaugurated with an epigraph from the classical father of medicine. And in Petersen's hands, this Hippocratic impulse had the effect of reducing human agency to the operations of natural law. In fact he had already been devoting his energies to retooling the Hippocratic corpus for the twentieth century. In 1938 he conceived the idea of producing a collection of Hippocrates's writings translated into the idioms of the modern medical lexicon. His aim was to change "archaic terms into what Hippocrates actually meant."[20] The book appeared in 1946 under the title *Hippocratic Wisdom*.[21] In the last analysis it amounted to little more than a sustained effort on the part of Petersen to insert himself into the text. It was less a work of exposition

FIGURE 9.2. "Weather, Mood and Body." Petersen, *Lincoln–Douglas*, 12. Illustration by Jean McConnell.

than an exercise in expropriation. Symbolic of the whole enterprise was the final chapter of the work, entitled "The Human as a Cosmic Resonator," which exemplified how Hippocrates was made to ventriloquize Petersen. As Sargent notes: "The physiology that Petersen found in the Hippocratic *Corpus* reflected his own knowledge and theories. Petersen could not think in the context of Greek medicine. He was thinking in the context of medicine in the twentieth century and read into the words of Hippocrates his own modern insights."[22]

Petersen's reputational legacy lies marooned somewhere between crank and sage. As one sympathetic commentator, Sallie Tisdale, put it in an appropriately entitled op-ed "Confessions of a Cosmic Resonator": "Petersen was the consummate crackpot scientist, a zealot who was either way ahead of or way behind his times, depending on your point of view. He devoted the last 20-some years of his life to a collection of anecdotes, statistical minutiae, and page upon page of graphs and tables analyzing everything from nasal pH to leucocyte counts in relation to changes in the weather." And indeed, while a 1943 reviewer of one of Petersen's writings for *Time Magazine* began with the quip, "Genius has had many odd explanations, but one of the oddest yet was advanced last week by an eminent Chicago pathologist," as recently as April 2013, an article by Stephen Rosen (author of *Weathering*) for the *East Hampton Star* insisted that "Landmark investigations by William F. Petersen, a pathologist," had "demonstrated that clinical symptoms, mental reactions, and abnormal behavior are conditioned by changing seasons and weather."[23] Rosen's book *Weathering*—a kind of biometeorological primer for a wider public that came out in 1979—contained more references to Petersen than to any other writer.[24] Whatever the ambivalence of these judgments, during the mid-twentieth century Petersen's perspective persisted in the endeavors of the International Society of Biometeorology (ISB), which came into being in 1956. The links to Petersen are readily discernible not only in the new society's entire focus but also in the fact that it established "the William F. Petersen Foundation" in 1963 "to commemorate the great biometeorologist."[25]

Petersen was not a lone voice. Far from it. In developing his Hippocratic medical-meteorology, he kept closely in touch with the work of the Yale geographer Ellsworth Huntington (1876–1947), whose love affair with environmental determinism more generally is well-known.[26] But perhaps less visible are the reports Huntington produced on human health in relation to climatic conditions and the way in which he marshaled a synthesis of evolutionary biology and neo-Hippocratic medicine for eugenic ends.

Huntington's interest in climate as an active health force initially surfaced during the writing of *Civilization and Climate*, published in 1915. A product of what he called "the new science of geography," this title stood self-consciously in a tradition reaching back to classical antiquity through Charles-Louis de Secondat, Baron de Montesquieu and Henry Thomas Buckle who, as Huntington reminded his readers, "believed that climate is the most important factor in determining the status of civilization." Huntington's interest was in the sources of civilization more generally, but now he inserted matters of health into his diagnosis using the neo-Hippocratic-sounding language of climate, blood, and circulation. Noting that "at high altitudes the number of red corpuscles in the blood increases enormously, and the capacity to absorb oxygen and to give out carbon dioxide is correspondingly modified," he believed the most likely explanation lay in climate's influence on the circulatory system.[27]

In the years to come Huntington devoted considerable energy toward elucidating causal connections between weather and health—research that received the commendation of Charles-Edward Amory Winslow who founded the Yale School of Public Health and his coauthor Lee Pierce Herrington of the State University of New York. In 1920 Huntington produced a statistical analysis intended to demonstrate "the control of pneumonia and influenza by the weather" in the wake of the "Spanish Flu" pandemic of 1918, one of the greatest medical catastrophes of the twentieth century. In 1923, in a separate analysis of the US experience, he concluded that "the favorable conditions of the air were the greatest factor yet detected in helping the people of the cities of the United States to ward off the influenza in the fall of 1918."[28]

And then in a much longer assessment, written in his capacity as chair of the National Research Council's Committee on the Atmosphere and Man, he delivered an assessment of the role of weather in explaining patterns of daily mortality in New York City. Here he emphasized the importance of diurnal temperature changes and insisted that a "drop of temperature is systematically accompanied by a relatively low death rate; a rise by a high death rate." Unsettled weather, it seemed, was "accompanied by a lower death rate than is uniform weather." Of crucial significance in his analysis was his idea that humankind flourished in what he called "optimum" weather conditions—a finding he explained on evolutionary grounds. And these circumstances formed the conceptual foundation on which his whole edifice was constructed. He concluded: "The whole thing may be summed up by saying that for each of the three climatic elements . . . investigated there appears to be a well-defined optimum. . . . Any appreciable departure from the optimum temperature either upward or downward is accompanied by an increased death rate."[29]

Beyond these relatively specialist inquiries, Huntington was keen to disseminate his climatic philosophy of the body much more widely. Thus in his 1933 general text *Economic and Social Geography*, cowritten with Frank E. Williams and Samuel van Valkenburg, he devoted a whole chapter to "Climate, Health, and the Distribution of Human Progress." Right upfront he announced, in confidently determinist mode: "A comparison of many kinds of maps shows that climate is directly or indirectly the main determinant of the geographical distribution of economic activity and civilization. This happens . . . primarily through its effects on man's health, energy, and mental alertness." His technique, though simple—if not simplistic—was visually striking: he used maps to correlate, for example, the distribution of what he called "climatic energy" in the United States with data on health derived from life insurance companies. The take-home message was crystal clear. Huntington wanted to segregate the globe into a hierarchy of spaces that he typecast on a spectrum from low to high, from inefficient to efficient, from unfavorable to "nearly perfect."[30] Thereby he could map the world's climatic regimes according to the biopolar categories of privileged and ill-favored, healthful and disease-ridden, stimulating and crippling. It was an imagined geography of enormous scope and power. Not only did he believe he could empirically map global biopower, he hoped his cartographic energetics could in turn be mobilized to manage human populations for eugenic ends.

Something of the same impulse also found expression in the writings of another key member of the biometeorological circle, Clarence Alonzo Mills (1891–1974), professor of experimental medicine at the University of Cincinnati and attending physician at the Cincinnati General Hospital.[31] Mills's interest in medical climatology had been stimulated during a visit to China in 1926 where he observed among Chinese diabetics a curious sensitivity to insulin and among a number of patients a functional insufficiency of the adrenal glands. As he pondered on such problems, Mills came to ascribe these syndromes to peaks of humidity and heat.[32] This led to a lifetime's work on the effects of climate on the body. The publication of *Medical Climatology: Climatic and Weather Influences in Health and Disease* in 1939 represents an early statement of Mills's developing outlook. Here, and in a host of particular case studies, he scrutinized the influence of climatic conditions on metabolic, cardiovascular, and infectious diseases; later he also devoted his attention to matters of air pollution and air conditioning.[33] The scope of his climatic diagnostics was extensive. Among the medical conditions he identified were links between weather patterns and thiamine deficiency;[34] causal connections between storminess and high blood pressure, resistance to infection, and respiratory problems;[35] the impact of atmospheric temperature on sexual cycles and the onset of puberty;[36] how the consumption of caffeine and alcohol was influenced by different climatic regimes;[37] and the role that prevailing weather played in the development of such disorders as diabetes, pernicious anaemia, leukaemia, tuberculosis, pneumonia, sclerosis, and failures of the vascular system.[38]

Huntington, in the main, liked Mills's effort, describing it as "a perfect mine of valuable suggestions" and as work of "high value" that "blazes a pioneer trail in its suggestion that the study of climate may be the clue which will lead ultimately to a reduction in metabolic or degenerative diseases."[39] Joseph J. Spengler, the distinguished American economic historian, was also enthusiastic. He was convinced that sociologists would have to revisit their modes of explanation in the light of Mills's work, which, to him, demonstrated "at length the importance of the effects of climate and weather upon the fundamentals of human existence."[40] By contrast, the reviewer for the American Medical Association was troubled by Mills's expansive claims and observed: "In spite of the author's statement in the preface that he does not desire to hold climate and weather responsible for all human reactions, the reading of the book demands the constant exercise of the critical faculties to avoid this conclusion."[41]

Like both Petersen and Huntington, moreover, Mills increasingly cultivated a more general climatic philosophy of history. Perhaps it was most prominently manifested in a report that appeared in the *Mason City Globe-Gazette* in March 1941. "Dr Mills believes," the author announced, "that the rise to power of Adolf Hitler in Germany and Benito Mussolini in Italy may be due in part to the gradually warming temperatures of the world. People are more docile and easily led in warm weather than in cold, Dr. Mills insists."[42] Evidently, in Mills's vision, the scope of climate's empire was vast indeed. So comprehensive was its power in fact that he told the readers of his 1942 volume *Climate Makes the Man* that "climatic influences . . . affect man's rate of growth, speed of development, resistance to infection, fertility of mind and body and the amount of energy available for thought or action." With climate's dominion extending over such an expansive territory, Mills concluded that the human race was nothing more than "a veritable pawn of the universe."[43]

A mere sampling of the sphere of political influence over which climate held sway reveals something of the scope Mills attributed to it. In cultural history he ascribed the triumphs of ancient Chinese civilization, the rise of early Greece, the advent of the "Dark Ages," the French Revolution, the modern imperial expansion of Japan, and such like to the vicissitudes of climate. In terms of economics, he connected business cycles with what he called "weather stimulation" and "human energy."[44] Indeed he was so convinced of his "energy theory of economic and social trends" and that the weather was "the basic dictator of man's energy and business activity," that he held out the hope that economists would "seek the help and advice of scientists in the fields of meteorology and physiology" in their aim of keeping "business on an even keel."[45]

Underlying Mills's climatic faith was the foundational conviction that weather and climate "determine the level of activity or 'pep,' the restlessness or complacency, bodily vigor or sluggishness, progressiveness or contentment with mere existence"—a view he believed he had substantiated from laboratory experiments on rats (two in particular, Ivan and Hilda) kept in temperature-controlled environments. His work on hot-room rats fed on diets deficient in thiamine, for example, apparently demonstrated "the tropical resignation and lack of initiative which keeps large sections of the human race content with a low level of existence." Plainly he found it remarkably easy to inflate Ivan's and Hilda's behavior in a hot box into a global geography of what he called climatic drive. As he explained to readers of his *Medical Climatology*: "It is most significant . . . that climatic environment . . . exert[s] its major and most direct effect on the ability of the body to produce energy."[46]

Climatological energetics became a dominating preoccupation shaping Mills's geographical imagination on a global scale. And so, like Huntington, whom he frequently cited, he did not hesitate to construct hierarchical catalogues of the world's regions arranged in "descending order of climatic energy." The results were both dramatic . . . and dramatically self-serving. First place, happily, went to "the storm belt of North America."[47] Particularly attractive was the region "around and just to the west of the Great Lakes." This zone possessed "one of the most invigorating climates the earth has to offer, and there human physical development shows clearly the effects of the intense climatic drive."[48]

By contrast, the "zone of lowest climatic drive" incorporated "Africa, India, South China, the Malay Peninsula, the East Indies, Philippines, northern Australia, much of South America east of the Andes, and low lying parts of Mexico and Central America." Here things were pretty grim. "Africa—'darkest Africa'—has little to offer for human development," he announced, while "Asia . . . is likewise damned by climate."[49] The tropical world was thus relegated to the margins of historical significance. Medically, its "depressive heat" meant a lowering of "bodily energy" with the result that "people there cannot meet physical emergencies with the vigor that temperate zone residents show."[50] Politically the tropics were impotent to govern themselves. As he starkly put it in 1963: "only under the most rigid dictatorship can tropical low-energy nations achieve and maintain today's ideal of an industrialized society."[51]

In his portrayal of the tropical psyche, gloomy verb was piled upon gloomy noun and adjective: "The heat of the tropics lulls people into a passive complacency and saps their vitality," he proclaimed.[52] Passivity, resignation, sluggishness, lassitude, monotony: these were diagnostic of the tropical body. Correspondingly, the temperate world, with its storminess and variability, was home to cyclonic civilizations where "the greatest wealth and activity" were to be found, as well as "the most advanced procedures in public health and the lowest death rates." Why? Stormy weather delivered to the occupants of the mid-latitude cyclonic zone "a wholesome and stimulating variety of life, inducing in people a restlessness which . . . drives them on to build skyscrapers, set up great factories, and pursue other energetic activities." Here, Mills asserted, there "is little monotony, either climatic, mental, or physical . . . health in general is most buoyant and life most interesting. The men dwelling amid such influences are the ones who have dominated the world in the past."[53]

During the middle decades of the twentieth century, the Hippocratic vision of the body and its functioning as being responsive to the imperatives of climate received renewed impetus from those intent on cultivating a science of biometeorology. In the hands of its advocates, moreover, this project frequently expanded into a universal philosophy of history as more and more dimensions of human life and culture were brought within the arc of climatic explanations. At the same time this latest Hippocratic intervention stimulated the production of worldwide imaginative geographies that segregated the globe into baleful and beneficial climatic regimes. In this respect biometeorology perpetuated a long tradition of post-Hippocratic writing extending from Jean Bodin to Montesquieu and well beyond. But in at least one crucial respect this latest incarnation of the medical-meteorological tradition marked a new departure. It sought to incorporate Darwinian modes of explanation within its agenda.

DARWINIZING HIPPOCRATES

The Darwinizing impulse in biometeorological inquiries surfaced in a number of different arenas. For instance, William Petersen devoted the twenty-second chapter of *Man–Weather–Sun* to the subject of "Genius"—a topic much beloved of eugenicists. Here, while acknowledging the influence of genetic background, Petersen drew on the writings of figures such as Ellsworth Huntington and the Italian physician and criminologist Cesare Lombroso to support the claim that there were direct links between birth month and intellectual accomplishment. Petersen's own feeling was that the season of conception was of even greater significance than birth date because of the plasticity of the embryo's cerebral tissues and thus its susceptibility to climatic stimuli. His own inquiries suggested that the conception of what he called "unusual human types" took place during late winter and spring when "meteorological variability" was at its height.[54]

The effects, Petersen was sure, could be generalized. What he referred to as "our western culture, primarily Hellenic in origin," for example, revealed "the interesting phenomenon of the sudden appearance on the world stage of the largest group of unusual individuals that ever appeared in a short time"—a circumstance that coincided with "the longest recorded period of the Aurora Borealis."[55] This association encouraged him to direct his attention to more recent European and American data on birth years and genius using information contained in biographical dictionaries and connecting these with sunspot occurrence. It all confirmed his judgment that there was probably "some relationship between differentiation of mentality and the meteorological environment which exists at the time of the individual's conception."[56]

Why did such a relationship exist? Darwinizing climate provided the answer. Prenatal and infant environmental stress arising from climatic conditions introduced a more rigorous "selection" of certain "human types at certain periods." Climate, to Petersen, was an active agent of natural selection, and this meant that "organic evolution" should not be thought of as "as bound by the (relatively) immutable laws of genetics" but, rather, driven by the "far greater organic plasticity arising from the interaction of the environment, the parental organisms, and the egg or the very young embryo." Evolutionary theory provided a vocabulary—variability, selection, adaptation—by which to make sense of the body as a biometeorological phenomenon.[57]

Huntington had also been devoting his energies to the whole topic of birth season and its geographical variation. Intrigued by the "effect of the seasons" as revealed in "the birthdays of men of eminence," he insisted in 1938 in *Season of Birth* that the most propitious temperatures for physical and mental development were due "to the selective effect of the climate in which the species *Homo sapiens* originated during the glacial period."[58] In the 450 or so pages that followed he elaborated on the geography of birth season by surveying conditions in Belgium, Russia, Japan, and the United States, while at the same time exploring the impact of the seasons on such eugenic themes as sex ratios, race, rank, longevity, leadership qualities, insanity, genius, and criminal behavior.

Several things stand out in this investigation. Primarily, Huntington's analysis was a conscious evolutionizing of the Hippocratic vision.[59] Right from the outset, he located his interest in birth season in the framework of early human evolution by seeking to ascertain the lingering significance of the climatic regime within which the human species had emerged. He called attention to the fundamental role in human affairs of the basic animal rhythm of reproduction. The seasonal patterns of human birth that he believed he had ascertained were thus rooted in evolutionary biology. The conspicuous "truncation of the mortality curves of both men and women throughout the reproductive period during the months when the basic animal rhythm of reproduction is at its height," he observed, "represents a deep-seated adaptation to seasonal fluctuations of the weather."[60] The reason had to do with the links between seasonality and spermatozoa that the Massachusetts state biologist, the parasitologist David Lawrence Belding MD (with whom he was in direct communication), had discerned in his research on fisheries. "The seasonal swing in the birth rate," Huntington noted, "appears to be the direct result of a physiological swing whereby the number of spermatozoa ebbs and flows with the seasons" (*Season of Birth*, 260). All of this was swept into a synthetic Hippocratic evolutionism the dimensions of which he spelled out in a chapter on "The Evolution of *Homo Sapiens*." The following excerpt gives a flavor of this vision:

> The first conclusion is that in man, just as in animals, an annual rhythm of reproduction is of great importance. . . . One of the most interesting problems . . . is the reason why a temperature in the neighbourhood of 62° is the optimum for reproduction. Why not 40° or 80°? We have come to the conclusion that the reason is found in a selective process that occurred among our primitive ancestors. In climates of an intermediate type, neither very cold nor so warm that they are monotonous, the children who are conceived when the average temperature rises to about 62°F. are born in the early spring when reasonably warm weather first arrives. That is the time when the supply of food begins to increase after the scarcity of winter. In other words, birth occurs at the season which is best for the survival of the infant. . . . A major feature of this book is the discovery that the optimum temperature for reproduction is also the optimum for physical health in general. (417, 420)

A second prominent fixture in Huntington's climatic edifice manifested itself in the opening pages of *Season of Birth*. Huntington announced it as a novel discovery, a further extension of the Hippocratic-evolutionary credo. The selective operations of climate in the evolution of Homo sapiens during the glacial period, he urged, constituted "a convincing explanation of the highly puzzling fact that man has two different optima of temperature—physical and mental. . . . The physical optimum apparently represents the season when conceptions were most likely to result in the birth of infants that were able to survive. The mental optimum represented the cooler season when births of this kind actually occurred" (v). The physical optimum occurred at 60–65°F, the mental optimum at 39–54°F. Why? And more particularly: "Why should metabolism and the glandular functions proceed most perfectly at that temperature?" (2).

Again the story of early human evolution provided the answer. The "conditions governing season of birth among our primitive ancestry apparently explain the mental as well as the physical optimum of temperature," he began. "Our bodies apparently function best," he explained, "at the temperature which prevailed in primitive times at the mating season, for children conceived at that season were best able to survive. This gives us a physical optimum." By contrast, human "minds function best at the temperature which prevailed in primitive times at the normal season of birth. At that time, more than any other, the survival of the new generation depended upon the alertness of the parents. Thus a mental optimum became established at a temperature lower than the physical optimum" (3).

Huntington had been toying with such ideas for well over twenty years. In his 1915 *Civilization and Climate*, for example, he had devoted a whole chapter to what he called "The Ideal Climate." Here he dwelt on the most suitable conditions for factory operatives and students to best perform their respective tasks, and he identified a "mental optimum of 38°" and a "physical optimum of 60° or possibly 65°."[61] Huntington seized on these numbers to sketch a global geography of "climatic excellence"; they were concentrated in a few "chief portions of the globe"—England, North America's Pacific northern coast, and New Zealand, with parts of South America being compromised on account of its rainfall deficiency.[62] Elsewhere the relative absence of annual and diurnal variability of atmospheric conditions condemned inhabitants to physical danger, mental lethargy, and moral peril. A few years later, in his *World-Power and Evolution* of 1919, he repeatedly deployed the idea of climatic optima in his discussion of such themes as "climate and health" and "the environment of mental evolution."[63] Thereafter the idea resurfaced periodically. In his 1920 analysis of how weather controlled pneumonia and influenza he characterized "ideal" atmospheric conditions for the "best treatment for all kinds of respiratory diseases" as air that is "variable in temperature," "fairly moist," and "not too warm."[64]

Huntington's mobilization of the idea of climate optima and birth season proved attractive to those interested in cultivating a science of biometeorology. His priority in identifying the significance of birth season was noted by Petersen. And Frederick Sargent II, another key figure in the evolution of biometeorology, paused in his survey of the Hippocratic heritage to expound Huntington's concept of climatic optima as a critically significant biometeorological finding. It was picked up too in Per Dalén's 1975 study of connections between birth season and schizophrenia and has continued to feature as a landmark intervention in a variety of medical arenas.[65]

The Darwinian motif surfaced conspicuously in the work of the influential German-born climatologist Helmut E. Landsberg (1906–1985) who achieved distinction for his pioneering work on the statistical analysis of climate, the biometeorological consequences of urbanization, and the effects of weather on human health and behavior. During his long career, Landsberg held many posts. He worked as an operations analyst with the US Army Air Corps, occupied the position of director of the US Weather Bureau's Office of Climatology, and was appointed to the faculties of the Universities of Chicago and Maryland.[66]

Among his numerous contributions was his 1969 text, *Weather and Health: An Introduction to Biometeorology.* Here he introduced his readers to a range of key topics such as the science of the sun's radiation and its effects on human skin, the impact on the body of air and altitude, and the medical effects of urban temperature, contaminants, air pollution, and sulphur dioxide. Landsberg's familiarity with contemporary meteorological science did not prevent him from resorting to the ancient writings of Hippocrates, and so, in his reflections on what he called "Weather Suffering," he found space to review the influence that Hippocrates believed wet years exerted on "fevers, gangrene, epilepsy, apoplexy, and quinsies," adding that some of these ideas "have stood both the test of time and scientific scrutiny."[67] The experience of pain, the flare-up of rheumatic conditions, the onset of asthma, and the occurrence of heart attacks all displayed seasonal patterns. At the same time Landsberg's Hippocratic sympathies were suffused with Darwinian sentiments. Thus his reflections on human responses to cold conditions were couched in the language of migration and "the race for survival in a cold environment" and were buttressed by Darwin's own observations drawn from the *Beagle* voyage. Similarly his comments on hot environments and heat stress were grounded in the human biology of evolution in a tropical climate. The close relationship the human species sustained with its climate thus stemmed from the operations of natural selection, which had ensured that early humans "were obviously adapted to an environment in which they could easily survive." This entire conception of the human body as a biometeorological form was thus rooted in the merging of Hippocratic medical geography and Darwinian evolution.[68]

In other ways too neo-Hippocratic biometeorology fitted with evolutionary ways of thinking. Central to a good deal of biometeorological thought was the image of the body as a mosaic of spaces, a variegated landscape responsive to the imperatives of weather and climate. This corporealized microgeography comes through clearly in the work of one of modern biometeorology's chief architects, Solco Walle Tromp (1909–1983), a Dutch geologist and geophysicist based at the Biometeorological Research Center at Leiden.[69] Tromp depicted the human body as a series of meteorotropic zones or "registration centres"—the skin, the nose, the lungs, the nervous system, and so on—which reacted in diverse ways to meteorological stimuli.[70] Of crucial importance was the hypothalamus, a section of the brain, through which, he insisted, changes in weather and climate affected the body on account of its thermoregulatry function. Not surprisingly this geographical reimagining of the human organism as a suite of somatic regions responding in distinctive ways to atmospheric forces drew advocates toward ecological modes of thought and to Darwinian adaptationism in particular.

A case in point was Frederick Sargent II (1920–1980), who took an undergraduate degree in meteorology before studying medicine. He was elected first president of the ISB when it was formed in 1956.[71] His early work at the Harvard Fatigue Laboratory and the US Army Medical Nutrition Laboratory in Chicago had confirmed his expertise in physiology, and he later held academic positions at the Universities of Illinois, Wisconsin, and the University of Texas School of Public Health. Besides his scientific labors, Sargent produced a lengthy historical treatise (the first history of human biometeorology, according to Helmut Landsberg) under the title *Hippocratic Heritage: A History of Weather and Public Health*, which appeared posthumously. In fact this was a work of apologetic, constructing an extended historical tradition for contemporary biometeorological science, and it was thus both Whiggish and presentist through and through. As he himself put it: "Although the first systematic record of meteorological influences on human health was made by Hippocrates some 2,400 years ago, it was not until the present century that this linkage was verified."[72]

If this work displayed the Hippocratic roots of Sargent's biometeorological vision, his most influential scientific publication, the collection of essays he edited in 1974 entitled *Human Ecology*, displayed to the full the ecological cast of his thinking. In his introductory chapter he emphasized the reciprocal bonds between "life-process and environment" but was insistent that the emergence of the human species in a tropical environment had so impressed itself on the human constitution that it limited its "capacity to cope with cold environments."[73] Thereby something of the human race's climatic destiny was sealed. Here the critical role of evolutionary adaptationism asserted itself. Elsewhere in the collection, other contributors drew on René Dubos's claim that "states of health or disease are the expressions of the success or failure experienced by the organism in its efforts to respond adaptively to environmental challenge" to confirm that "disease" was fundamentally synonymous with "maladjustment."[74] The observation made by these authors that those with diseases were "unfit people" raises the specter of eugenics—an issue that Sargent hinted at when reflecting on what he called "Malthusian limits." "In a world of prejudice," he mused, "the problems of eugenics . . . can scarcely be broached."[75]

EUGENIC ENTICEMENTS

If, by the mid-1970s, Sargent felt that eugenics could only be alluded to in passing, several of his predecessors were decidedly more vocal in their recruiting of the biometerological body to precisely that purpose. During the earlier decades of the twentieth century this proclivity manifested itself in several interrelated concerns: the birth of genius, the season of conception, and climate and migration. And in each of these arenas it was precisely the fusion of Hippocratic and Darwinian modes of thought that facilitated a eugenic turn.

Inquiring into the origin, nature, and future of "genius" had long been a pet preoccupation of eugenic enthusiasts. The pioneer of eugenics, Francis Galton (Darwin's half cousin), for example, famously set out to show that human ability was inheritable, and he sought to test this intuition by tracing the descendants of eminent individuals over several generations. His findings, published in 1869 in *Hereditary Genius*, led him to lament the decline in renown between first-, second-, and third-degree relatives and later, in his 1883 *Inquiries into Human Faculty and Its Development*, to seek for ways of encouraging early marriage among families of high rank.[76] The subject of eugenics and genius was controverted, as Havelock Ellis (1859–1939)—physician, social reformer, and president of the Galton Institute—explained to readers of his 1916 essay on "Eugenics and Genius." Already the author of *A Study of British Genius* in 1904, Ellis addressed the complaint that eugenicists were "stamping out the germs of genius" inasmuch as geniuses were often thought to be "poor citizens, physical degenerates the prey of all manner of constitutional diseases, sometimes candidates for the lunatic asylum" and therefore dysgenic.[77] His own view was that such characterization was unwarranted, at least when geniuses were treated as a group. But in the course of reviewing the entire subject he highlighted the different positions that could be taken on the whole subject.[78]

Now, in the middle decades of the twentieth century, neo-Hippocratic enthusiasts worked to bring the subject of genius within the sweep of climate's
realm. Petersen's Darwinian construal of the connections between intellectual
accomplishment and the season of conception has already been noted. But it is
worth adding that, like many others, Petersen was at pains to convince his readers
that the conception period favoring the birth of genius—"the meteorological
variability characteristic of the late winter and spring"—was no less causally
active in producing other "unusual human types," including what he called the
insane, schizophrenics, and various "asocial groups."[79] As he further explained:
"Under the same unfavourable situation we produce more unusual types. The
greater the number the greater the possibility that some will have a better chance
for survival. Under this same environmental situation we produce more genius."
Generalizing yet further he concluded: "The greater the environmental variability at the time of conception, the greater the variability in the human types that
develop from such conception. The more stable and quiescent the environment
at the time of conception, the greater the possibility that the course of perfectly undisturbed genetic developmental trends can continue and the greater the
possibility that the product will be normal, stable, and more homogeneous."[80]

Convictions of this stripe certainly had eugenic possibilities as the writings
of Huntington and Mills amply confirm. Huntington had long been enticed
by the promises of eugenics. As early as 1910 he was in touch with a number of
intellectuals whose minds ran along similar lines. He was in communication
with John H. Kellogg of the Health and Efficiency League of America; the
Cambridge University professor of theology W. R. Inge wrote him expressing
his pleasure at the implementation of new immigration policies since they would
raise the birth rate among old stock American families; Henry Fairfield Osborn,
celebrated paleontologist, eugenicist, anti-immigrationist, and president of the
American Museum of Natural History, invited him to speak in 1910 at the Half
Moon Club, a leading Progressive Era social group.[81] Besides this Huntington's
concerns over unrestricted immigration merged with the eugenic outlook of
publicists such as Edward Ross, Madison Grant, and Lothrop Stoddart, and
he willingly participated in the Second International Congress of Eugenics,
which Osborn orchestrated in 1921 at the American Museum of Natural History. Later when the American Eugenics Society was formed in 1922, he closely
associated himself with it, eventually becoming president in 1934. During his
tenure of the presidency, moreover, he did much to shift the society's focus
from positive to negative eugenics—the move from encouraging "superior"
families to have more children to seeking ways of preventing "degenerates"
producing offspring.

It is not surprising then that Huntington directed his gaze toward subjects—genius, insanity, criminality, suicide, idiocy—that were the favored fixations of fellow eugenicists. But he gave these subjects a decidedly biometeorological twist by insisting that the "season of birth" bore a "close relation to genius and eminence."[82] Geniuses, apparently, were more likely to be born during the winter months, suggesting that "intellectuality may be fostered by temperatures below those best for bodily vigor alone" (*Season of Birth*, 348). The birth seasons of other groups—"mental defectives," "idiots," "imbeciles," "morons," and the "insane," to use his terms—similarly displayed their own patterns. The enticements toward eugenics here were manifold. Huntington was certainly convinced that breeding management was necessary to curb the reproductive activities of the "intellectually weak" and the emotionally impoverished, both of whom were "especially likely to yield to the sexual stimulation which marks the chief season of reproduction." After all they were largely "responsible for a large percentage of the persons who become criminals or suffer insanity" (18). But what he wanted was to insert neo-Hippocratic modes of thinking into the eugenic agenda. Because none of the "weak types of people" showed "the normal tendency toward increased conceptions when the temperature falls to the physical and then to the mental optimum in the autumn," it was obvious that matters of climate and season of birth should fall within the remit of eugenic oversight (408). As he put it: "It has been supposed in the past that eugenics is concerned only with hereditary qualities, and such is undoubtedly the case. Nevertheless, if eugenics means the applied science in which the objective is that children shall be well born, the prenatal and preconceptual influences of climate and weather come close to being included. This does not mean that there is any confusion between heredity and environment. What it does mean is that an unfavourable physical environment may depress a given group of people for generation after generation regardless of the people's genetic constitution" (445).

The implications were far-reaching. The spatial distribution of climatic optima for mental development explained the intellectual flair manifested in "the scientific tendencies of the Swedes, and the philosophical tendencies of the Scotch" (16). But it also meant that the people of the tropics were forever condemned to a low position on his hierarchy of human worth. As he had already explained in a textbook on social and economic geography: "No matter what races we deal with, we can scarcely expect the same degree of activity and progress among tropical people as among people in more favorable climates. This same reasoning applies with diminished force to other relatively unfavorable climates such as are found in China, Central Asia, and the far north."[83]

Here was a way of imagining global geography that could feed the eugenic craving to manage national biopower. For, if climate stamped its indelible mark of inferiority on the peoples of certain climatic zones, then this fact had major eugenic repercussions for immigration policy, and Huntington did all in his power to advance a restrictionist agenda. As he put it in a letter to Madison Grant in 1925: "I believe that by rigid restriction of immigration and the application of the best eugenic practices, America will not only benefit enormously but will do infinitely more good for the world than it can do in any other way."[84] In doing so Americans would simply be continuing by other means what Darwinian evolution had already achieved. As he explained in *Builders of America*, which he coauthored with the then secretary of the American Eugenics Society, Leon Whitney, in 1927, natural selection had ensured that the first Europeans to settle New England—the original "builders of America"—were strong in body and mind and possessed of "fine temperament, fine intelligence, and fine health." They were "one of the most highly selected groups the earth has ever seen." After all, they had long experience of one of the world's most stimulating climates. The only way to preserve this rich bioheritage was to "diminish the birth rate among the less valuable parts of our society and increase the birth rate among the more valuable parts."[85]

Later, in 1935, similar sentiments were enshrined in his catechetical *Tomorrow's Children: The Goal of Eugenics*, an expanded and revised version of Whitney's 1923 *Eugenics Catechism*. Published in conjunction with the directors of the American Eugenics Society and widely distributed to learned societies, schools, and various church groups, this book again displayed its author's Darwinian inclinations. "Immigration is one of the main agencies in changing the innate quality of a population," he began. "Most authorities agree that it is a selective process," inasmuch as those "who migrate far face great difficulties and dangers, and for the sake of high ideals, are generally of unusual value eugenically." At the same time "an adverse or dysgenic selection" often took place where the excellence of settlers was not monitored. "The eugenist believes that if we permit immigration, we ought to take stringent measures to make sure of the quality of the immigrants."[86]

To Huntington, it is clear, evolutionary biometeorology, the politics of birth rate, and the connections between immigration and biopower were intimately intertwined. It was much the same for Clarence Mills. Mills's biometeorological creed readily translated into the same eugenic fixations: immigration and reproduction. In 1939 Mills warned the users of his *Medical Climatology* textbook of the troubles attending certain kinds of migration: "Too great emphasis cannot be placed on the dangers encountered by people who migrate from regions of low

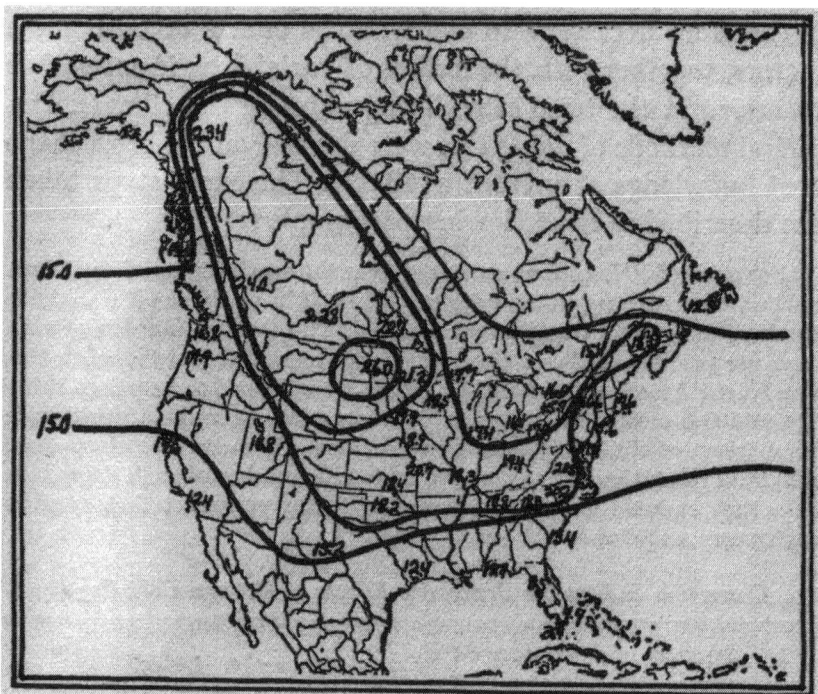

FIGURE 9.3. Mills's map of "Climatic Stimulation over North America." Mills, *Medical Climatology.*

climatic vigor into our northern storm belt where climatic stimulation is probably the most intense to be found anywhere on earth" (see figure 9.3).[87]

A few years earlier he had devoted a whole chapter in *Living with the Weather* to the subject "Climatic Stimulation and the Immigration Question." Resorting to the tedious tropes of tropical denigration and temperate admiration, which were grounded in the assurance that "the energy level and vitality of a people depend greatly on the character of the climate in which they live," he concluded his assessment with the following declaration:

> The day of unrestricted migration of large masses of people should not be permitted to return. With the knowledge we now have of climatic effects, due foresight would prevent a repetition of past errors. Migration from a more to a less stimulating region is not so likely to lead to trouble as is the reverse. . . . With immigration from a less to a more stimulating climate, however, it is the new-found energy of the new arrival which causes him to be such a misfit in his surroundings, affecting not only himself but

others about him. If by such study we can in future avoid past blunders in dealing with world problems of migration, any effort expended will be well worth while. Far too long has immigration been considered from the economic aspect,—what it now needs is a concern for its biologic results and possibilities.[88]

Or as he summarized it yet more pointedly: "Hand in hand with the low energy existence goes a lack of inhibition, a free and easy life. . . . On such a basis as this probably lies much of our immigration troubles with peoples from less stimulating lands."[89] In Mills's eyes, declining biopower was the net result of immigration from climatically undesirable regions of the world. And at the same time he seemed troubled by the thought that people from regions of low climatic vigor would acquire newfound energy when they arrived in a more stimulating climate.

A second prong of Mills's eugenic philosophy centered on matters of sex and reproduction. His combination of melancholic tropicality and procreative seasonality enabled him to offer the following eugenic advice to colonial settlers: "In real tropical heat there is no optimal period—vitality is low at all times of the year. Prospective parents living in the tropics who desire to practice the highest type of eugenics and give their children all possible benefits should spend several months in northern cold before conception takes place." He held out the hope that such advice would meld with "new marriage laws" compelling "young people to undergo medical examination before a [marriage] licence is issued."[90] The message that there were decided advantages in "cold-weather conception" needed widespread circulation, and he hoped that the "facts" he had accumulated would soon "find expression in high school or college eugenics courses."[91]

This concern for greater oversight of reproduction was only one manifestation of Mills's fascination with the biopolitics of climate and sex. In his *Medical Climatology*, for example, he devoted a whole chapter to "Variation in Fertility and Sexual Functions" dwelling on the onset of menarche and sexual maturity, and the relations these "bear to climatic stimulation in different regions of the earth." Here he urged that "tropical children lag far behind those of the stormy temperate regions, and that the average tropical girl of 15 years appears no nearer maturity than the 12-year old girl of cool temperate regions."[92] Elsewhere he used the authority of Hippocrates to counter "the fallacy of early tropical maturity."[93] Such perceptions, of course, had immediate implications for immigration too. As he later explained, "among the millions of migrants who came to this stormy stimulating continent [the United States] much earlier maturity was noted in the children who grew up here." This climatically determined change in the onset of puberty created problems, he believed, on account of the "more volatile nature of sub-tropical natives" compared with "more reserved northerners."[94]

To Mills, then, research on the season of conception, racial reproduction differentials, and the role of tropical-to-temperate migration in lowering the age of sexual maturity was of massive eugenic significance. And so, calling on the support of Huntington, he concluded his analysis of birth season with the following cri de coeur: "The days of leaving human reproduction to blind chance in its timing should no longer exist for intelligent couples. Optimal parental health and proper season of conception mean much to the existence of the next generation, and no parent should wish his children in future years to look back and feel that perhaps he did not do all he might have done for their welfare. To appreciate in full measure the importance of these matters, every prospective parent should read Huntington's *Season of Birth: Its Relation to Human Abilities*."[95]

CONCLUSION

The reinvigoration of Hippocratic modes of thinking during the first half of the twentieth century owed a good deal to the endeavors of a number of key individuals who reconceptualized the human body as a biometeorological entity. Much of this work dwelt on the ways in which atmospheric conditions and changing climate shaped human health both directly and indirectly. What facilitated this neo-Hippocratic revival was a range of themes that were intrinsic to Darwinian evolution—natural selection, adaptation to environment, and a growing awareness of the prevailing climatic circumstances during the period of the emergence of Homo sapiens. Taken together these impulses fostered a distinctive geographical imaginary that traded in the bipolar juxtaposition of invigorating and debilitating climates and thus of fit and feeble people. But these practitioners of medical climatology did not rest content with probing matters of health and disease and with empirically mapping biopower across the globe. Rather, they routinely looked to a broader horizon in their cultivation of a general climatic philosophy of historical explanation that incorporated much of human culture. The rise and fall of civilization, economic cycles, moral precepts, patterns of warfare, political biography, and many more aspects of human culture were all brought within the domain of climate's empire. These preoccupations frequently had the effect of pathologizing large portions of the globe as climatologically deficient and thus condemning the inhabitants of these regions to a bleak future. With such proclivities, some of those working on the interface between medical practice and climate science would yield to the allure of eugenics and would seek for ways to insert biometeorological concerns into a eugenic agenda to regulate national biopower. Chief among these were the implications of season of birth for reproductive policies and the significance of global patterns of climatic optima for immigration programs. At least among these advocates, the practice of biometeorology and the desire to engage in the politics of biopower were intimately intertwined. Considered in this light, Jim Fleming's astute observation on the legacy of Huntington's climatic determinism contains a cautionary warning: "his categorical errors seem destined to be repeated by those who make overly dramatic claims for weather and climatic influences."[96]

10

CIVILIZATION, CLIMATE, AND OZONE

Ellsworth Huntington's "Big" Views on Biophysics,
Biocosmics, and Biocracy

James Rodger Fleming

> The climate of many countries seems to be one of the great
> reasons why idleness, dishonesty, immorality, stupidity,
> and weakness of will prevail. If we can conquer climate,
> the whole world will become stronger and nobler.
>
> **—Ellsworth Huntington**

ELLSWORTH HUNTINGTON (1876-1947) wrote extensively on the effects of
weather on individuals and the dominant influence of climate on world civilizations. He worked outside of the nascent community of climatologists and became
an outsider in geographical circles, relying for income mainly on royalties from
his many popular books. His musings on environmental determinism mapped
regions of high "climatic energy" with "high civilization"—all located exclusively
within Earth's temperate zones. He further argued for the stimulating effects of
the regular passage of storms, locating swaths of exceedingly high energy and
correspondingly high civilization along the "storm tracks"—the most influential
being the track between his workplaces in Cambridge and New Haven, conve-
niently passing over his family home in Milton, Massachusetts. In the 1930s he
developed a reductionist "ozone hypothesis" of climate and history based on the
stimulating influence of ozonated fresh air on mental and physical activity, especially during the passage of storms. The aim in this paper is to review medical

and environmental knowledge of ozone in the early decades of the twentieth
century and examine Huntington's appropriation of the ozone molecule for his
climatic determinism. His hypothesis and his involvement in an international
conference in 1939 on biophysics, biocosmics, and biocracy form the context for
examining the physiological and political aspects of his (very creative) geographical imagination, which was in essence a geographical imaginary. Huntington's
case opens up a larger conversation about the convergence of scientists, the spaces
of science, and the geographies of atmospheric knowledge—of agency, place, and
time. Why Huntington? Why New England? Why the 1930s? Such questions
situate the seemingly perennial nature of climatic determinism and reductionism
within a very situated and contingent geographical and temporal context.

BEGINNINGS

Ellsworth Huntington was born in Galesburg, Illinois, on September 16, 1876,
the third child and eldest son of a Congregationalist minister, Henry Strong
Huntington, and Mary Lawrence Herbert.[1] The Huntingtons were proud of their
Puritan ancestry, which they traced to 1633. Following the call of the ministry,
the family moved to Gorham, Maine, in 1877 and then to Milton, Massachusetts,
a wealthy suburb of Boston, in 1889. Ellsworth attended the public high school,
where he excelled in athletics and academics. His biographers have called him
reclusive, but his brother suggested that perhaps he was humble rather than shy.

Huntington passed the Harvard entrance examinations, but family finances
precluded his enrollment there. Instead, he attended Beloit College where he
studied both classics and geology and graduated in 1897. His church connections
landed him his first job—in Turkey, where he served as an instructor and assistant to the president of Euphrates College, a small missionary school in Harpoot.
There he mapped the area, investigated the culture of the region, and established
a local weather station. In 1901 he shot the rapids of the upper Euphrates River
in a raft made of inflated goatskins, a feat that earned him the Gill Memorial
Award of the Royal Geographical Society.

Huntington returned to the United States to study physiography at Harvard
under William Morris Davis, who impressed upon him the importance of climatic influences. He received Masters degree in geography 1902 and spent that
summer doing fieldwork with Davis in Utah and Arizona. After graduation,
Huntington received two offers to return to Turkey—one from Euphrates College
and a second to serve as the US consul at Erzerum. He rejected both. Instead, he
joined the 1903 Carnegie-sponsored expedition to Central Asia led by Raphael
Pumpelly.[2] The expedition was seeking the remains of a lost Aryan civilization
that was thought to have inhabited a Mediterranean-sized sea left by the melt

waters of the retreating glaciers of Eurasia. All that remained now, however, was a desiccated land, nomadic inhabitants, and the Caspian and Aral Seas. In 1905–1906 Huntington joined an expedition led by Robert L. Barrett, a founding member of the Association of American Geographers, through the Himalayas into the Tarim Basin of China. Huntington's early views on climate, culture, and civilization were undoubtedly shaped by these experiences, which impressed upon him the importance of the geographic basis of the study of anthropology and "the immense influence which changes of climate have exerted on history." Among the ruins of ancient civilizations, Huntington found clear evidence of progressive desiccation and historical climatic changes. The result was his first book, *The Pulse of Asia* (1907), written while he was a Hooper Fellow at Harvard.[3]

In this book Huntington emphasized three major themes: (1) the power of climate to mold the habits and character of peoples; (2) the variation of climates over historic times, a phenomena he called "pulsations"; and (3) the fact that climatic changes cause corresponding economic, physiological, behavioral, and even mental changes in people and populations. Each climatic "pulse" could also be an ominous "throb" that threatened to dislocate the center of civilization. "Each throb has sent pain and decay to the lands whose day was done, life and vigor to those whose day was yet to be." This led him to conclude that "geography . . . is the basis of history in a way that is not generally recognized; and that climatic changes have been one of the greatest factors in determining the course of human progress." Beginning with evidence from dried-up lakes in Central Asia, Huntington universalized his results to include moral judgments about all the nations of the world:

> To-day, the strongest nations of the world live where the climatic condi-
> tions are most propitious. Japan and north China in Asia; Russia, Austria,
> Germany, France, and England in Europe; and the United States and
> Canada in America, all occupy regions where the climate is of the kind
> which we have defined as most favorable to the progress of mankind. Much
> as these nations differ in race, in ideals, and in type of civilization, they all
> agree in possessing a high degree of willpower and energy, and a capacity
> for making progress and for dominating other races.[4]

These themes, and the circular reasoning behind them, would characterize Huntington's writings for the next four decades.

A CHECKERED ACADEMIC CAREER

In 1907 Huntington passed his preliminary examination at Harvard for the PhD, but he failed to sustain his final PhD examination by a committee vote of four to two, with one abstention. Ironically, his committee found he was "deficient in his knowledge of climatology and showed great weakness in historical geology."[5] They found his work "uncritical" and lacking in focus, with a tendency to adjust the facts to fit his theories. The weaknesses pointed out by Huntington's professors reappear in later critiques of his work.

Huntington returned home to Milton, Massachusetts, to regroup, and he soon began a lifelong (albeit tenuous) relationship with Yale University. Although he was hired as an instructor in geography, he considered himself first and foremost an investigator. He admitted he "liked" teaching but "loved research . . . [and] let the research crowd [his] teaching to the wall," filling his time with travel, exploration, and writing.[6] He served on no committees at Yale and became known as somewhat of a recluse, with obvious problems filling a classroom. Students found it inconceivable that such an unimpressive person could be an accomplished world traveler and geographer. Still, in 1909, Yale did award Huntington a PhD degree and, a year later, promoted him to the rank of assistant professor. He continued his travels and explorations, returning to Asia in 1909 and visiting California, Mexico, and Guatemala between 1910 and 1915. His traveling and writing left little time for course preparation.

Yale viewed Huntington's 1912 application for promotion to full professor as "problematical." A letter of evaluation written by R. S. Woodward of the Carnegie Institution of Washington pointed out that Huntington's work lacked focus; "[he] is here and there in his writings and in his methods of procedure quite naive if not immature." Eduard Brückner of the University of Vienna thought Huntington was industrious but also uncritical in his use of data and sources. Brückner's most severe criticism was that Huntington had "shown several times the desire to fit the facts to his theory." Albrecht Penck of the University of Berlin concurred. Professor of geology E. S. Dana, who had solicited the letters, was of the opinion that Huntington had "brilliancy" but was "immature" (at the age of thirty-six) and had not yet settled down to produce "steady, sound work." Dana also thought that Huntington had the disadvantage "of an enormous overestimate of his own importance to the University" and recommended that Yale make no commitment to his future.[7] Rather than a promotion, Huntington received a two-year extension of his appointment. His request for reconsideration was not successful, and in 1915 Huntington left Yale to work from his parents' home in Milton. Huntington was a prolific author whose publications sounded the repetitive themes of climatic determinism and racial eugenics. In addition to numerous articles, he published *The Climatic Factor as Illustrated in Arid America* (1914) and *Civilization and Climate* (1915).

During the First World War, Huntington worked for the Military Intelligence

Division of the National Research Council (NRC), where he helped produce *The Geography of Europe* (1918). Here he illustrated his theory of environmental determinism with an example from the Italian peninsula:

> Few countries of equal size are more diverse than Italy. The north is among the most prosperous and progressive parts of the world, but the south is decidedly backward. In the north not only is agriculture highly developed, but manufacturing is well advanced in proportion to the resources of coal and metals, and science and art are well established. In the south, on the contrary, there is almost nothing in the way of manufacturing, in spite of the fact that water transportation is easier than in the north. Instead of science and art there is the most widespread illiteracy. Farming is the chief occupation, but it is carried on much more carelessly than in the north. The population is much less dense than in the north, but the people are poorer, in spite of the larger amount of land at their disposal. Many are undernourished, and this combines with other things to cause disease. In order to understand this contrast between the north and south the first necessity is to compare the climatic differences.[8]

Huntington turned down several opportunities to teach at universities in the Midwest and returned to Yale in 1919 as a research assistant with a nominal salary of two hundred dollars a year. He continued to write and teach courses there, sometimes as much as a half load at far less than half salary.[9] *World Power and Evolution* (1919) and *Red Man's Continent* (1919) followed in quick succession.

As chair of the NRC Committee on the Atmosphere and Man (CAM), Huntington focused his efforts on the interrelations of weather, health, and productivity. From its inception in 1921 to its demise in 1929, CAM members focused on four broadly biophysical projects: an investigation of the influence of meteorological conditions on factory productivity, physiological experiments under laboratory conditions, experiments in hospitals, and an investigation of mortality caused by influenza in New York City. In conjunction with this work, Huntington developed a sort of "meteorological Taylorism," named for the famous efficiency studies of Frederick Taylor. As he wrote to Davis:

- Children tested for speed and accuracy on the typewriter for a very brief period work fastest at 67°.

- Girls working for the whole day at mechanical work do best at about 60°.

- Adult men doing mechanical work which also demands a little brainpower, but not much, are at their best at about 50–55°.

- Students at Annapolis, 1300 of them, doing prolonged and concentrated brainwork are at their best at about 40°.[10]

The repetitive themes of climatic determinism and racial hygiene echoed through a prolific number of Huntington's texts: *Principles of Human Geography* (1920), *Business Geography* (1922), *Climatic Changes* (1922), *Earth and Sun* (1923), *The Character of Races* (1924), *West of the Pacific* (1925), *The Pulse of Progress* (1926), *The Human Habitat* (1927), *The Builders of America* (1927), and *Mainsprings of Civilization* (1945). The academic tone of Huntington's work led many to believe it was rigorous and authoritative, but in fact it was neither. Harlan Barrows, a professor of geography at the University of Chicago was of the opinion that "Ellsworth [should] write less and think more." Isaiah Bowman suggested that if Huntington did not shave one morning he could write an article in the time saved.[11] Concerning the ostensibly scientific topic in the book *Climatic Changes*, US Weather Bureau scientist William Jackson Humphreys wrote that "its broader conceptions are mere fantasies, while its details show little regard for facts and none for physics. . . . [I]t is as far from being scientific as *Alice in Wonderland*."[12]

TEMPERATE ZONES AND HIGH CIVILIZATION

Huntington argued that the cool temperatures and variable weather conditions found in temperate zones, combined with climate pulsations, promote the highest expression of civilization by providing inhabitants with more physical and mental stimulus than stimuli experienced in tropical or polar climates. He thought he had discerned a biocosmical relationship between climate and civilization. In 1913 he sent out a long questionnaire to over two hundred intellectuals and public figures—anthropologists, geographers, soldiers, and statesmen—mostly from the United States, Great Britain, and Europe, asking their cooperation in the "preparation of a map showing the distribution of the higher elements of civilization throughout the world." His purpose was to map the world's civilizations from highest to lowest and show that regions of "high civilization" were correlated with regions of high "climatic energy."[13] Based on this survey, he produced a set of world maps depicting what he called the distribution of civilization (see figure 10.1 top) and the distribution of climatic energy (see figure 10.1 bottom), with the highest levels located exclusively in temperate zones. The optimum for civilization, as defined by Caucasian standards, was a climate with decided seasonal swings but without great extremes of heat, cold, aridity or humidity, and with frequent moderate changes due to cyclonic storms. By civilization, Huntington meant *Western* civilization:

> The power of initiative, the capacity for formulating new ideas and for carrying them into effect, the power of self-control, high standards of honesty and morality, the power to lead and control other races, the capacity for

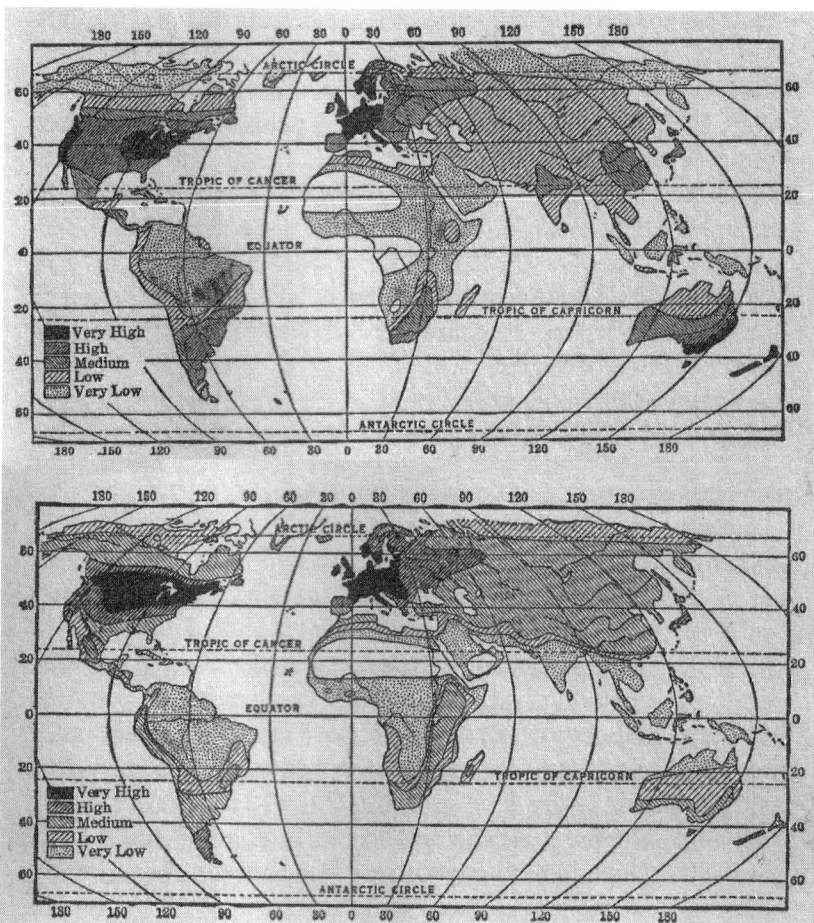

FIGURE 10.1. Huntington's maps of distribution of civilization (*top*) and climatic energy (*bottom*) Reproduced from Huntington, *Human Habitat*, 139, 145.

disseminating ideas, and other similar qualities which will readily suggest themselves. These qualities find expression in high ideals, respect for law, inventiveness, ability to develop philosophical systems, stability and honesty of government, a highly developed system of education, the capacity to dominate the less civilized parts of the world, and the ability to carry out far-reaching enterprises covering long periods of time and great areas of the Earth's surface.[14]

Huntington was seeking a consensus of expert opinion ranking the relative civilization of 185 regions of the world. Apparently, he believed (or hoped) that

FIGURE 10.2. "An energetic family of children on a fruit farm in the United States (making cider). Scenes like these are almost unknown except where the climate is healthful and stimulating." Reproduced from Huntington, *Human Habitat*, following 136.

FIGURE 10.3. "Anaemic Indian coolie trying to sleep and at the same time cool his British rulers by working a punka or fan. This man's feeling of lassitude, anaemia, and sleepiness is typical of steadily warm moist climates." Reproduced from Huntington, *Human Habitat*, following 144.

by averaging the opinions of his correspondents, individual idiosyncrasies would disappear. Not surprisingly, instead of mapping civilization and climate, Huntington measured the collective bias of his fifty respondents—twenty-five Americans, sixteen northern Europeans, eight southern Europeans, six East Asians, and one Russian. There were no Africans, Latin Americans, or Southern Asians (see figures 10.2 and 10.3).

STORM TRACKS

"The people of the cyclonic regions rank so far above those of other parts of the world that they are the natural leaders. For instance, the form of democratic government which was worked out in France and England but which was first really tried in the United States, is the form which every country in the world is gradually trying to adopt."[15] In late December 1938 Huntington spoke on "mass outbreaks of human genius" at the Association of American Geographers meeting in Chicago—he gave examples of ancient Greece and thirteenth-century Italy. Huntington argued that climate changes and cultural changes were strongly correlated and proposed that both individual genius and creative eras in general were the result of increased mental and physical stimulation, mainly in the temperate zone, caused by increased storminess.[16] He claimed that the glories of Palestine and Alexandria in the first century and the creative influence of individuals such as Jesus of Nazareth, Saint Paul, Philo, Origen, Polonius, Strabo, Clement of Alexandria, Polybius, and Cleopatra were connected to the assumed path of cyclonic storms in ancient times. Huntington further speculated that New England might experience a cultural renaissance as a result of the frequent passage and intensity of its storms.

Charles J. Kullmer, a professor at Syracuse University, influenced Huntington's ideas when he suggested there could be correlations between high barometric pressure and peaks in library circulation records for nonfiction books. (Lighter fiction was more likely to circulate during stormy periods.) Kullmer also suggested that Huntington should consider the psychological effects of atmospheric electricity: "The field that seems to me most interesting is to investigate the atmospheric electrical potential of a barometric depression to ascertain if there is a gradient. . . . If there is such a gradient, it would be a fine field to conduct psychological experiments with varying electrical potential of the atmosphere. We already know that vegetation is strongly stimulated by high potential wires strung over the field; perhaps the mental operations are similarly affected." Kullmer's research indicated that financial institutions lying along what he called the typical storm track supposedly received larger deposits than banks elsewhere. He generalized this to the speculative proposition that climatic change—especially shifts in location of the storm track—may ultimately account for shifts in the location of civilizations.[17]

AN OZONE HYPOTHESIS

But for Huntington something still was missing. The typical weather elements, temperature, rainfall, barometric pressure, even the location of storm tracks, could not explain everything to his satisfaction. A hurricane that passed through Massachusetts in 1938 changed that.

On September 21 the "Great New England Hurricane" of 1938 devastated Long Island, coastal Connecticut, and Rhode Island before passing over Amherst, Massachusetts, in its rapid passage northward.[18] The freshman class of the Massachusetts State College was taking a battery of intelligence tests at the time. Howling eighty-mile-an-hour winds felled trees and drove the rain in horizontal sheets; the college lost electrical power. Yet the examination went on in the late afternoon gloom, and when the results were in, a class deemed by other measures to be no better than average had produced test scores that were nearly off the charts—some 20 percent above average. How did this happen? Huntington, who noted that his own productivity had soared during the hurricane, speculated that the storm had exposed the students to a stimulating wave of ozone that heightened their mental abilities. This hypothesis was reinforced by the observations of two college chemists who detected the pungent smell of ozone during and immediately after the passage of the storm.[19] Was ozone a determinant of health and intelligence? Was it the key to civilization, culture and climate? Huntington thought it might be and began to promote a hypothesis of molecular reductionism linking ozone, storm tracks, and temperate climates to high civilization and racial superiority.

Early in 1939 Huntington was invited to participate in a prominent meeting where he could highlight his ideas, especially his ozone hypothesis. The International Congress of Biophysics, Biocosmics, and Biocracy, chaired by Louis C. Barail, president of the American Society of Biophysics and Cosmobiology was to be held in mid-September in New York City in conjunction with the World's Fair.[20] According to a press release in the *New York Times*: "*Biophysics* studies such matters as the circulation of the blood, breathing, the effect of radium on the body's cells, and the adjustment of the bodily organs to light, heat, moisture, electricity, and gravity; *Biocosmics* considers, among other things, the effect of weather, the seasons and electricity on man and lower life forms; *Biocracy* covers the general improvement of mankind, through heredity, training of children, racial origins, geography, medicine and political science."[21] Huntington chaired the section on biocosmics, read a paper on "Ozone Weather and Health," and

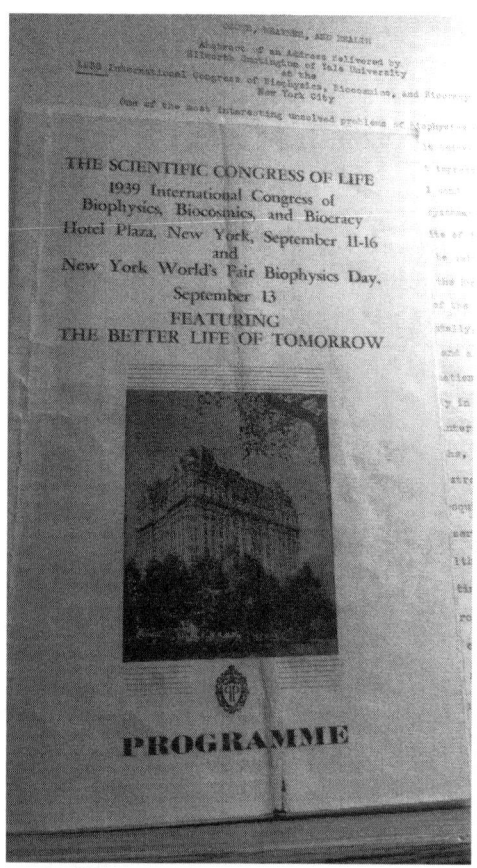

Figure 10.4. Cover of Huntington's personal copy of the Programme of the 1939 International Congress of Biophysics, Biocosmics, and Biocracy. Reproduced, with permission, from the Papers of Ellsworth Huntington, Yale University Library.

gave a popular lecture on "What the Weather Does to Human Beings" for Biophysics Day, September 13, 1939, at the New York World's Fair (see figure 10.4).[22]

The first day of the congress, September 11, was the kind of day that convinced Huntington of the truth of his theories. The weather was bright and clear with gusty winds out of the northwest and a high of only 65 degrees—bracing and stimulating conditions for intellectual challenges. Unfortunately for the organizers, many European delegates were not able to attend since war had just broken out in Europe.[23]

At the conference Huntington argued that the standard meteorological variables—temperature, pressure, and humidity—were not sufficient to account for the cultural differences between climate zones. The location of storm tracks helped, especially since Huntington thought that the passage of storms brought to the surface trace amounts of ozone, which was a mental and physical stimulant. He argued that ozone in concentrations of 1:3 million was stimulating—less than that was ineffective; more than that was toxic. He also surveyed the circulation records of some forty-three public libraries to see if he could correlate them with ozone levels.[24]

AFTERMATH

Immediately after the congress, Huntington prepared a long typescript, meant to become a book, on "Ozone, Health and Progress," but it was never published.[25] Here Huntington reviewed the physical factors that help or hinder the progress of civilization and included his arguments for the positive influence of temperate zones and the benefits of living along the storm track. He then constructed an ozone hypothesis of history, working, as usual, mainly from secondary sources and loose correlations. As a founding member of the Society for the Study of Cycles, he further attempted wild correlations between fluctuating ozone levels in Paris and London and disparate phenomena such as deaths by heart disease in the northeastern states and the populations of Canadian lynx, New Brunswick salmon, and tent caterpillars in New Jersey!

> From all this we come to the conclusion that the best air is found where mountains, seacoasts, storms, high latitude, and outdoor life in the country are most happily combined. Among these various factors storms appear to have a special significance. They give rain at all seasons, which is of special importance to agriculture. More important, however, may be the fact that they apparently give people health and energy through the waves of ozone that arrive when masses of polar air press forward with westerly winds in the wake of almost every storm. The world's storm belts in the United States, Europe, and Japan, and again in the middle latitudes of the southern hemisphere are the regions which get the most benefit in this way. Being constantly stimulated by waves of ozone, as well as by the changes of temperature and fluctuations from clear skies to clouds, their people are endowed with much more energy than those in other parts of the world.[26]

Ozone was identified by its pungent odor and named by Christian Friedrich Schönbein in 1840. Originally thought to be an elemental substance, it was soon identified as a molecule containing three oxygen atoms (O_3). It is typically produced by an electrical discharge—near heavy motors or in thunderstorms, for example. Ozone has long been known as a powerful oxidant and disinfectant. In Huntington's era, ozone was considered to be a stimulant, synonymous with "fresh air." Ozone was at a maximum out-of-doors, in fresh mountain air and along coastlines. There was thought to be no ozone in large towns, inhabited rooms, or near decomposing organic matter.[27] Ozone generators were sold for home use. Lindstrom and Company of Chicago, for example sold units that produced "medical electricity, mechanical vibration, high frequency violet rays and ozone: nature's great helpers in regaining health, power, and beauty."[28]

Was ozone the missing key to explaining the link between civilization and climate?—a "neglected climate factor," a magic determinant of health, mental activity, and ultimately, culture? Huntington thought so. Temperature, humidity, wind, and so forth did not fully explain the full range of human reactions to changes of weather. "Ozone may account for the peculiarly stimulating quality of the air at such places as Berkeley where the U. Cal. is located. My own experiences, like that of many others, is that the air there makes me feel keyed up to intense activity at the time. In fact, one almost longs for a chance to be let down."[29]

Huntington further generalized his ozone theory, moving it from biophysics to biocosmics, hinting at the geographical improvements possible through biocracy, and linking the theory to his earlier mapping of high civilization onto temperature zones along the storm tracks:

> [There are] well known contrasts between good air and bad air which are closely associated with differences in the amount of ozone. Under ordinary circumstances indoor air has what is often called a "dead" feeling, whereas outdoor air has a vital and invigorating quality. A similar contrast makes city air a very poor substitute for that of the country. In both cases the poorer air has lost much of its ozone through contact with walls, dust, and so forth [pre-auto smog era]. Ventilation engineers have been forced to acknowledge that window ventilation somehow satisfies people much better than the most perfect system where "conditioned" air is blown into rooms. The reason is probably found in the slightly greater amount of ozone in the untreated air.

One of his close colleagues who read his article on ozone was not convinced: "Personally I feel that you have taken a shot in the dark regarding ozone as a responsible factor, with your arguments based more upon logic than fact and with even the logic most tenuous throughout. This is quite unfortunate since it will encourage many people to doubt the validity of your earlier writings. What you say of ozone effect *may* be true, but it was unworthy of the dignity of publication in such length until supported by convincing facts."[30]

DENOUEMENT AND DENOUNCEMENT

The ghost of Ellsworth Huntington still haunts the geographical imagination.[31] No matter how odd his ideas appear today, climate determinism is still a pervasive ideology. The US Department of Education advises that, "*climate* very much affects the character of a place. The amount of sun or rain, heat or cold, the direction and strength of the wind, all *determine* such things as how people dress, how well crops grow, and the extent to which people will want to live in a particular spot." The US Central Intelligence Agency uses the pejorative term *"enervating"* to describe the effect of high temperatures and humidity on visitors to equatorial Africa. Did they really mean to impugn an entire nation like the Congo with the multiple negative connotations attached to the term (physical weakness, nervous impairment, indolence, wanting in strength of character, spiritless, unmanly, effeminate)?[32] As recently as 1990, the UN Environmental Program convened a workshop, "On Assessing Winners and Losers in the Context of Global Warming," where discussion centered on possible social impacts of climate change. The assumption was that climate change would impact different cultures in different ways, and the losers would be predominantly the Southern tier or developing nations, with border conflicts and environmental refugees likely. All people would likely suffer physiological and psychological stress in part due to erratic weather patterns.[33]

The author of a widely discussed book makes the following surprising claims in a chapter titled "Climate and Civilization: A Short History":

> • Five million years ago *Australopithecines* left the forests and stood on hind legs because of global cooling.

> • Ice Ages led to the emergence of *Homo sapiens* because "incredible ecological change put a premium on the larger brains needed to adapt to rapidly changing climate conditions."

> • 40,000 years ago a "cultural explosion of tools and jewelry may have coincided with an unusually warm millennium in Europe."

> • 1816, "the year without a summer" led to food riots, the fall of governments, unprecedented crime epidemics, and a dramatic increase in the number of suicides and executions. "The bureaucratic, administrative tendencies of the modern state were given great impetus."

> • "As the climate pattern begins to change, so too do . . . the feasts and famines, the seasons of peace and war."

Here both human evolution and social stability are directly linked to climate change. The source of these generalizations?—Al Gore's book, *Earth in the Balance!*[34]

Nor is the link between "molecular reductionism" and determinism any weaker today than it was in Huntington's time. If ozone was his "key" to civilization and climate in the 1930s, uranium and plutonium symbolized the dawn of a nuclear age, while DNA promised to serve as the "code of codes" for life itself. Returning to weather and climate issues, silver iodide (AgI) was considered to be a "trigger" mechanism that would allow widespread modification or even control of the weather in the 1950s and 1960s; protecting the global environment as symbolized by stratospheric ozone became the goal of the Montreal Protocol and its successors; and everyone's environmental molecule of choice, carbon dioxide (CO_2), has become an international symbol of human intervention in the climate system, somehow codifying both affluence and apprehension.

And what of biocracy? Although the term does not appear in the *Oxford English Dictionary*, perhaps it should. Here are three distinct historical meanings of the term "biocracy":

> 1. The rule of life, wisdom of the body, or a society governed in conformity with the laws of life. The term was used by Walter B. Cannon during the Great Depression. It is structurally similar to "technocracy" and was, as discussed, a major component of "The International Congress of Biophysics, Biocosmics, and Biocracy" in 1939.
>
> 2. Racial cleansing, eugenics, genocide of the sort practiced by the Nazis: "National Socialism is nothing but applied biology" (Rudolph Hess, 1934).
>
> 3. More recently, the ecopolitics of bioethics or the rights of nature, as used by the environmental movement: "This large biocracy, this large open democracy which involves all other beings in our planetarian adventure" (Earth Charter Continental Meeting, Brazil, 1998).[35]

Throughout his life, Huntington openly expressed his disdain for what he called "the plain straightforward work of careful men who are competent to develop little ideas but not big ones." He admired macrohistorians such as Arnold Toynbee, who accepted the pulsatory hypothesis of climate change to gain a great advantage over their more conservative colleagues. Toynbee said he was "enormously influenced" by Huntington's ideas about the relation between human beings and their physical environments. It was Toynbee's opinion that "students of human affairs may agree or disagree with Huntington, but in either case they will be influenced by him, so it is better that they should be aware of him."[36] Although Huntington's thought was indeed influential in its time, since then his racial bias and crude determinisms have been largely rejected. Nonetheless, his categorical errors seem destined to be repeated by those who make overly dramatic claims for weather and climatic influences.

Huntington critics, and even his friends, characterized him as a "brilliant, erratic, and rather careless man" whose conclusions were frequently "called into question."[37] His enthusiastic but basically unscientific contributions to climate studies cast a long shadow over the field and contributed greatly to the perception, early in the twentieth century, that climatology involved a healthy population of kooks and cranks and was not yet ready for an intellectual synthesis. One lesson we can take from Huntington is that our understanding of climate and climatic change is not a story of inevitable progress. There were detours and dead ends. His link to the past—to earlier environmental determinists—is obvious. His influence on his own generation was indeed significant and should serve as a warning—that environmental determinism is not dead and is again rearing its ugly head in the form of excessive claims. Toynbee's practice of macrohistory is now reemerging as "Big History," whose practitioners claim to be providing a framework for all knowledge by incorporating cosmological, geological, biological, anthropological, and historical insights into a blended narrative of the universe and humanity's place within it. Such historians think telling all-inclusive "just so" stories will increase the reach and relevance of history. However, most historians, trained to eschew grand narratives, see such practices as little more than interdisciplinary and rather undisciplined forays into the secondary literature of other fields, potentially doing more harm than good. Grand narratives based on what we now know and value are fragile constructions at best.[38] Imagine a historian/geographer of a much earlier era, say 1542, setting out to write a big history. Ptolemaic astronomy and the providential history of the Bible would be center stage, and the chosen theme might very well be God's providence for a geocentric universe. The focus would be on alchemy, not chemistry, and the three elements (as most everyone then knew) would be salt, mercury, and sulfur. We need to remember that today's science will soon be tomorrow's history of science and that universal geographies are still written by people in particular places.

Huntington was called "the most notable exponent of environmentalism in the English-speaking world in the twentieth century." His obsessive promotion of hard climatic determinism and racial eugenics had a devastating effect on the nascent field of human geography and further complicated the already tangled connections between history and geography. In this sense David Hackett Fischer's opinion, echoed by Harold Dorn not that long ago, does not seem viable: "Huntington's thesis was never really refuted. It was merely ridiculed, because it failed to fit the metaphysical framework of social science at the time. Perhaps it is time for those issues to be reopened."[39] Perhaps not. Toynbee's advice is more nuanced. He wrote that we should be aware (or *beware*) of Huntington's influence. Knowing something about him, or at least knowing enough to avoid his categorical errors, will help contemporary students avoid the pitfalls of environmental determinism and its associated excesses that are lurking in today's literature on climate change, history, and geography.[40]

The convergence of place and temporality is evident in Huntington's climate work and in his overheated geographic imagination. He set out to construct a grand synthesis of climate and civilization for the entire world and for all time, invoking both global and local factors to construct his geographical imaginary. His eugenic theories fit well with his predilections favoring peoples of temperate climes. His early work on climatic pulsations led him to assume that the rainfall patterns in the Middle East were linked to the rise and fall of both cities and civilizations. He adopted a cosmic view of the world, but was, in actuality, a dyed-in-the wool New Englander. He lived most of his life near Cambridge, Massachusetts, and New Haven, Connecticut, which he assumed boasted the most stimulating climate and highest "mental energy" on the planet. He valued living along a storm track, where he experienced mental stimulation from the vicissitudes of the weather, especially the passage of storms. A singular event, the passage of the 1938 great New England hurricane, led him to his speculative ozone hypothesis, in which a trace molecule holds the key to both mental energy and high civilization. The ozone layer in the stratosphere had been discovered in his adult lifetime, and ozone in the countryside, mountains, and seashore was widely regarded as beneficial "fresh air." Such influences led him to focus his big ideas about environmental determinism—about biophysics, biocosmics, and biocracy—on reductionist claims involving temperate latitudes, storm tracks, and the influence of the ozone molecule on climate and civilization. It is an understatement to say that Huntington had a big imagination. He thought he had captured the big picture of the course of world climate and civilization. But like today's practitioners of Big History, his geographical imagination was very much a product of his situatedness.

11

THE SHADED MODERNISM OF
THE GLOBAL INTERIOR

Climate and Risk in the Architecture of MMM Roberto,
Rio de Janeiro, 1936–1955

Daniel A. Barber

IN THE PERIOD SURROUNDING the Second World War—from the 1930s
to the 1950s—mechanical systems of heating, ventilation, and air conditioning
(HVAC) became increasingly available and affordable. The so-called thermal
comfort that HVAC systems provide is today a thoroughly regulated and standardized condition in many buildings, especially office buildings, commercial
institutions, and other spaces of global culture—what Peter Sloterdijk refers to as
"the World Interior of Capital."[1] As Sloterdijk also suggests, such systems contain
a range of political and economic consequences: "Air conditioning, in the literal
sense," Sloterdijk wrote in 1999, "will establish itself as the main space-political
theme of the coming era."[2] Indeed, in the face of concerns over anthropogenic
climate change, it can be difficult to overestimate the role of HVAC systems in
buildings, relative to carbon emissions, to prospects for industrialization and
global economic development, and to transformations of living standards and
quality of life, albeit unevenly, around the globe.[3] Building systems operate at the
intersection—or, better, at the interface—of economic systems and ecological
knowledge.[4]

Local climate has played a role in architectural decision-making for centuries.
Before the rise of HVAC, the relationship between a given design and surrounding climatic conditions played out through attention to site, orientation to the

sun, and coordination of materials to heat and humidity. With the emergence of modern architectural techniques beginning in the mid-nineteenth century, the relationship of a building to its climate and the means to understand and engage this relationship underwent a significant transformation. Subsequent to this shift, climate design methods became more scientific, in relationship to contemporaneous developments in the atmospheric sciences. The period of the 1930s and 1940s was one of transition, where the general parameters of thermal knowledge came to be a field of creative engagement for architects.

This chapter is part of a larger project exploring how, from the late 1930s to the early 1960s (as HVAC was also developing), numerous designers and engineers around the world explored methods to manage seasonal climatic change by architectural, rather than mechanical, means. These climate-focused methods developed as an aspect of the design principles of architectural modernism—especially in the influential work and writings of the Swiss French architect Le Corbusier—and were, I will argue, essential to the spread of modern techniques outside Euro-American metropoles. Indeed, modern design strategies saw much development in the so-called periphery, as these regions were the site for climate design experiments and also experiments in postcolonial forms of governance. British architects in West Africa, researchers exploring climate-design fundamentals in the American suburbs, Le Corbusier's projects in India, and émigré architects designing American embassies across the Middle East and Africa are some of the better-known examples of this entanglement of design methods, colonial science, and the changing aspirations and techniques of governance. Elaborate and effective means to temper the thermal conditions of the interior without the use of mechanical strategies—or, at least, using architectural means to reduce reliance on expensive air-conditioning systems—was essential to the growth of modernism as a style, and to the prospects of economic development in many areas of the Global South, for better or worse.

The focus of this essay is the development of climate design strategies in Rio de Janeiro, Brazil, at the beginning of this mid-century period, considered in relationship to the emergence of an architectural imaginary that attempted to bridge the gap between global knowledge of climate systems and the design imperatives of local conditions. This imaginary was articulated in the context of transformations to the economic and social infrastructure of Brazil, a process in which these ideas and buildings participated. As one of the first examples of the broader interest in climatic design, Brazilian architects—those in Rio in particular—absorbed and transformed the principles of modern architecture according to climatic dynamics and through the design element of the shading system. These projects were essential in indicating the wider possibility of adapting the so-called International Style to the specifics of a given region—in terms of climate, materials, political environment, and forms of inhabitation. They are also essential to understanding the role of architectural ideas and imaginaries in the economic and political systems that followed.

These three factors—climate design methods, climate science, and the anticipation of social, industrial, and economic modernization—come together in the work of the architectural firm MMM Roberto. Three brothers, Marcelo, Milton, and Mauricio, the firm won commissions for a number of new building types essential to the integrated process of social and economic modernization, including airports, government agencies, speculative office buildings, apartment complexes, and buildings for insurance corporations.[5] They became specialists in buildings that represented, materially and symbolically, the modernization of Brazil and the transformation of architectural methods. Their ideas also help to clarify how architectural modernity became more attentive to its role in facilitating economic development. Essential to this facilitating role was the effective management of the thermal interior. MMM Roberto's work, and in particular the 1942 design of the headquarters of the Brazilian Reinsurance Agency (Instituto de Resseguros do Brasil [IRB]) will form the core of the narrative that follows—albeit in a broad context of other projects and built work that clarify the resonance of climate methods across familiar histories of modern architecture and across a broader history of the relationship between modernization, architectural design, and the unanticipated consequences of reliance on fossil fuels.

THE GLOBALIZATION OF THE "INTERNATIONAL STYLE"

The history of modern architecture is beginning to take into account the importance of climate considerations in the development of design strategies. This is symptomatic of a broader transformation of the field—away from narratives focused on masters and their acolytes and toward a more profound engagement in the local and global complexities of the built environment. Other pertinent historiographic symptoms include a restructuring of the center/periphery model of regional influence, an exploration of the role of technological innovations, and an analysis of the deep imbrication of architectural modernism across colonial and postcolonial regimes of governance.[6] Climate—as a material reality, a scientific concept, and an aspect of modernization projects—offers a compelling window onto the complex relationship between the technicity of the built environment and these sociopolitical transformations.

The period of the most intense activity in the development of modern architectural ideas (the interwar period generally considered) could also be read as the moment when the global transition to fossil fuels had an impact on buildings.

The factory-like single-paned glass façade of the main building at the Bauhaus in Dessau, to take one prominent example, could not have been imagined without increased availability of coal for heating. Fossil fuels facilitate design innovations. However, there was little access to affordable mechanical means of cooling, and the role of design strategies in adapting a building to its climatic conditions was an essential aspect of the global proliferation of modernist innovations.

The work of Le Corbusier is especially significant.[7] Among the most important architects of the twentieth century, Le Corbusier was quick to recognize the significance of his own interventions. He is known not only for his buildings but also for his writings on architecture and his promotion of the tenets of modernism that he and others were developing. Indeed, his 1914 diagram of concrete floor plates supported by steel beams—the so-called dom-ino diagram, a play on domicile and the modular, repeatable design condition a domino tile suggests—is a landmark in the production of an architectural imaginary filled with aspirations for how new materials could transform approaches to design and, as a result, transform ways of living (see figures 11.1A and 11.1B).[8]

The dom-ino, on these diagrammatic terms, articulated a straightforward proposal. The use of structural steel beams meant that architects were no longer reliant on a wall of brick or stone to hold the building up. Concrete and steel meant that both the façade and the arrangement of the interior were no longer burdened by tectonic requirements and were now available for creative architectural expression.[9]

As the façade became liberated from structural demands, it was filled with glass. A light-filled interior was not only more pleasant throughout the day, because of the natural effects of daylight, it also allowed for solar radiation to play a more pronounced role in the thermal experience of the building. For Le Corbusier and others, sunlight spoke to a modern experience of a healthy interior, which would be clean, sanitary, and filled with light.[10] "Teach your children," Le Corbusier stated in a 1929 lecture in Buenos Aires, "that architecture is about sunlight on floors"—a cryptic phrase that suggests a healthier relationship to natural patterns was an aspect of the modern architectural imaginary.[11]

In regions outside of the temperate climates of Europe and North America, where architectural modernism's collusion with colonial, postcolonial, and neocolonial practices was intensifying and producing opportunities for building, this new openness to the sun was a bit more complicated. A further design strategy was needed. Much as Le Corbusier was essential to articulating and promoting the principle of modern structure with the *dom-ino*, he sketched *brise-soleil* on buildings in Bracelona, Algiers, Geneva, and elsewhere in the 1920s and 1930s (see figure 11.2). "Brise-soleil" (French for "sun-break") became a general term

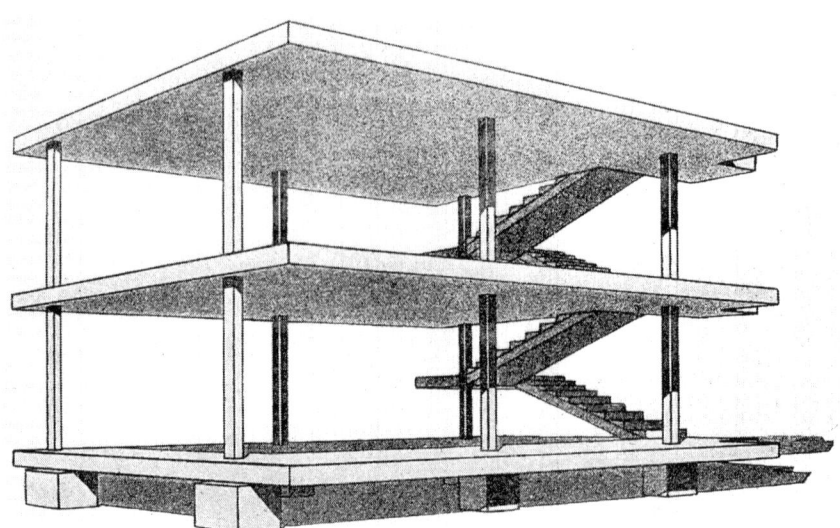

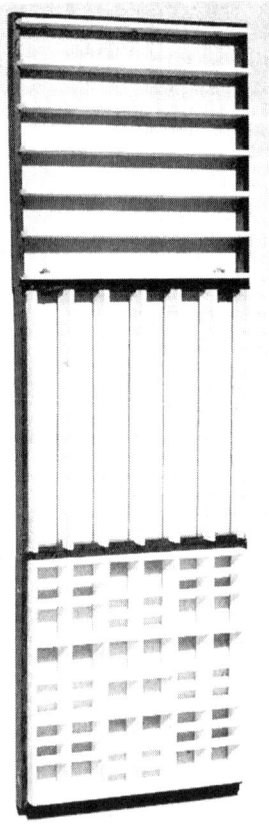

Figure 11.1A (*above*) and B (*left*). Le Corbusier, Dom-ino diagram, 1914, from the *Oeuvre Complète, 1910–1929*. © FLC/ADAGP, Paris / Artists Rights Society (ARS), New York, 2019. Model of different louver orientations for brise-soleil façade attachments, as an illustration to Olgyay and Olgyay, *Solar Control and Shading Devices*. Reproduced with permission.

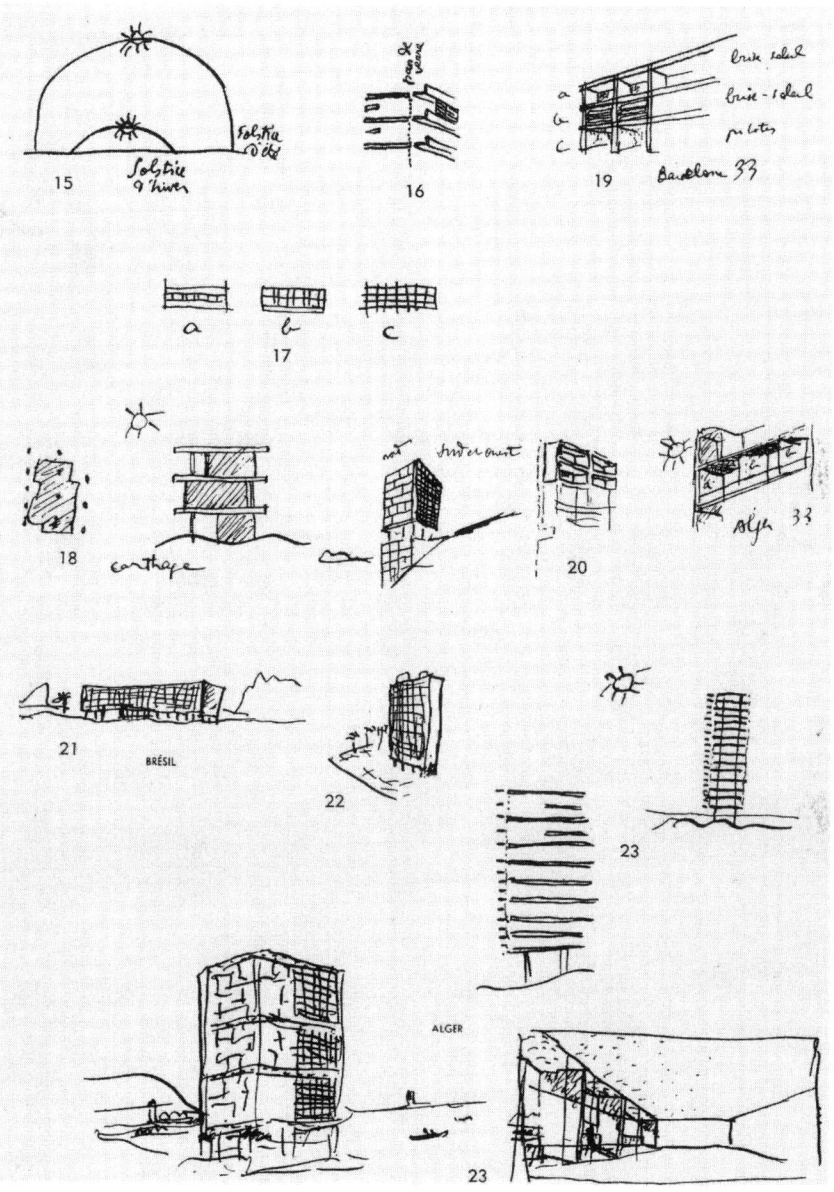

FIGURE 11.2. Le Corbusier, sketches for "Une petite histoire du brise soleil chez Le Corbusier," which was published at the beginning of the *L'Architecture d'Aujourd'Hui* special issue on "Architecture au Bresil," August and September, 1947. © FLC/ ADAGP, Paris / Artists Rights Society (ARS), New York, 2019.

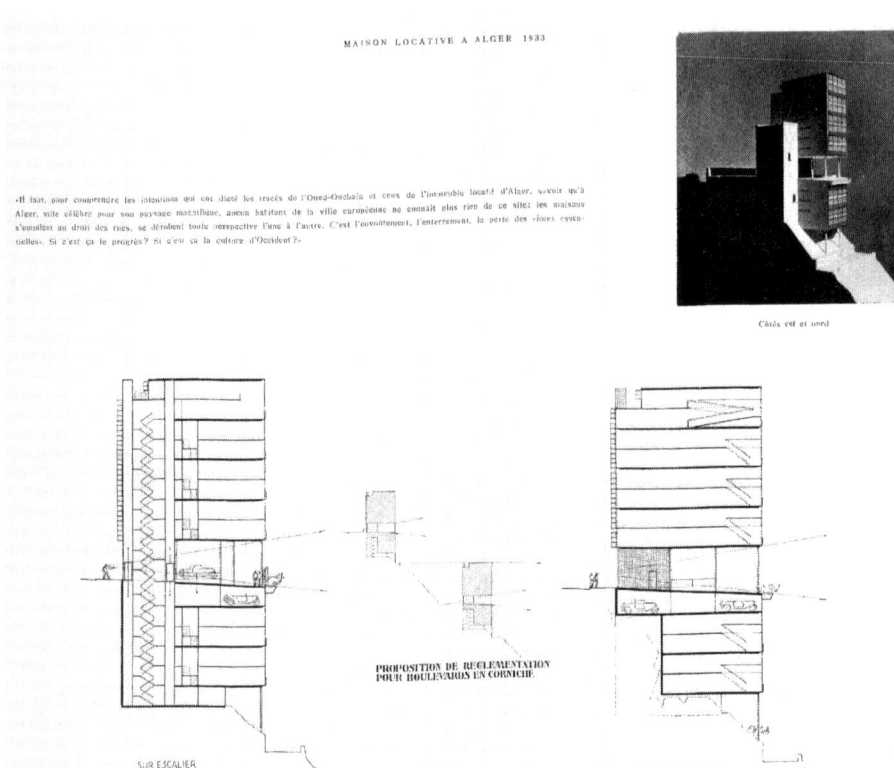

MAISON LOCATIVE A ALGER 1933

FIGURE 11.3 (*left and right*). Le Corbusier, Maison Locative, Algiers, 1933. From the
Oeuvre Complète © FLC/ADAGP, Paris / Artists Rights Society (ARS), New York, 2019.

for louvers, eaves, or fins, often moveable, carefully placed on the façade with
the intent of precisely modulating the relationship of the sun, according to its
seasonal and daily solar path and relative to radiation and lighting conditions
and the desired experience of the interior.[12] In an important sense, the brise-soleil
was not so much an invention as it was the integration of traditional strategies
into the terms of modernist architectural developments, applicable across regions
and building types.[13] Shading systems, generally speaking, have been essential to
modulating solar effects on a building's interior for centuries.

One of Le Corbusier's most influential proposals was the Maison Locative,
which was designed for a site in Algiers in 1933 (figure 11.3). This unbuilt project,
a twelve-story hillside tower, included a concrete grid of protruding shading
elements on the sun-exposed facades. Following the dom-ino idea, the building

sets up a diagrammatic approach to the tall office building built with steel
structure, concrete floor slabs, and a glass façade: it is open to the sun on the
southern façade and covered in a second skin—a shading grid—on the northern
sun-exposed façade.[14] The shading system was, in principle, calculated so as
to allow the sun in during cooler months but to block it during the summer.
Le Corbusier proposed the project as part of a broader set of plans intended to
facilitate the economic modernization of Algiers as a central conduit for North
African commerce with France—suggestive, albeit here only vaguely, of the importance of these new design techniques and their accompanying imaginaries as
both a response to and a facilitator of the political and economic transformations
of the post-colonizing period to follow.

   The dom-ino and the brise-soleil can be seen as essentially paired, the one
requiring the other, especially in the Global South. Together, they were important to the development of modern architecture as a global building strategy and

to claims of regional adaptability—to the various articulations of the promise of modernism as an "International Style," as made explicit in the exhibition of that name at New York's Museum of Modern Art (MoMA) in 1932 or, as indicated more broadly in the series of meetings, partly instigated by Le Corbusier, of the International Congresses of Modern Architecture (CIAM) from 1928 to 1959.[15] This paired elaboration of the dom-ino and the brise-soleil was a substantive intervention of modern architecture, as a material and symbolic emergence that transformed building practices as part of development regimes in the Global South. These strategies also led to the conceptualization of the thermal interior as an object of architectural analysis.

ARCHITECTURE IN THE ESTADO NOVO

The best-known example of shaded modernism in Brazil is the Ministry of Education and Health (Ministério da Educação e Saúde [MES]; see figure 11.4).[16] Designed by Lúcio Costa, Carlos Leão, Jorge Moreira, Oscar Niemeyer, Affonso Reidy, and Ernani Vasconcelos beginning in 1936, the building is a tall narrow structure with a more amorphous form, an auditorium, intersecting at the base. The MES established the basic *parti* of the shaded office building, produced throughout Brazil and the Southern Hemisphere more generally—it also prefigures Neimeyer's design for the UN Headquarters in New York.

The Ministry Building in Rio approached climate along the lines of the Corbusier plan for Algiers. It is the dom-ino diagram repeated up to the height of fifteen stories, with a steel frame and concrete slabs and the façade available for climatic modulation. The north sun-facing exposure is carefully mediated by banks of louvers nested in an egg-crate façade; the south façade is all glass, without solar protection. The egg-crate modules hold the façade together as a visual piece, as a monumental screen, while the variation in each module is both visually dynamic and effective as a thermal and daylight modulation device. In this case, unlike that diagrammed at Algiers (see figure 11.4), the inhabitant could adjust the shading louver with a lever system and thus alter the light and heat conditions of the interior. This was an interactive device, activating and thematizing, as cultural practice, the relationship between the path of the sun and the inhabitants' experience of the interior.

When Le Corbusier visited Brazil in 1929, he exerted a strong influence on Lúcio Costa, the lead architect on the MES project. Costa was briefly the head of the Escola Nacional de Belas Artes (ENBA) while two of the Roberto brothers were studying there.[17] Le Corbusier returned to Rio in 1936 by invitation from Gustavo Capanema, the minister of education and health, who had commissioned Costa to design the new building; Costa had requested a large Brazilian team and Le Corbusier as consultant. When Le Corbusier visited, he proposed to move the building from the Distrito Federal in central Rio to a site by Guanabara Bay, which was not then available.[18]

The MES design process is often seen as the moment when Le Corbusier brought the brise-soleil to the architectural culture of Rio, though even the slightest glance at the chronology of related structures belies this—Niemeyer's Obra do Berço, in Lagoa, with movable vertical louvers, was under construction by the time Le Corbusier returned to Brazil. The Roberto brothers' Brazilian Press Association (Associação Brasileira de Imprensa [ABI]) was also near completion in the Distrito Federal. According to Brazilian historian Gustavo Rocha-Peixoto, one of these two buildings should be seen as the first brise-soleil structure to be built worldwide; the first example of shaded modernism.[19] As the American historian and curator Barry Bergdoll recently wrote, these and other important buildings of early South American modernism "are not the belated reflections of examples set in Europe, but previsions of a modernization to come," a significant historiographic inversion that will be further explored below.[20]

The Ministry of Education and Health itself—that is, the government agency, not the building—was created ten days into the Vargas administration, in November 1930. Getúlio Vargas had come to power through what was effectively a bloodless coup; Vargas, with the army beside him, took advantage of alarm over a supposed communist plot.[21] On the one hand, the Vargas regime—after 1937 referred to as the Estado Novo (the New State)—was forward looking; the creation of the Ministry of Education and Health suggests the regime's approach to managing and providing for the body politic. On the other hand, Vargas was an authoritarian leader, and between 1930 and 1945 he concentrated power in his federal administration, his confidants, and the agencies, including the MES and the departments for labor, resources, and finance that were under his control.

The ministry was, in significant and straightforward ways, the face, or façade, of the Estado Novo. It both symbolized and facilitated a modern regime of health management, including collective insurance, clinics and vaccines, care around pregnancy and birth, and early education, as part of a rich and complex history of public health in the country.[22] It represented hope, and a trajectory toward, a prosperous Brazilian future.

The ministry, and the Estado Novo, articulated a very specific form of bureaucratized governmental activity that can be glossed as the conceptualization of citizens as a population to be optimized, directed toward economic growth and territorial modernization. Forms of labor, economic activity, insurance, and growth framed this approach to the population, alongside media and processing systems through which demographic, educational, health, and other population

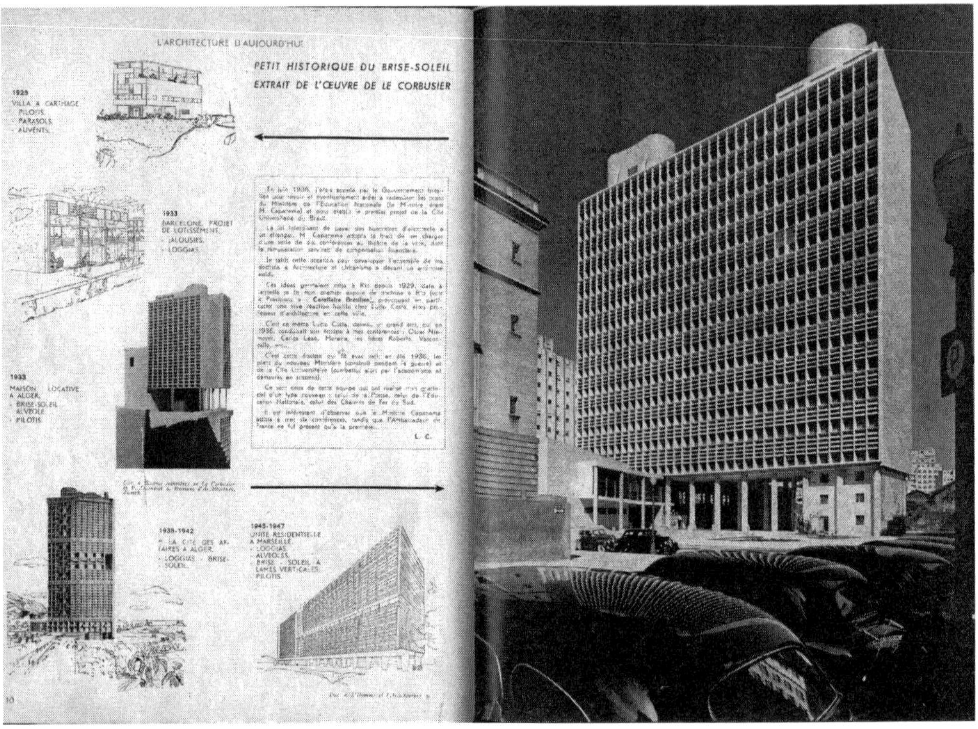
L'ARCHITECTURE D'AUJOURD'HUI

PETIT HISTORIQUE DU BRISE-SOLEIL
EXTRAIT DE L'ŒUVRE DE LE CORBUSIER

1929
VILLA A CARTHAGE.
- PILOTIS.
- PARASOLS.
- AUVENTS.

1933
BARCELONE. PROJET DE LOTISSEMENT.
- JALOUSIES.
- LOGGIAS.

1933
MAISON LOCATIVE A ALGER.
- BRISE-SOLEIL ALVEOLE.
- PILOTIS.

1938-1942
« LA CITE DES AFFAIRES A ALGER.
- LOGGIAS - BRISE-SOLEIL.

1945-1947
UNITE RESIDENTIELLE A MARSEILLE.
- LOGGIAS.
- ALVEOLES.
- BRISE - SOLEIL A LAMES VERTICALES.
- PILOTIS.

L. C.

10

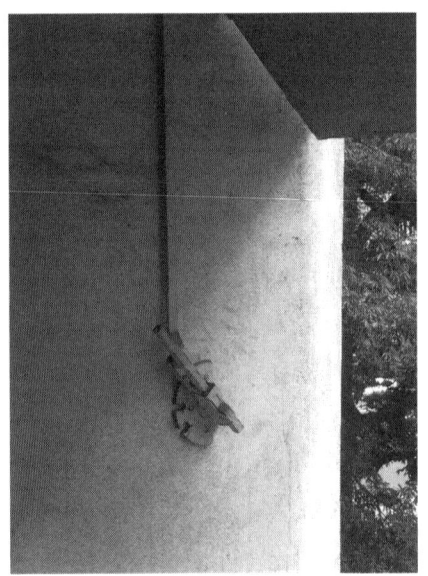

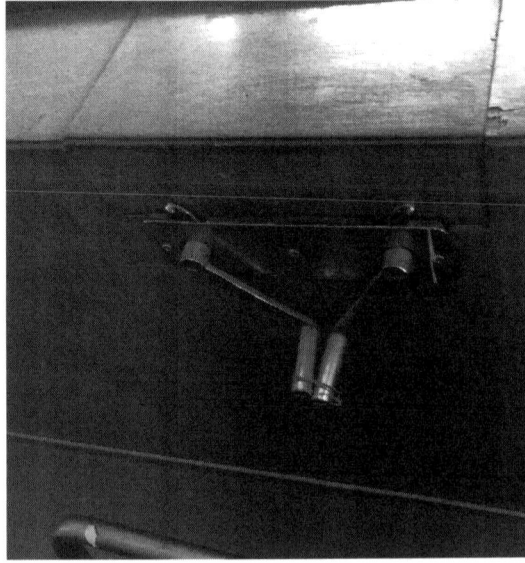

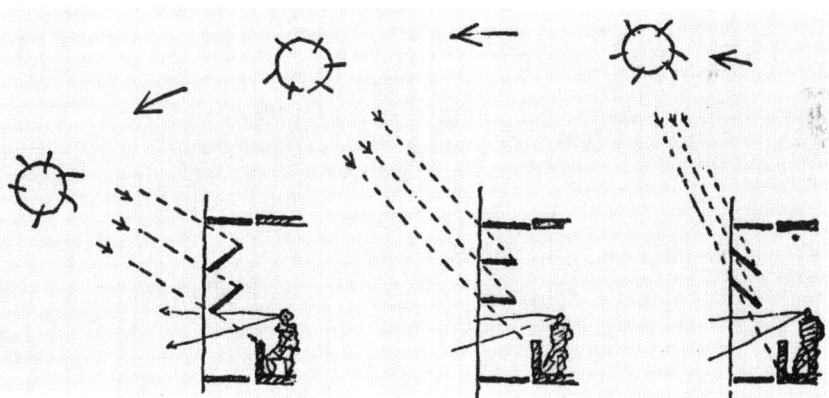

FIGURE 11.4. Lucio Costa, Oscar Niemeyer, Carlos Leão, Affonso Eduardo Reidy, Ernani Vasconcelos, et al, Ministerio da Educação e da Saúde (MES), Rio de Janeiro, 1936-1943; from *L'Architecture d'Aujourd'Hui*, 1947 and *Architectural Forum*, August 1944. Louvers and louver-adjustment system at the MES. The building was undergoing renovation at the time of these photographs, and the system was not yet working; Lucio Costa's drawing of the daily path of the sun in relationship to the louver system and the daylit conditions of the interior of the MES.

data could be best put to work. The distant Brazilian interior in particular came to be subject to the work of agencies such as the MES, as a means of exercising state power over a population seen to require integration into modern ways of life. The population of the interior was subject to assessment through risk mechanisms and monetization as a labor force and was placed in relationship to natural resources and other territorial opportunities according to an emergent logic of globalized capital.[23] MES Radio, in concert with the Press Association (housed in the Roberto brothers' ABI Building just a block away from the Ministry Building) brought entertainment and educational programs deep into the interior. Health bulletins and training programs, it was hoped, could shepherd the wider population toward a new quality of life standard—including better education, improved life expectancy, and the eradication of communicable diseases.

Climate was a crucial aspect of these modernization efforts, both symbolically and materially. The first minister of education and health, Francisco Campos, saw the project of the ministry as that of "rejecting four decades of social engineering informed by pessimistic theories of racial and climatological degeneracy."[24] Campos's interest in overcoming the purported obstacle of climate was relative to a general notion of climatic determinism that was characteristic of the environmental sciences emergent in this period. A number of European and American physiologists assessing the conditions of the colonies and the Southern Hemisphere saw climatic conditions as determinant of a region's potential role on the world political and economic stage (see Livingstone and Fleming, this volume). "One of the reasons," as one such imperialist wrote in 1942, "for the rise of [one] nation [rather than others] in modern times *is its control over climatic conditions*: that nation which has led the world, leads the world, and will lead the world, is that nation that lives in a climate, indoor and outdoor, nearest the ideal."[25] The purported ideal climate was that of the temperate zones of Europe and North America. Climate knowledge, in this sense, emerged as part of racially stratified forms of colonial reason, also resonant across the thermal interior.

Management of this built thermal interior was a small but significant aspect of the goal of conservative revolutionaries to modernize Brazil through social and economic reforms, and one that implied a convergence of these reforms both with the promise of modern architecture and with the hope for a more substantive role of Brazil in global economic systems. One also hears echoes of Le Corbusier's dictum, first offered in a 1929 speech in Buenos Aires, that someday soon "every building, around the globe, will be 18 degrees." The universalist notion of a world market, a consistent norm for the expansion of social and economic possibility, was embedded simultaneously in political and architectural imaginaries of the period.[26]

The entanglement of climate and architecture with geopolitics and geophysics was of course not exclusive to Brazil. In 1944, for example, Austro-American architect Richard Neutra was commissioned by the US government to build schools and hospitals around Puerto Rico and developed a series of prototype designs and specific climatic methods, mostly focused on ventilation, that were intended to improve the life conditions of the island's inhabitants. The work would be published, by a Brazilian press, as *Architecture of Social Concern for Regions of Mild Climate* in 1948.[27] In the meeting in San Francisco to establish the United Nations in 1945 (at which Neutra was the CIAM representative), he referred to the project as a "Planetary Test" for the efficacy of modern architecture to address conditions outside of Euro-American urban centers.[28] With correlates in West Africa, India, Singapore, and elsewhere, Neutra's "Planetary Test" intended to find out how modern architecture could, through climate methods, improve the quality of life in what were beginning to be called developing economies.[29] This was a fundamental shift from prewar functionalism toward a postwar design operationalism, largely through architectural strategies of climatic adaptability. Neocolonial projects across West Africa by Jane Drew and Maxwell Fry and other so-called Tropical Architects, climatic modernisms in Singapore and Malaysia, experiments across Australia, among many other examples, begin to suggest the wider dissemination of these strategies in the period just before air conditioning took over the building industry and the designers who worked with it.[30] This wider proliferation of strategies articulates, in some ways, the new world that this modernist imaginary had promised—framing knowledge of the population in concert with understanding of global climatic patterns, the complexities of regional identity, and the changing technological prospects of the built environment. The façade, in this sense, becomes an interface for understanding and manipulating a cultural approach to climate in a given time and space.

KNOWLEDGE OF CLIMATE

The period from the mid-1930s to the mid-1950s was one of expanded research into the relationship between architecture and climate. American and European meteorologists began to understand architecturally relevant low-level atmospheric conditions in new ways. By the end of the Second World War, projects had been seeded that allowed for an increased rate of knowledge development. John von Neumann's well-known Meteorology Project at the Princeton Institute for Advanced Studies, using the ENIAC computer at the Aberdeen Proving Ground in Maryland, was just down the street from the Princeton Architectural Laboratory, where Victor and Aladar Olgyay experimented with climate methods, heliodons, thermoheliodons, and other modeling devices—though there was no substantive interaction between the two groups.[31] The Olgyays research focused on the developments in Brazil as a realization of modernism's climatic promise; their book *Solar Control and Shading Devices* in 1957 started with Le Corbusier and ended with a ninety-seven-page section of "Architectural Examples" that included thirty-eight buildings from Brazil.

The Olgyays also contributed to the Climate Control Project, the dynamics of which allow for an indication of the endgame that the Brazilian efforts also aspired to. This project was as a collaboration between the American Institute of Architects and *House Beautiful* magazine. Helmut Landsberg, an Austrian émigré climatologist in the United States since 1936, who would continue to write on urban climate matters, had trained James Marston Fitch, a young architectural historian, in meteorology during the war—Fitch would later write *American Building and the Forces that Shape It*.[32] Fitch was a central protagonist on the Climate Control effort.

The phenomenal increase in weather data after the war focused on upper atmospheric phenomena rather than on near-ground observation and the minute details of microclimatic conditions relevant to architectural matters.[33] As the macro-climate became of increasing interest, observation stations moved to airports and other sites at a distance, relatively speaking, from population centers, leading to data that was not directly applicable to architectural concerns. Architects and others involved in the built environment saw a need to employ specialists who could address lower atmospheric phenomena—"cooperation between architects and climatologists," Fitch wrote, "will yield designs better adapted to their environment."[34]

The Climate Control Project sought to estimate the climatic conditions for thirteen growing metropolitan regions in the United States. Landsberg—working with Paul Siple, from the US Army Corps of Engineers and well-known as an explorer of the South Pole—originally intended to demarcate four general regions across the continent and to commission an architect to develop a typical design for each region. It quickly became clear that there would have had to be "hundreds of zones," because any simplified analysis according to temperature and precipitation ignored too many variables that were significant to design issues (see figure 11.5). Siple "could find no assistance from individuals in the building field who, [he] had hoped, could give criteria that would assure that the zones were significant."[35] And here was the central issue confronting Brazilian experiments as well—gathering data was important, but of even more concern was how to effectively communicate climatic knowledge to the architectural profession. "What I visualize," Siple later wrote, "are diagrams so clearly depicting [climatic issues] that texts will not be necessary for further interpretation."[36]

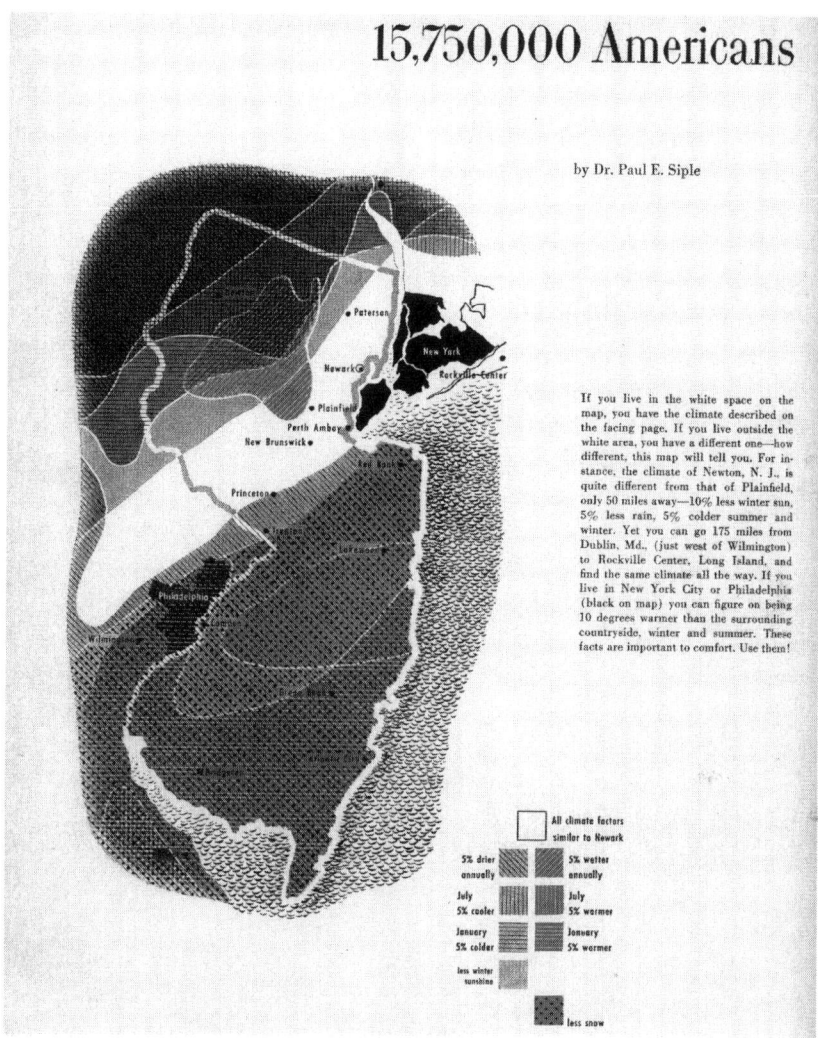

FIGURE 11.5. Paul Siple, Climatic Map of Metropolitan New York and New Jersey for "15,750,000 Americans Live in This Climate," House Beautiful (November 1949).

This brief synopsis of a slightly later architectural-climate modeling experiment clarifies the concerns that were on the table in interactions between meteorologists, engineers, and architects in Brazil in the 1930s and 1940s. Even basic efforts made in mapping the sun path required interpretation, visualization, and translation into an architectural context. In 1936 journals around the world—*Architectural Forum* in the United States, *RIBA Journal* in the United Kingdom,

Arquiteto e Urbanismo in Brazil, and others in France, Spain, Australia, Chile, and elsewhere—began publishing sun-path diagrams and other climate data.

This broad architectural interest was met with a scientific and governmental interest in carefully understanding specific local conditions—those of the coastal plain on which Rio stood, for example—in order to optimize development. In designing the IRB, the Robertos refer to "computation of Weather Data done at the Instituto Nacional de Tecnologia" (the institute was another Vargas agency). Such analyses were being done in both Rio and São Paulo, and the focus was on mapping the specific characteristics of Rio's complex climate. Rio had been subject to numerous urban restructuring schemes, from the Plan Agache in 1928 to the Boulevard Vargas over the 1940s and into the 1950s. The municipality had also, for some decades, been flattening hills and rerouting rivers so as to maximize development potential, especially in the central parts of the city. Knowledge of Rio's geography was undergoing transformation, and it was increasingly crucial to the calculations of the state. Interest in Rio as a body of data, available for scenarios of optimization, further encouraged precise analysis by both architects and climatologists in the 1930s.[37]

With Rio nestled between coast and mountains, wind and wave conditions have a great effect on its microclimate. In one foundational article from 1936, Ibrahim Carone asked "What Is Rio's Healthiest Location?" an obvious question for targeted urban development (see figure 11.6). For Carone, controlling humidity and solar radiation was the key to the experience of a comfortable temperature. Specific areas of the city were more appealing in terms of general air quality—relative to a sense of "freshness" through proximity to the ocean rather than to a concern about pollution. Carone discussed the neighborhoods of Tijuca, Copacabana, and Jacarepaguá, and then looked at the Distrito Federal, in the center, which he describes as being "ventilated by the south winds" largely because of the stark elevation drop from one side of the occupied coastal plain to the other.[38]

If the challenge of the global interior was some semblance of consistency across varied regions, in Brazil the obstacle was the experience of heat, which could be mitigated, even in sweltering weather, by any persistent breeze that could be induced through strategic openings in a building. The engineer Paolo Sá, quoted by the Robertos below, was an expert in understanding the induced experience of ventilation as cooling—"the importance," as he referred to it, "of eliminating extra heat produced by the human machine." Sá further summarized, "When the air is in motion, the layer that was in contact with the body, heated and moistened, is replaced by a fresher and drier one, which increases the feeling of comfort."[39] A breeze drawn out of prevailing winds and facilitated by ventilation was an essential adaptation of the building to different kind of thermal experience.

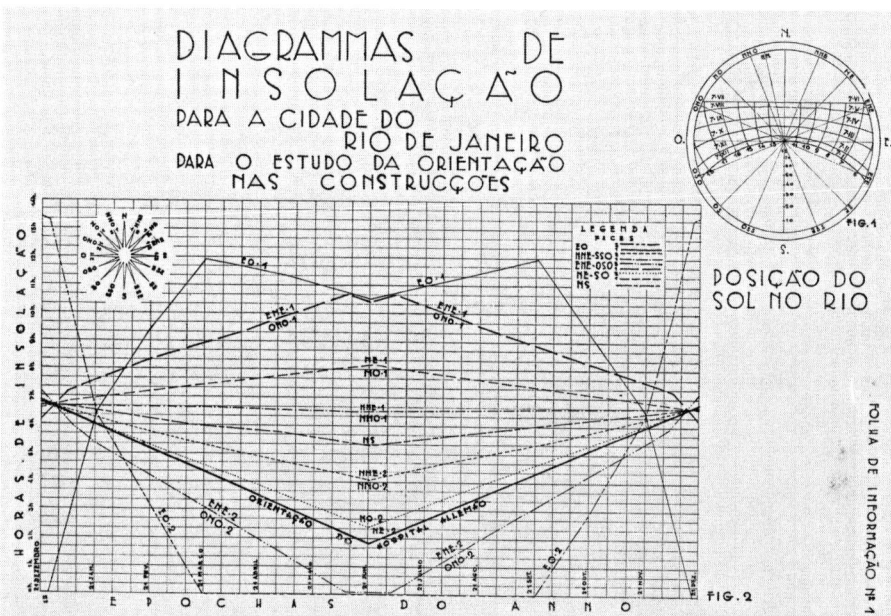

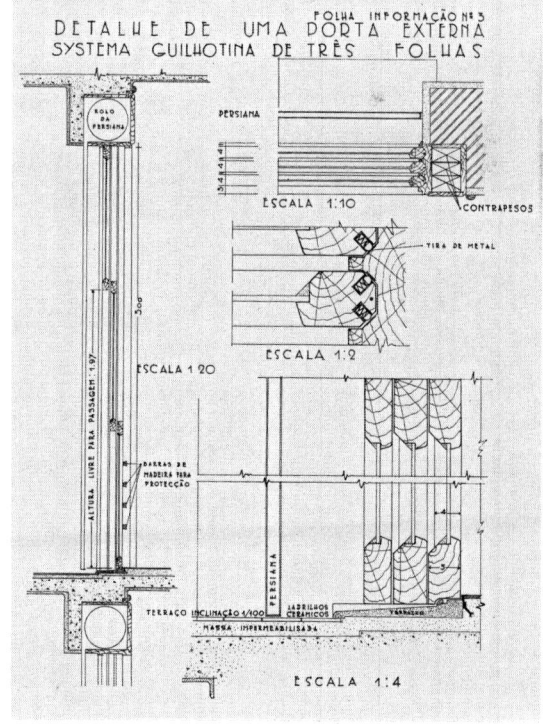

Figure 11.6. Paolo Sá (Instituto Nacional de Tecnologia), Insolation Diagrams and Systema Guilhotina in *Arquitetura e Urbanismo* (May–June 1936).

Consideration of the thermal needs of the body and the satisfaction of these needs through design means was an important thread among architectural developments in this period. In the first part of the twentieth century, factories, large public areas, homes, and office buildings were being analyzed and reconsidered according to their comfort conditions.[40] Drawing on the work of the French meteorologist André Missenard, who worked closely with Le Corbusier in the development of climate management systems, Sá proposed the adaptation of a coefficient of heat, humidity, and air speed so that this sensation of freshness amid tropical pressure could be codified and regulated.[41] Such a process was not only effective for Rio but can be extended to "other parts of the territory" and according to "other genres of activities."[42]

As part of this general expansion of climatological knowledge, the scientists in the AIA/*House Beautiful* project and the engineers in São Paulo and Rio were struggling to articulate specific visual means to facilitate architectural experimentation with a radically site-specific notion of climate. These investigations aimed to consider how systematic knowledge could be brought to bear on the details of a given location; and to consider how this knowledge could be applied through a given set of devices and methods that collectively sought to bridge global and local knowledge of climate patterns. Techniques of data gathering and analysis were brought into a regulatory regime as standards for thermal comfort conditions. The initial medium of this transformation, before HVAC, in Brazil and elsewhere, was the façade shading device.

TOWARD AN ACCLIMATIZED MODERNITY

Climate design methods helped to reflect local characteristics of the modernization process, at the same time that climate data proliferated in a number of different intellectual and managerial environments: in engineering schools and architecture firms and in designs for government agencies and insurance corporations. The shading device was an essential aspect of the insertion of Brazil into global flows of knowledge, as part of a broader reconfiguration of economic relations.[43] From this perspective, the promise of architectural modernity is less its implicit progressivism and more a capacity for *adaptability*—to different climates and different political regimes and to different regimes of behavior, habit, and maintenance.

The first building designed by the Roberto brothers was the ABI Building (figure 11.7). It was started in 1936, a year after Vargas won a second term in a democratic election and then dissolved the electoral process. The center of Vargas's newly centralized government was the Distrito Federal, in central Rio, where construction on the ABI was getting underway just as the design process for the MES Building began.[44] The ABI was a commission for a quasi-government agency; the ABI was concerned not only with press but also educational films and rural radio programs, on which it collaborated with the Ministry of Education and Health.

At the ABI, fixed louvers sit at a distance from the interior wall, across a "heat dispersion zone" that also operated as a corridor for auxiliary circulation (seen in the left of figure 11.7). The shading panels are thick slabs of asbestos cement, almost monumental in appearance. They are angled to block the sun most effectively at summer solstice.[45] Unlike those at the slightly later MES Building and most that follow, these louvers were fixed and could not be modulated according to seasonal conditions.

The Robertos' second government commission was for the Aeroporto Sántos Dumont, begun on a site to the southeast of the Distrito Federal in 1936 (figure 11.8). It is just across the street, literally, from where the Instituto de Resseguros do Brasil would be built starting in 1938. The airport is visible from the open entryway to the IRB, and from its Roberto Burle-Marx–designed roof terrace. Today the secondary, regional airport in downtown Rio, Sántos Dumont, was at the time a central piece of Vargas's modernization project, connecting the growing city to Europe and the United States, to São Paulo, Salvador, and the interior. At Sántos Dumont, Rio's modernity was imagined as one of openness and selective engagement with the sun. The airport terminal's large movable louvers on the glass façade helped to acclimatize the open, high-columned interior. This shading skin was essential—thermally, experientially, symbolically, politically—to catapult Brazil onto the world stage.

The liminal condition of the shaded office tower is clearly expressed in the IRB (figure 11.9).[46] It is liminal for its capacity to straddle and integrate a range of significant historical passages—from low- to high-rise modern office buildings, from climatic determinism to architectural experimentation, from the social project of the masses to the calculational matrix of the population. The IRB itself (the institute, not the building) was begun as part of Vargas's initial rise to power in 1930. It was a fund—60 percent of government money, 40 percent from major global insurers and reinsurers operating in Brazil—to stabilize the property market and provide capital and coverage for investment in Brazil.[47] The building was the front door for international capital arriving at Sántos Dumont and participating in the economic development of the Estado Novo.

The IRB Building had an elaborate set of shading mechanisms—it was carefully designed for dynamic interaction with its microclimate. The different facades had different treatments, according to their solar exposure—what the Olgyays' would later term "bio-climatic," and indeed, despite the nonoperable

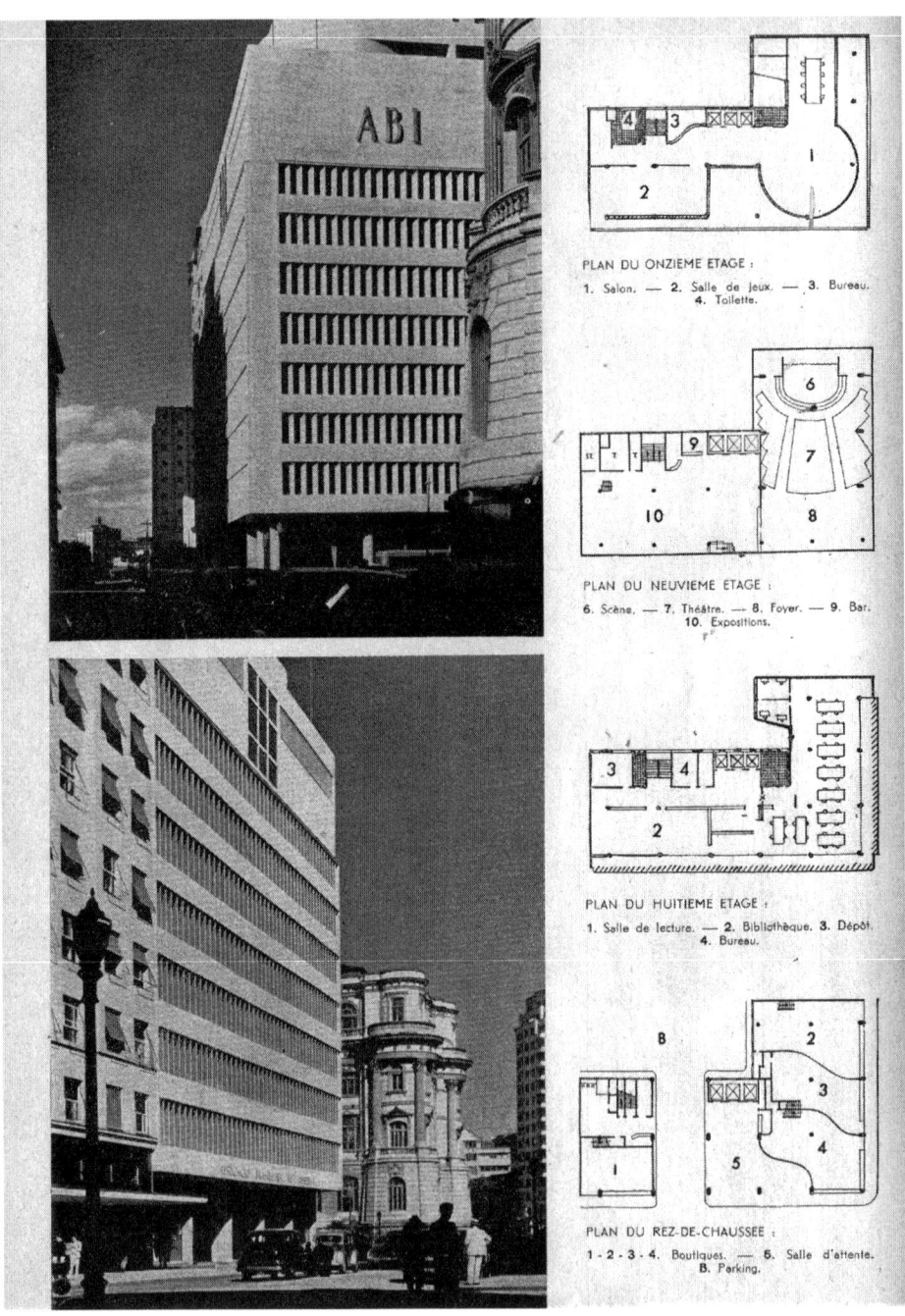

FIGURE 11.7 (*left and right*). MM Roberto, Associação Brasileira de Imprensa (ABI), Rio de Janeiro, 1936, in *L'Architecture d'Aujourd'Hui*, 1947.

BUILDING "A.B.I."

RIO DE JANEIRO

MARCELO ET MILTON ROBERTO. ARCHITECTES

La construction de l'A.B.I. (Association de la Presse Brésilienne), un des plus importants buildings parmi les édifices modernes du Brésil, fut confiée aux frères Roberto, alors que deux d'entre eux étaient encore étudiants.

Cet édifice, dont le parti se rapproche par certains côtés de celui du Ministère de la Santé, se présente sous un aspect très différent. L'utilisation de bandeaux horizontaux et d'une façade sans ouvertures à l'étage de la salle de spectacles, ont abouti à un volume compact et un peu sévère d'un caractère solide et assez monumental.

Cet aspect est encore renforcé par le revêtement des façades en dalles de granit brésilien et de travertin argentin.

Le brise-soleil sur les deux faces, également exposées au soleil, est du type fixe en éléments de béton armé coulé au ciment et au sable blancs. Cette grille protectrice est séparée de la paroi vitrée des bureaux par un étroit couloir qui forme une circulation extérieure.

L'isolation thermique des murs pleins est assurée par des matelas d'air entre la paroi béton et un mur intérieur en briques creuses.

Le bâtiment contient aux étages supérieurs des locaux spacieux de club : salle de spectacle, salles d'exposition, restaurant et une terrasse-jardin aménagée par Burle Marx.

1 et **2**. TERRASSES AVEC JARDIN AUX DEUX ETAGES SUPERIEURS.

3. VUE INTERIEURE DE L'AUDITORIUM AU 9ᵉ étage.

4. VUE SUR LE CORRIDOR EXTERIEUR SEPARANT LES BUREAUX DES BRISE-SOLEIL FIXES.

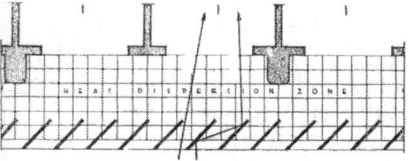

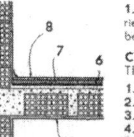

CI-DESSUS : DETAIL DU BRISE-SOLEIL.
1. Bureaux donnant sur un couloir extérieur. — Brise-soleil en éléments de béton fixes.

CI-CONTRE : DETAIL DE CONSTRUCTION DES FAÇADES.
1. Travertin.
2. Béton armé.
3. Vide d'air.
4. Briques creuses.
5. Enduit.
6. Asphalte sur chape.
7. Mâchefer.
8. Parquet.

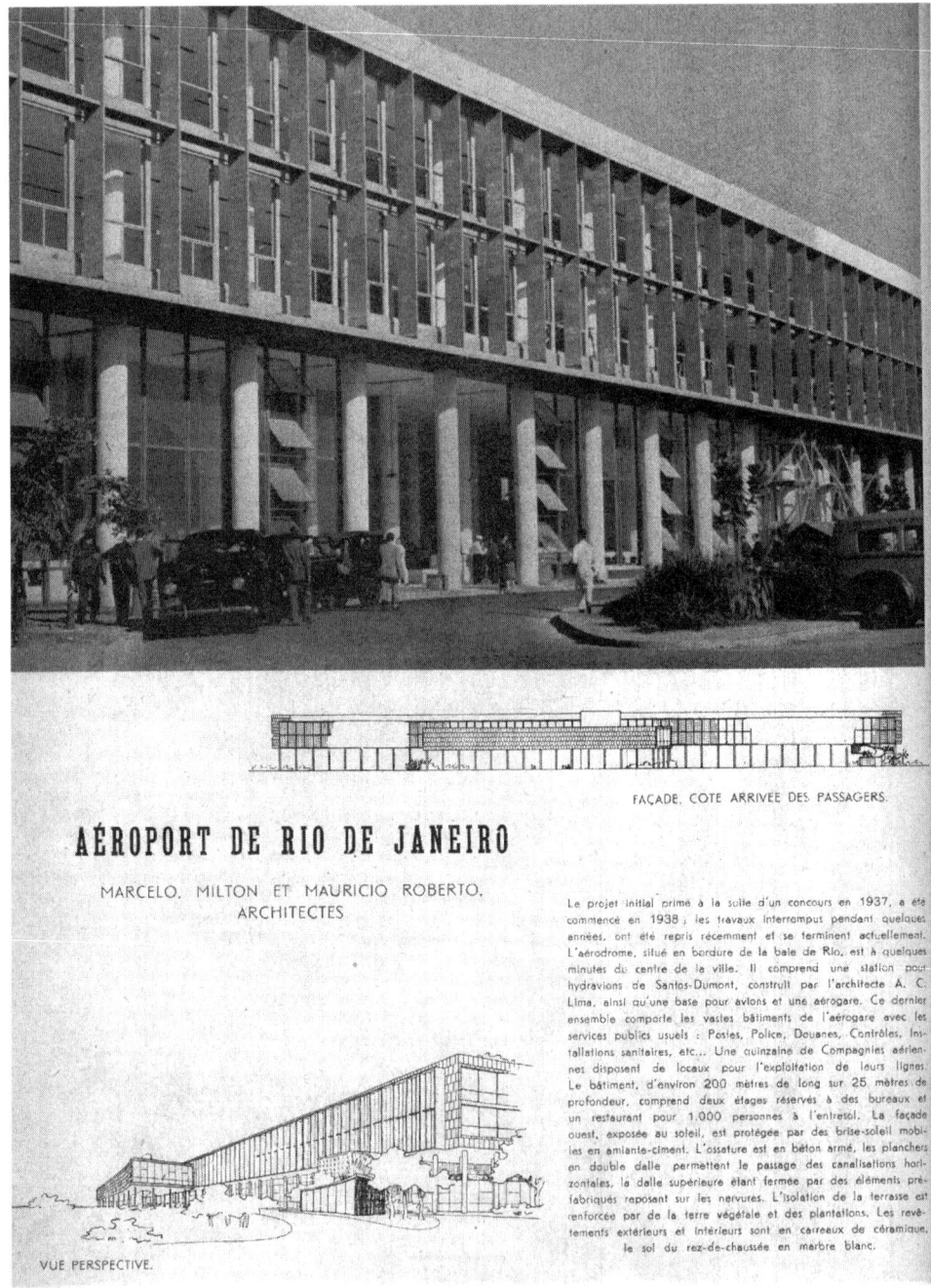

AÉROPORT DE RIO DE JANEIRO

MARCELO, MILTON ET MAURICIO ROBERTO,
ARCHITECTES

FAÇADE, COTE ARRIVÉE DES PASSAGERS.

Le projet initial primé à la suite d'un concours en 1937, a été commencé en 1938 ; les travaux interrompus pendant quelques années, ont été repris récemment et se terminent actuellement. L'aérodrome, situé en bordure de la baie de Rio, est à quelques minutes du centre de la ville. Il comprend une station pour hydravions de Santos-Dumont, construit par l'architecte A. C. Lima, ainsi qu'une base pour avions et une aérogare. Ce dernier ensemble comporte les vastes bâtiments de l'aérogare avec les services publics usuels : Postes, Police, Douanes, Contrôles, Installations sanitaires, etc... Une quinzaine de Compagnies aériennes disposent de locaux pour l'exploitation de leurs lignes. Le bâtiment, d'environ 200 mètres de long sur 25 mètres de profondeur, comprend deux étages réservés à des bureaux et un restaurant pour 1.000 personnes à l'entresol. La façade ouest, exposée au soleil, est protégée par des brise-soleil mobiles en amiante-ciment. L'ossature est en béton armé, les planchers en double dalle permettent le passage des canalisations horizontales, la dalle supérieure étant fermée par des éléments préfabriqués reposant sur les nervures. L'isolation de la terrasse est renforcée par de la terre végétale et des plantations. Les revêtements extérieurs et intérieurs sont en carreaux de céramique, le sol du rez-de-chaussée en marbre blanc.

VUE PERSPECTIVE.

FIGURE 11.8 (*left and right*). MM Roberto, Aeroporto Santos Dumont, Rio de Janeiro, 1936-1944, from *L'Architecture d'Aujourd'Hui*, 1950.

AEROGARE DE RIO DE JANEIRO
DÉTAILS DE STRUCTURE

1. UN ESCALIER TOURNANT
EN BETON ARME AVEC LIMON
DANS L'AXE ET MARCHES EN
CONSOLES ; 2. UNE VUE DU
GRAND HALL PUBLIC EN
COURS D'EXECUTION ; 3. DE-
TAIL DE FAÇADE ; ON APER-
ÇOIT, A GAUCHE, LES BRISE-
SOLEIL MOULES ; A DROITE,
LA GRILLE EN BETON ARME
QUI LEUR SERT DE SUPPORT ;
4. UNE VUE DEPUIS LE
PREMIER ETAGE.

1   2

3   4

BATIMENT D'ADMINISTRATION " I.R.B. "

MARCELO, MILTON ET MAURICIO ROBERTO, ARCHITECTES

DIRECTION DES ASSURANCES D'ETAT A RIO DE JANEIRO

Cet édifice abrite un organisme officiel d'assurances, créé par le Gouvernement brésilien et financé pour 30 % par des compagnies privées et pour 70 % par des organisations de sécurité sociale de l'Etat. Le bâtiment est particulièrement intéressant par les méthodes de construction employées. L'ossature en béton armé est du type classique, mais la constitution et le montage des façades ont été réalisées selon des procédés nouveaux mis en jeu par les architectes. Le remplissage de chaque travée est constitué par un élément préfabriqué monté à sec, en atelier et posé comme une menuiserie. Cet élément, dont on trouvera plus loin le détail, a permis de monter la totalité des façades en dix-neuf jours. La recherche des détails et la finition, d'une précision dont la qualité a étonné les techniciens américains, confèrent à l'ensemble une finesse remarquable. L'utilisation d'éléments de remplissage, conformément à la destination des locaux a eu pour résultante un jeu de textures et d'ombres qui compensent la sécheresse inhérente à toute architecture de buildings pour bureaux. On remarquera la clarté des plans, la séparation des circulations verticales du public et des employés. Les services sociaux pour les employés sont remarquablement développés. Le neuvième étage et la terrasse leur sont entièrement consacrés. On peut affirmer qu'ici, pour la première fois à notre connaissance, a été créé un précédent de cette importance.

FIGURE 11.9. MMM Roberto, Instituto de Resseguros do Brasil [IRB], Rio de Janeiro, 1942. North façade (*top*) and the south and east façades (*bottom*), from *L'Architecture d'Aujourd'Hui*, special issue on "Architecture au Bresil" (August–September 1947).

louvers, it was likely the first self-consciously bioclimatic building to be made in collaboration with scientific knowledge about climate.[48] The north, sun-facing façade (see figure 11.9) had fixed shading louvers framed by vertical areas of glass bricks—these provided daylight for two distinct circulation systems, one for the public and one for government and corporate access. The north façade shading system consisted of a double wall: the interior wall was about two-thirds glazing, with a second, shading façade hung at a slight distance.

As the section drawing in Figure 11.10 indicates the exterior, second skin, which was prefabricated and attached on-site in a mere two weeks, contained fixed brise-soleil elements. These were reinforced concrete louvers formed in a shallow S-curve in plan—the outer face of the louver was a "heat-deflecting surface," to block the penetrating rays of the summer sun, and the inner face was "light reflecting," to increase the daylight transmitted to the interior. There was also a "heat dispersion space"—the louvers, as noted, hung at a short distance from the glazed wall. This space contained a ventilating draw from above (in the photograph, the upper set of semiclosed louvers) to draw heat away from the glazing. This ventilation element was itself not simply open but also contained louvers, so as to be manipulable and able to seasonally protect the interior from solar radiation. The interior facade (evident in the section) was divided between a thick storage block, on the bottom, and a glass window that could be partially opened.

The southern and eastern façades were also activated for their microclimatic positioning. The banding of the exterior was attuned to solar incidence, allowing the building to further reduce reliance on mechanical systems, even amid seasonal extremes. Milton Roberto played out this intervention in a series of diagrams (figure 11.11). Next to a schematic plan, section, and elevation of an unshaded glass wall, he notes: "The exterior walls could all have been thus, very modern, of course. But in Rio there's much sun and much light." Therefore, he continued, "according to scientific computing done with data from the Instituto Nacional de Tecnologia, Rio de Janeiro, Research by Paolo Sá, we concluded that the walls should be thus." He continued with another set of drawings, of the same façade but with a rectangular window set amid a masonry or concrete wall. "Nevertheless," Milton Roberto continues, with reference to debates in Chicago during the early development of the steel and concrete skyscraper, "everyone knows since [Louis] Sullivan about the advantages of a window with full horizontal development. Therefore we have done thus." The third and final diagram shows a series of drawings that match the condition of the façade as built, in which bands of windows alternate with concrete—a thin window at the top, to allow light into the room both directly and by reflection from the ceiling, and a wider

VIEW UPWARD THROUGH BAFFLE SYSTEM

Tools, hardware and some building material, including pipe and electric wiring, were imported from the U. S., as were the majority of the business machines and typewriters. On some of the floors, walls and columns, the architects, who confess to a liking for the color and play of materials, used marble, granite, stone and ceramics—not arranged as decoration but to emphasize architectural intent. The native granite of Brazil is pinkish-grey, the sandstone orange or brown; marble is usually imported. Cement comes from a large modern plant just outside Rio. The Robertos, like all Brazilian architects, work closely with the building industry. They are deliberately free with materials, being concerned to break down established practice and regulations which prevent the development of an experimental architecture.

The reinforced concrete floor construction has an over-all thickness of 12 inches, made up of a reinforced concrete slab, standard beam construction and a ceiling of precast concrete slabs, with an air space between for piping and conduits. A detail of the floor is shown below.

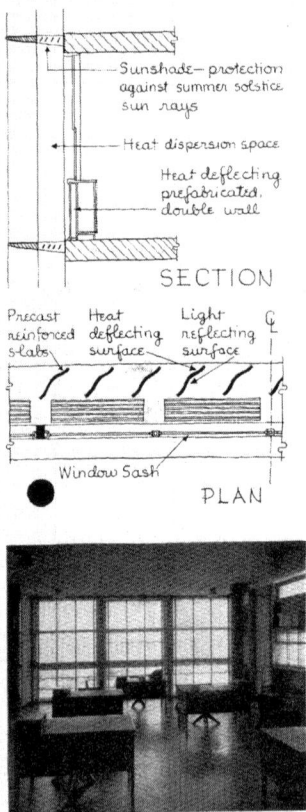

SECTION

Sunshade—protection against summer solstice sun rays

Heat dispersion space

Heat deflecting prefabricated, double wall

Precast reinforced s-labs

Heat deflecting surface

Light reflecting surface

Window Sash

PLAN

Projecting windows on east side have floor-to-ceiling fenestration and are equipped with interior sun blinds for occasional use.

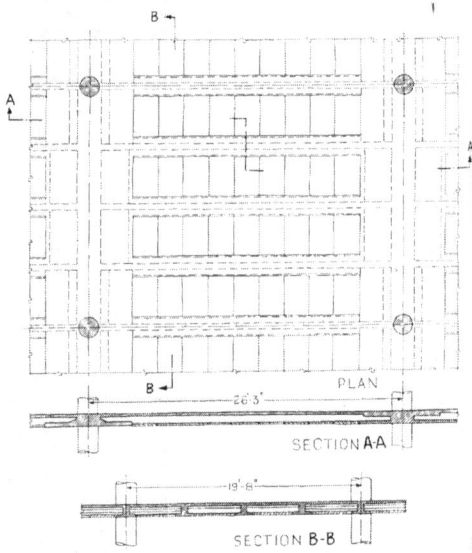

PLAN

SECTION A-A

SECTION B-B

FIGURE 11.10. MMM Roberto, Instituto de Resseguros do Brasil. North façade details of the IRB, from *Architectural Forum*, August 1944.

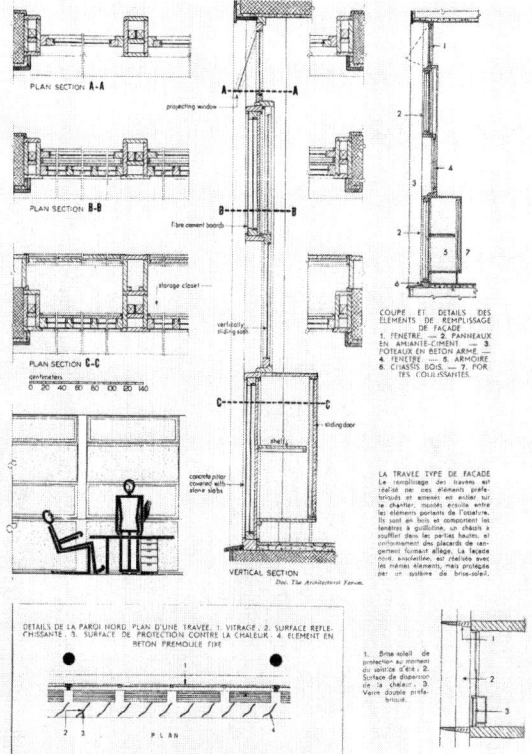

FIGURE 11.11. South and east façade of the IRB, sections of the façades collected in *L'Architecture d'Aujourd'Hui*, special issue on "Architecture au Bresil" (August–September 1947), and an explanatory diagram (*below*) drawn by Marcelo Roberto. Courtesy of the Núcleo de Pesquisa e Documentação, Faculdade de Arquitetura e Urbanismo, Univesidade Federal do Rio de Janeiro.

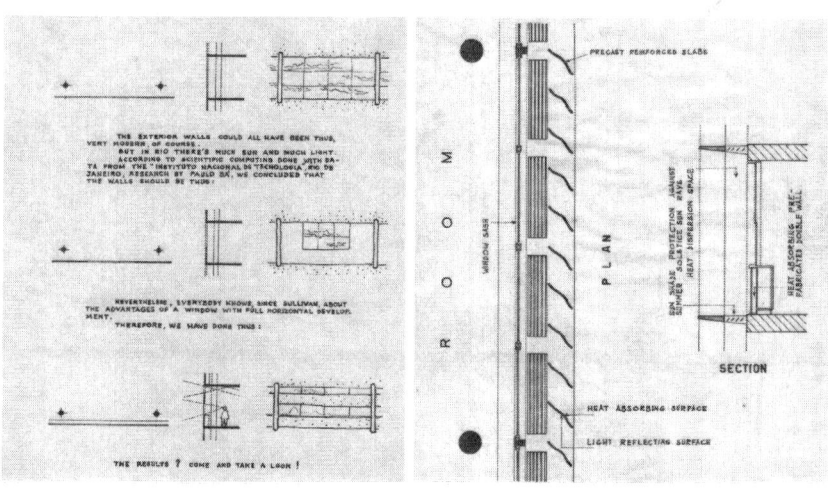

window at more or less the height of the inhabitant, able to gaze out comfortably while also allowing light in. "The results?" Roberto concludes, "Come and take a look!"[49] That these drawings were annotated in English is significant—they were published in *Architectural Forum* and translated into French for *Architecture d'Aujourd'hui*, in small articles discussing the buildings in the larger context of Brazilian modernism.[50] Detailed sections also show how top windows operate and, incidentally, the use of filing cabinets as an additional thermal buffer. The façade opened up the interior to extensive daylight, for both open office space and more intimate meeting spaces, minimizing costs while allowing for programmatic flexibility.

Like much of downtown Rio, the IRB sits on a flattened hill. It was something of an outpost, along with the airport, an extension of the central federal district toward the bay. Built first, it sat in a field of construction sites and earth removal projects as central Rio shaped itself toward global modernity—shaped around, in some ways, the climate-mediated interiors of the reinsurance agency. Other MMM Roberto projects across the 1950s included a headquarters and factory for the Caterpillar company, industrial, insurance, and finance headquarters, and a series of technical institutes around Rio de Janeiro to train workers for these factories and office buildings.[51] They were also well-known designers of shaded middle-class apartment complexes around Copacabana, Ipanema, and Botafogo. More generally, Brazil is an acknowledged center for the development of architectural modernism in this period, with too many examples to mention; many such examples were built with their climate engagement and use of shading devices in mind.[52]

RISK, HABIT, CLIMATE

These carefully shaded buildings for the insurance industry, in their planimetric arrangements and technological details, are resonant with what Francois Ewald has termed an *insurational imaginary*. For this imaginary, insurance is "a series of de-centerings, disruptions, destabilizations in the life-world, a sort of pulling through of the people into a calculational matrix of population, of the way one considered people, things, and their relationships." The rise of insurance marks the moment of an approach to life—to housing, to public health, to securing future possibilities—focused on optimization.[53] Shading devices and insurance both operated as means to process data toward the care of the population, mediating between social patterns and their potential for aggregate, and unexpected, consequences.

The designed façade mechanism similarly marks the moment of an approach to life that is rendered manageable through risk assessment.[54] Risk in these buildings is the quantitative management of possible futures—the capacity for knowledge,

FIGURE 11.12. MMM Roberto, Marques do Herval, Rio de Janeiro, 1953. Photographs of the façade. Courtesy of the Núcleo de Pesquisa e Documentação, Faculdade de Arquitetura e Urbanismo, Univesidade Federal do Rio de Janeiro.

and near-term simulation, to shape those futures. Attention to the variability of the shading device, the premise of interactivity, indicates a different sort of trajectory of the physiological, the bioclimatic, and the biopolitical than that afforded by HVAC. The representation of a bodily engagement with climatic conditions occurred both through the comfortable experience of the interior and the physical manipulation of the parts of the complex system. Two of the Roberto brothers, at the Marques do Herval of 1952, demonstrate this physical interaction between the interior and the exterior (figure 11.12). The body becomes the object of climatic analysis, increasingly prevalent as an architectural means of defining comfort.[55]

MMM Roberto also designed a weekend retreat for IRB workers (figure 11.13).[56] Nestled in the hills west of Rio, the retreat represents an intensification in the sense of integrated, systemic attention to the life conditions of the individual—as worker, as asset, as citizen, as growth opportunity, as site of optimization. This bodily aspect was functional and experiential. The experience of self-care is further intensified through more commissions for insurance, and even a house for the extended Roberto family. The weekend colony is a delicate, extended

VUE DU BATIMENT DEPUIS LES JARDINS

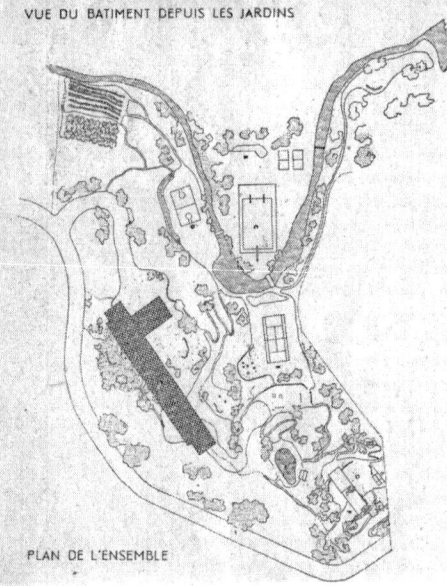

PLAN DE L'ENSEMBLE

Edifiée pour un personnel de salaires modestes, cette colonie est située dans la forêt, mais à quelques minutes de la ville. Elle est destinée : partie aux vacances proprement dites, partie au repos de fin de semaine, et peut recevoir au total 83 personnes. Elle comprend des dortoirs pour jeunes gens des deux sexes et des chambres pour les couples avec et sans enfants. Les salles et le restaurant sont de dimensions suffisantes pour recevoir les visiteurs dominicaux.

Le bâtiment principal est construit sur un terrain en pente. Les pilotis ont été utilisés pour le rez-de-chaussée ouvert sur le parc et qui comprend le garage, la gérance, le coiffeur, un hall de service avec vestiaire, les meubles des employés et la buanderie. Au niveau de la route, et directement accessible, le premier étage comporte le restaurant et ses services, les salles de vie commune, ainsi qu'une grande terrasse donnant sur le parc. Les salles de vie commune sont entièrement ouvertes sur le parc. Leur façade est composée de trois éléments : fenêtres coulissantes, volets coulissants à lames brise-soleil horizontales, et, à la partie supérieure, des châssis fixes à lames horizontales assurant la ventilation des locaux. Une subdivision individuelle a été obtenue dans les dortoirs des jeunes filles par l'utilisation d'éléments composés d'une cloison légère, lit et armoire.

L'ossature du bâtiment est en béton armé. Le granit gris local a fourni les murs du rez-de-chaussée. La charpente et les rampes d'appui des balcons au deuxième étage sont en bois, la toiture en fibro-ciment ondulé. Les balcons et certaines parties des vérandas sont fermés aux extrémités par des treillis losangés en bois très légers, peints en beige rosé. A l'intérieur, les murs sont peints en crème, les menuiseries en bleu clair ou en crème. Les étages sont desservis par un escalier central en béton armé, en spirale.

FIGURE 11.13 (*left and right*). MMM Roberto, Colonia de Ferias do Instituto de Resseguros do Brasil (IRB), Boa Vista, Rio de Janeiro, 1943, from *L'Architecture d'Aujourd'Hui* (1950).

PLAN DU DEUXIEME ETAGE
1. PARTIE DU BATIMENT POUR LES VA-
CANCES (31 personnes); 2. PARTIE DU BATI-
MENT POUR LES WEEK-END (52 personnes);
A. DORTOIR DES JEUNES FILLES; B. DOR-
TOIR DES GARÇONS; C. JEUNES FILLES;
D. JEUNES GENS; E. INSTALLATIONS SANI-
TAIRES; F. CHAMBRES; G. VERANDAS;
A. HALL

PLAN DU PREMIER ETAGE
1. RESTAURANT; 2. CUISINE; 3. BAR;
4. VERANDA; 6. SEJOUR; 7. JEUX;
8. TERRASSE

PLAN DU REZ-DE-CHAUSSEE
1. ENTREE; 2. GARAGE; 3. GERANCE; 4.
COIFFEUR; 5. SALLE DES ENFANTS; 6 ET
9. VESTIAIRE; 7. HALL DE SERVICE; 8.
BUANDERIE; 10. EMPLOYES

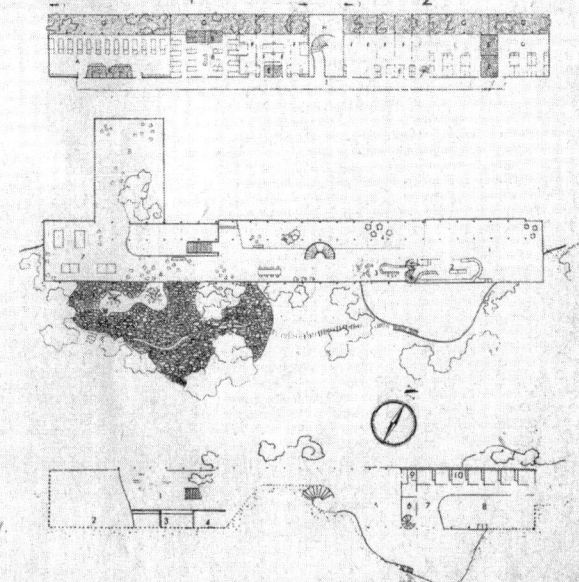

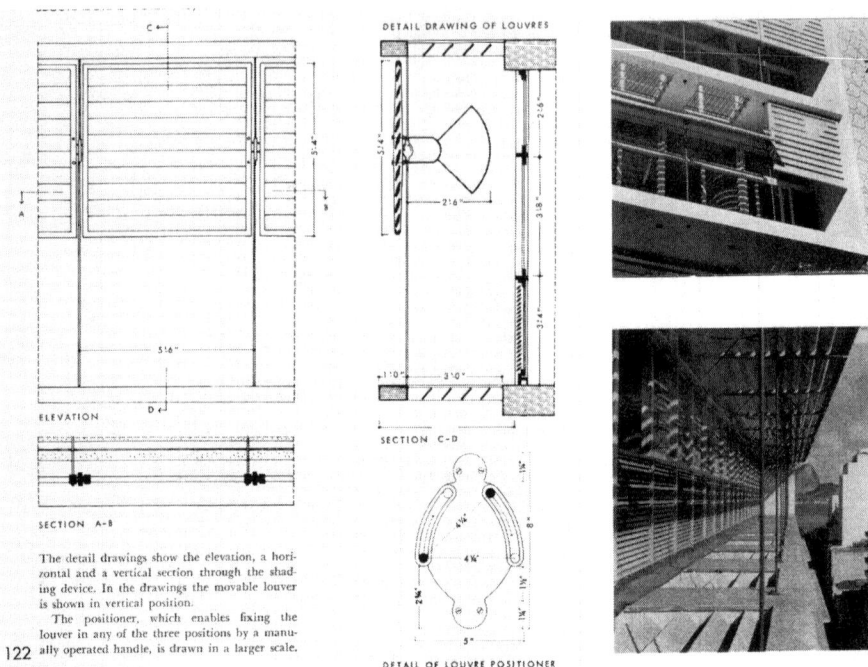

The detail drawings show the elevation, a horizontal and a vertical section through the shading device. In the drawings the movable louver is shown in vertical position.

The positioner, which enables fixing the louver in any of the three positions by a manually operated handle, is drawn in a larger scale.

FIGURE 11.14 (*above and top right*). Roberto, Edifício Seguradoras, Rio de Janeiro, 1949, spread from Olgyay and Olgyay, *Solar Control and Shading Devices*, 1957. Reproduced with permission.

Figure 11.15A (*bottom right*). Wurdeman and Beckett Prudential Insurance Building, Los Angeles, California, 1949, from Olgyay and Olgyay, *Solar Control and Shading Devices*, 1957. Reproduced with permission.

building, traversing a small ravine in a larger valley, sited to take advantage of prevailing breezes. Rio has since become such a beach city that the retreat to the hills for comfort seems anachronistic—though Copacabana, at the time of the IRB Building's construction, was barely developed.[57] This opportunity to escape the city into the countryside can be seen relative to a general transformation of the concept of leisure as another precondition for modernity.[58]

The Robertos also designed the Edifício Seguradoras in 1949, a speculative office building for the property insurance industry, to encourage more foreign investment. It could be placed next to Neutra's Northwestern Mutual Insurance in Los Angeles (1950), Skidmore, Owings, and Merril's building for Pan American Insurance in New Orleans (1952), and Wurdeman and Beckett's Prudential Building in Los Angeles (1949), showing adaptations to this now

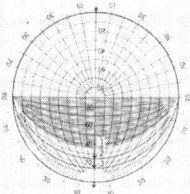

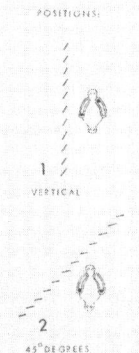

1

VERTICAL

2

45° DEGREES

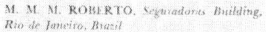

3

HORIZONTAL

M. M. M. ROBERTO, *Seguradoras Building, Rio de Janeiro, Brazil*

The elevation of the building which receives the most sunshine is evolved in practically two layers; one is the inner side with windows, which is the actual enclosure, and the other three feet before it, which is the visual one, serving as a filter against the sun's heat. The changeable position of the movable louvers results in a variable and playful character throughout the day. The horizontal shade is the continuation of the floor slab made of reinforced concrete. It is formed as a frame with inserted asbestos louvers, so that the heat accumulated outside the wall can escape upward. The manually operated movable louver can be fixed in three positions securing different shading characteristics, as shown on the mask.

123

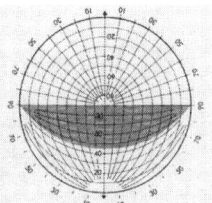

WURDEMAN & BECKET, *Prudential Building, Los Angeles, California*

This elongated office building in Los Angeles faces south, parallel to Wilshire Boulevard. The photograph shows the sun protection afforded the clerical offices, at right angles to the blank service tower wall. The sun-shades are of aluminum and consist of fixed slanted louvers which contrast to the pink terra-cotta spandrels. Despite the sunshades, glare has proved a problem at certain times, so draw curtains were installed on the south windows.

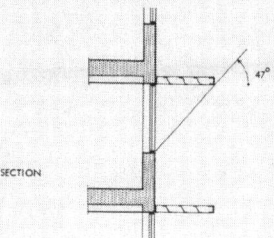

SECTION

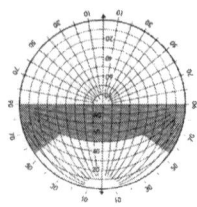

SKIDMORE, OWINGS & MERRILL AND C. E. HOONTON, *Pan American Life Insurance Building, New Orleans, Louisiana*
The main rectangular unit of the building is covered on all four sides with fixed vertical aluminum louvers, which in combination with the extended horizontal concrete slabs result in an eggcrate mask. The use of the same type of device around the building gives an architectural unity, but can not be justified functionally. The same effect could have been achieved with different dimensioning toward the various orientations.

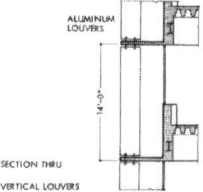

FIGURE 11.15B. Skidmore, Owings, and Merrill with C.E. Hoonton, Pan American Life Insurance Building New Orleans, Los Angeles, 1952, spread from Olgyay and Olgyay, *Solar Control and Shading Devices*, 1957. Reproduced with permission.

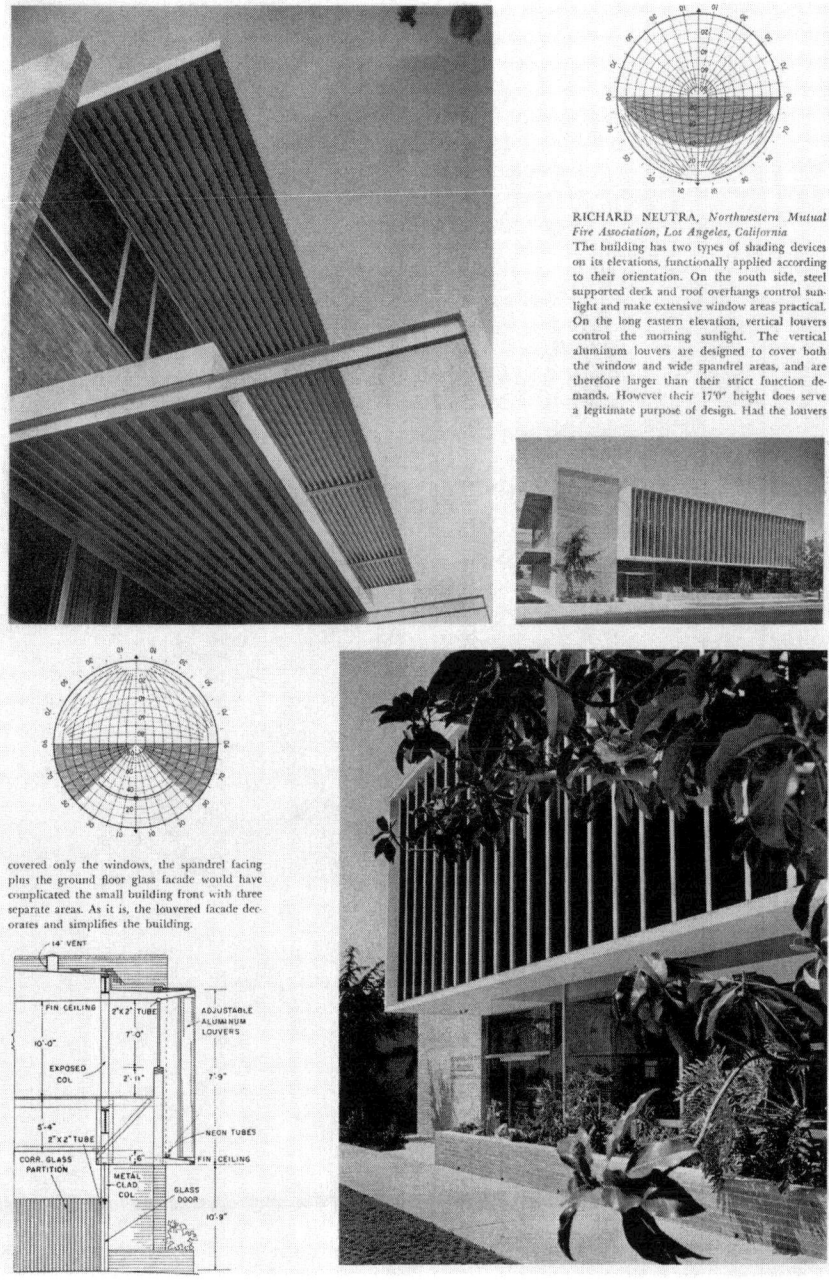

RICHARD NEUTRA, *Northwestern Mutual Fire Association, Los Angeles, California*
The building has two types of shading devices on its elevations, functionally applied according to their orientation. On the south side, steel supported deck and roof overhangs control sunlight and make extensive window areas practical. On the long eastern elevation, vertical louvers control the morning sunlight. The vertical aluminum louvers are designed to cover both the window and wide spandrel areas, and are therefore larger than their strict function demands. However their 170° height does serve a legitimate purpose of design. Had the louvers covered only the windows, the spandrel facing plus the ground floor glass facade would have complicated the small building front with three separate areas. As it is, the louvered facade decorates and simplifies the building.

FIGURE 11.15C. Richard Neutra, Northwestern Mutual Fire Insurance Building, Los Angeles, 1950, spread from Olgyay and Olgyay, *Solar Control and Shading Devices*, 1957. Reproduced with permission.

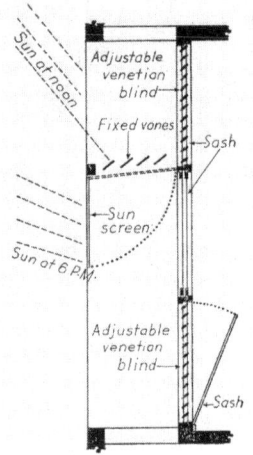

FIGURE 11.16 (*above, left, and right*). MMM
Roberto, Edifício Mamãe, Rio de Janeiro,
1945. Courtesy of the Núcleo de Pesquisa e
Documentação, Faculdade de Arquitetura e
Urbanismo, Universidade Federal do Rio de
Janeiro.

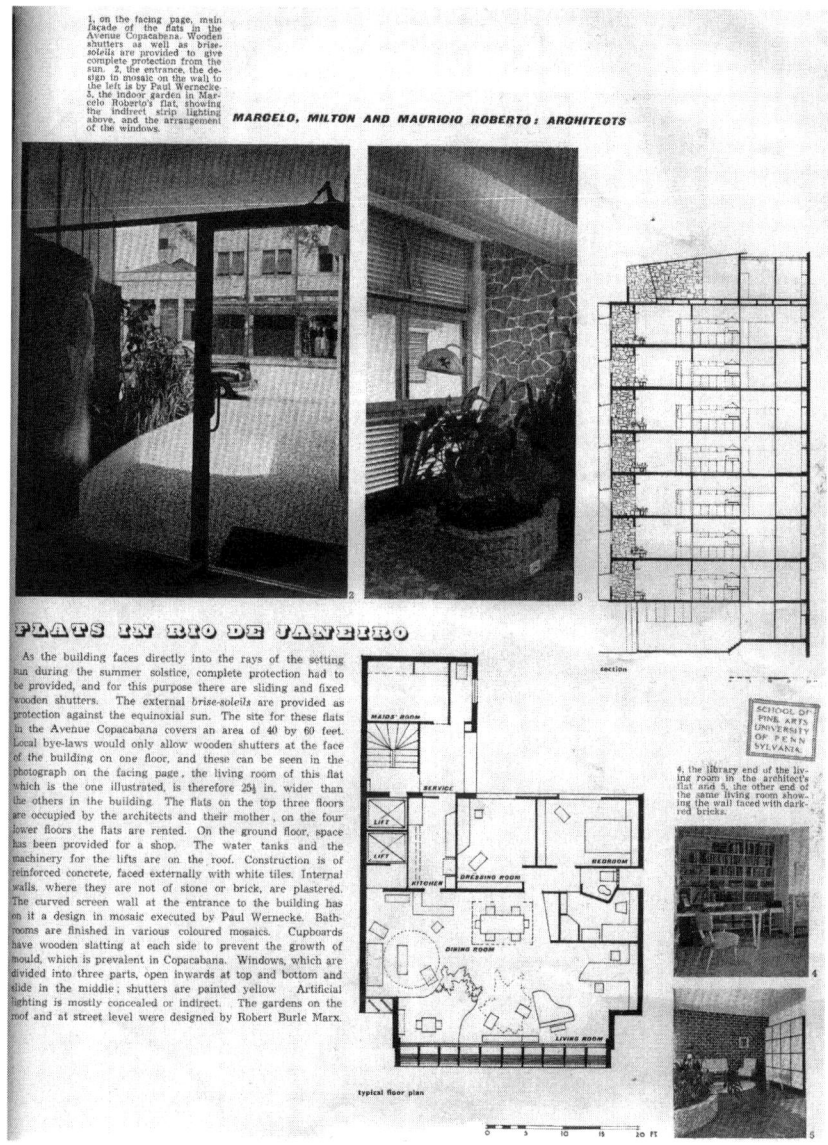

1, on the facing page, main façade of the flats in the Avenue Copacabana. Wooden shutters as well as brise-soleils are provided to give complete protection from the sun. 2, the entrance, the design in mosaic on the wall to the left is by Paul Wernecke. 3, the indoor garden in Marcelo Roberto's flat, showing the indirect strip lighting above, and the arrangement of the windows.

MARCELO, MILTON AND MAURICIO ROBERTO: ARCHITECTS

FLATS IN RIO DE JANEIRO

As the building faces directly into the rays of the setting sun during the summer solstice, complete protection had to be provided, and for this purpose there are sliding and fixed wooden shutters. The external brise-soleils are provided as protection against the equinoxial sun. The site for these flats in the Avenue Copacabana covers an area of 40 by 60 feet. Local bye-laws would only allow wooden shutters at the face of the building on one floor, and these can be seen in the photograph on the facing page, the living room of this flat which is the one illustrated, is therefore 25½ in. wider than the others in the building. The flats on the top three floors are occupied by the architects and their mother, on the four lower floors the flats are rented. On the ground floor, space has been provided for a shop. The water tanks and the machinery for the lifts are on the roof. Construction is of reinforced concrete, faced externally with white tiles. Internal walls, where they are not of stone or brick, are plastered. The curved screen wall at the entrance to the building has on it a design in mosaic executed by Paul Wernecke. Bathrooms are finished in various coloured mosaics. Cupboards have wooden slatting at each side to prevent the growth of mould, which is prevalent in Copacabana. Windows, which are divided into three parts, open inwards at top and bottom and slide in the middle; shutters are painted yellow. Artificial lighting is mostly concealed or indirect. The gardens on the roof and at street level were designed by Robert Burle Marx.

4, the library end of the living room in the architect's flat and 5, the other end of the same living room show-ing the wall faced with dark-red bricks.

SCHOOL OF FINE ARTS UNIVERSITY OF PENNSYLVANIA

section

typical floor plan

quite refined shading style (figures 11.14 and 11.15).[59] A global proliferation of buildings saw modern architecture as especially effective for its sensitivity to climatic conditions.

A final compelling trajectory is represented by MMM Roberto's Edificio Mamãe, built in 1945 (figure 11.16). It is an apartment building that housed their extended family—each of the brothers had at least one floor, and their mother, by then widowed, occupied the seventh story.[60] The façade of the Edificio Mamãe demanded elaborate attention from the inhabitant, with a number of adjustable and fixed elements, as seen in the section. As at the Marques do Herval, the emphasis is on the shading systems of the façade as a designed membrane that simultaneously represents and activates cultural desires for a specific type of habitation. The many façade manipulations allowed for varied engagement across diurnal and seasonal patterns. The inhabitant, the family, attended to its needs through these interactions, and through manipulating the effects of the façade on the interior. More likely, in this case, domestic help was involved in the manipulation of the façade system and thus the regulator of thermostasis. This was an architecture of habits and practices, with complex consequences across concerns of climate and agency.

CONCLUSION

The climatically dynamic buildings of mid-century Rio are events in the history of a future yet to come, liminal moments of engaging bodies directly in regimes of modernization. These buildings seem less like a past, since overcome by the forces of progress and economic growth, and more like a future, a diagram for a new approach to cultural and climatic contingencies, a new kind of physiomaterial substrate for processing distinctions between interiors and exteriors—as the façade, increasingly sealed and impenetrable, has become filled with mechanical HVAC systems and has come to represent a different sort of relation in the crisis-ridden present.[61]

The insurational imaginary in Brazil is compelling historiographically both for the specifics of the case and for the schematic causal diagram it presents—a broader narrative that offers, instead of the teleological march of reason, the circular, repetitive patterns of climate. And an unexpected corollary in the importance of habit, as both subject and object of design, as a sort of rejoinder to the everywhere-ness of petroleum and the optimization of the interior. The relevance of the past to the present is not as a prevision but as a premediation, allowing for a consideration of cultural techniques that encourage habits of self-care on a societal scale, an agglomerative process of casual behaviors that counters risk. Aspects of the future are embedded in this past, not nostalgically but open to engagement in the present, as a decarbonized method for climate management.[62] "The building" becomes an object of study as a space for habitual engagements with climate, at the intersection of individual actions and corporate or governmental practices. A counterinsurational imaginary embedded in design methods and their potential effects opens up the possibility of climate as an experiential field for political engagement.

FIGURE 11.17. The IRB building had window air conditioners placed in various places in the façade, reportedly in the 1990s; since then, as the three images on the right show, the window units have been removed and a low-velocity system has been installed on the interior. Photograph by the author.

Two final images help to clarify these points: the first is a recent image of the façade of the IRB, now filled with in-window air-conditioning units (figure 11.17)—one can find similar images of almost all of the Brazilian buildings discussed herein, built with shading systems at mid-century, since reconfigured for ad hoc, fuel-dependent, inefficient, mechanical conditioning. A second image, taken about the same time, is of the renovation to the housing block of Pedregulho (figure 11.18). Designed in 1947 by Affonso Reidy, it was celebrated, even by Le Corbusier, as a model of public housing. Designed for city workers, the building was officially called the Conjunto Residencial Prefeito Mendes de Moraes, after one of Vargas's allies. The curving hillside arrangement was carefully organized for minimizing summer solar incidence. Pedregulho fell into decay in the 1970s, largely because of an absence of public funding for maintenance. The on-site health center closed and the pool sits empty, though the elementary school remains. As it was needing renovation, the residents and the city reached an agreement that allowed for participation of residents in the renovation process.[63] Part of this involved a careful refurbishment, rather than replacement, of the shading system—in concert, it must be noted, with upgraded in-window air-conditioning units. The sanded, stained, almost shining

FIGURE 11.18 (*above and right*). Affonso Eduardo Reidy, Conjunto Residencial Prefeito Mendes de Moraes (Pedregulho), Rio de Janeiro, 1947–1952. Undergoing renovation, 2015, photographs by the author.

wood brise-soleil both express a capacity to help rethink architectural historical patterns and also can be considered, on technical-efficiency terms, indications of a renewed interest in shading devices among architects and their clients, and the renewed viability of spatial and climatic accommodation as a comfort management strategy.

The shaded modernism of mid-century Brazil helps to place in relief a new history of architecture, one that takes into account its profound relationship to climate. The climate design strategies of MMM Roberto can be seen in the context of developments in modern architecture, in the climate sciences, and in the complex conditions of governance that accompany Brazil's entry into global modernity in this period. Climate design, climate science, and economic modernization come together in these projects to form a powerful imaginary, which in turn coevolved with forms of practice and lived experience to create enduring patterns in the built environment. Architecture becomes an interesting discursive space to explore these interconnected histories across the twentieth century and up to the present. These examples begin to suggest new analytic methods and historiographic contours for a history of architecture that attends to the increasing reliance on fossil fuels, a reliance that presents a challenge to ways of life in the present.

AFTERWORD

Historiographies and Geographies of Climate

Mike Hulme

IN THIS BRIEF AFTERWORD I consider the motivations behind this volume and reflect on the wider public value of the types of historically and geographically sensitive climate scholarship contained in the preceding pages. In particular, I make three observations. First, given how pervasive the idea of climate change has become in the contemporary world, it is important to challenge simplistic and historically emaciated accounts of what climate is understood to be. Second, richer accounts as offered here of the multifaceted idea of climate are emancipatory for contemporary politics. They challenge the dangerous hegemony of a naturalistic climate science—"every society needs a cohort of intellectuals to check the dominance of a single perspective when its ideological hand becomes too heavy"—and open up new ways of framing climate–society relations.[1] Third, these case studies illustrate the multiple ways in which the idea of climate is performative; *how* one comes to know climate, and the account one gives of its changes, is never politically neutral nor without effect on the social ordering of the world.

KNOWING CLIMATE HISTORICALLY

Successive generations of university students, especially those studying geography, have learned much of what they know about the science of the atmosphere, weather, and climate from Barry and Chorley's eponymous textbook, originally published half a century ago.[2] It has been phenomenally successful, going through nine fully revised editions, the most recent in 2009. Although the words "atmosphere," "weather," and "climate" also feature in the title of this current volume, the editors and contributors offer a very different account of the knowledge that Barry and Chorley describe. The clue is in the four other words of the title adopted by the editors: "geographical imagination" and "placing knowledges." Rather than a systemized account of the science of meteorology à la Barry and Chorley, this volume seeks to understand when, where, and how atmospheric knowledge came to be made and what effects these forms of scientific knowledge of weather and climate have had on social ordering and political life. The case studies contained here are disciplined by being historical and geographical rather than by being merely scientific.

Another remarkable book of recent decades with similarly wide readership has been Sir John Houghton's *Global Warming: The Complete Briefing*.[3] This too has gone through multiple editions, the fifth and latest edition published twenty-one years after the original work appeared in 1994. Houghton offers a popular account of anthropogenic climate change, translating the reports of the Intergovernmental Panel on Climate Change (IPCC) for a wider audience.

I mention these two popular academic books about weather and climate because they both offer relatively thin historical treatment of the understanding of meteorological and climatic knowledge and of the origins of the idea of climate change. Barry and Chorley's book emerged shortly after the International Geophysical Year (IGY) of 1957 and the birth of the satellite era of the 1960s. Houghton's book, as one might expect of one relying heavily upon the IPCC, reinforces indirectly the notion that the phenomenon of climate change was only discovered in the 1980s. In contrast, geographical knowledge of climates and their interaction with diverse social worlds is much older and draws upon wider epistemic traditions.[4]

This collection of studies therefore brings a welcome and necessary thickening of the idea of climate and its changes. It offers an antidote to the deficient notion that global atmospheric science started only after the IGY in 1957 or the idea that climate change was discovered in 1988. The thrust of the eleven studies reported here is to challenge any Whiggish histories of climate science, climate scientists, and their cumulative impact on society and politics.[5] How we come to believe what we do today about climate is *not* an inevitable outcome of ever-accumulating scientific knowledge about the atmosphere. History, context, and contingency matter profoundly for the scientific enterprise. Jim Fleming's book *Fixing the Sky: The Checkered History of Weather and Climate Control* illustrates this wonderfully in relation to the idea of geoengineering and dispels the careless idea that climate engineering started in 2006 with Paul Crutzen's essay on albedo enhancement through stratospheric sulfur injection.[6] This volume does something similar for the idea of climate. It undermines the dominant account of the last thirty years that climate only "really" became known with the arrival of Earth System science and the IPCC in the 1980s. The truth is that humans have always sought to make sense of their weather in imaginative ways. Our ancestors embarked on many elaborate enterprises for constructing knowledge about the atmosphere's behavior, inventing the idea of climate along the way. "The geographical imagination" is a useful lens through which to view this story.

THE VALUE OF HISTORIES AND GEOGRAPHIES OF CLIMATE

Studying diverse climatic knowledges using the approach realized in this volume is important. Much can be learned from the cases reported here, which are of benefit for the contemporary epistemic practices, discourses, and politics of climate change. It is not just a matter of intellectual curiosity to study, for example, how currents in the eastern Indian Ocean came to be named in the early twentieth century (Morgan, this volume) or how nineteenth-century life assurance companies constructed cartographies of climatic risk (Kneale and Randalls, this volume). One doesn't have to be a fully signed up Foucauldian to recognize that there is value in showing how ideas—about climate as much as about madness—are socially constructed. In other words (and here to borrow directly from Foucault), to recognize that an archaeological and genealogical approach to the study of climate can illuminate some of the wider philosophical, cultural, social, and political contingencies that shape understanding of climate, and its changes, in the present day. The discursive formation of climate change that is dominant today is not the inevitable outcome of rational practices. It emerged as it did in the (largely) Anglophone world of the 1970s and 1980s for contingent reasons. It did not have to be thus.[7]

Histories and geographies of climate are inevitably subversive, as much as were Foucault's archaeologies and genealogies of madness and clinics. Or at least they are subversive of claims about the matter-of-factness of what is known today about climate change or about the self-evidence of the "climate crisis." What this "crisis" is really about depends upon what you think climate to be, what forces act upon it, and how climate itself becomes a force. Let me illustrate the subversive nature of this approach with three examples, each of which have been illuminated by studies contained in this volume.

First is the question of scale. At what scale *are* climates understood? At what scales *should* climate be understood? Even if the contemporary impulse is to think of climate as global, it is nevertheless still commonplace to talk about microclimates, urban climates, regional climates, or even national climates.[8] Several of the studies in earlier chapters examine how empires, observatories, metrology, and communication technologies all played their part in enlarging the geographical reach of climatic knowledge in past eras. Yet such collapsing of epistemic distances did not necessarily result in understanding "climate-as-global" as conceived today. Climatic knowledge interacted with distinct geographical imaginations in different ways, such that Australian, Indian, British, and American climates maintained epistemic distinctiveness and different cultural resonances. This reminds us that there are other possibilities of thinking about human interdependencies with climate on scales other than the global. Urban, oceanic, or polar climates might offer greater cultural, political, and technological traction for effective interventions than maintaining the fiction that it is only global climate that requires governing.

A second example concerns the origins of taken-for-granted climatic phenomena. Like any science, climate science is in the business of simplifying complex realities through reduction and classification. Climatic concepts—which end up seeming timeless and placeless, essential rather than constructed—have origins; they emerge in specific times and places. Studies in this book examine the origins of the Southern Oscillation (Adamson, this volume) and the Leeuwin Current (Morgan, this volume), and other work has shown how phenomena like El Niño and La Niña become reified, to the extent of being endowed with agency.[9] There are many such climatic phenomena—for example, the North Atlantic Oscillation, the Arctic Oscillation, the Madden-Julian Oscillation, the Multi-Decadal Atlantic Oscillation, or the Pacific Decadal Oscillation (PDO). Understanding how these climatic "personalities" come into being is important, since, as I argued above, they "perform"; agency is granted to them. This is illustrated nicely in the case of the PDO, which is now regarded by many to be partly responsible for the global warming hiatus of 1998–2013.[10]

My third example concerns the aspiration for detachment. Can scientific accounts of climate and its changes ever be detached (or "purified") from various forms of subjectivity?[11] In relation to climate change this question is highly charged given the tussles over the claimed authority of the IPCC to give scientific accounts of climate change that are "policy-relevant, but policy neutral."[12] What several of the studies in this volume show historically (Livingstone, Kneale and Randalls, McKittrick, and Endfield) is how this separation of scientific knowledge from cultural, ethical, or political values is a chimera. Deepening an appreciation of why this is in the context of climate is important simply because there are so many contemporary instances of where climatic knowledge is "impure." For example, Orlove, Lazrus, Hovelsrud, and Giannini show how the attention given to the regional effects of a changing climate varies as a consequence of those regions' colonial histories and cultural imaginaries.[13] When it comes to regional climates, not all are equal. And then in the context of climate modeling, far from being an objective universal pursuit, Mahony and Hulme have shown how distinct national styles of institutional science direct and guide the design and application of global climate models.[14]

HOW CLIMATE KNOWLEDGE PERFORMS

It is well established that the idea of climate means different things to different people in different cultures in different eras. And that the knowledges of climate constructed by these people emerge with particular political and cultural attachments, which condition the authority of such knowledge in public life. I have offered a brief evidential history of these realities in my book *Weathered: Cultures of Climate.* The studies reported in this current volume offer a further set of lenses through which not just to understand how different climate knowledges are made but also to learn about their performativity.[15] Knowledge constructed about the world is constitutive of the social order by which the world is governed, orders that in turn constrain or enable the possibilities of further knowledge-making. Science and society are coproduced; they make and remake each other.[16] Climatic knowledge can therefore never be politically neutral; neither is it impotent. Climate knowledge performs; it has material and social effects. This is why the notion that the IPCC produces "policy-neutral" knowledge is a fiction and why it matters who has the right to claim that they have "seen" climate change.[17] The studies in this volume illustrate different realizations of how climate knowledge coproduces the social worlds in which that knowledge resides. I will briefly summarize a couple of examples.

The first example concerns how climatic knowledge is bound up in constructions of what constitutes dangerous changes in climate. Although contemporary fears of climate are usually associated with in situ changes in climate, examples in this volume (Endfield, Livingstone, and Kneale and Randalls) show how fears about ex situ changes in climate (moving location) were constructed through different geographical imaginations. This is a theme also picked up in a contemporary setting by Strengers and Maller, who studied how climates become "dangerous" when international migrants move from one climatic regime to another—in their case, Australia.[18] Different climatic knowledges, whether embodied or abstract and whether historical or future, open up different possibilities for climates to be deemed "safe" or "dangerous" and condition the types of responses to climatic risk made by people in particular social settings.

A second example of how climatic knowledge "performs" concerns the recurrent aspiration of humans to "control" or engineer their climate. This ambitious goal requires equally ambitious knowledge—about the factors that control climate, how these factors might be modified and with what degree of certainty. The long history of climate engineering is littered with partial or hubristic knowledge claims that have inevitably resulted in futile and abandoned efforts to engineer the climate.[19] The case study offered by Mahony in this volume demonstrates this nicely in the case of colonial East Africa; and climate knowledge "performs" rather badly. McKittrick's study of South Africa's white settler rainmaking ambitions shows how scientific claims to be able to make "artificial" rain served larger racial purposes. The chapter by Barber offers a different perspective on the relationship between climatic knowledge and its performativity in the context of modernist architecture. This offers a different, more pragmatic and humble, way of thinking about climate control in which scale-limited climate knowledge leads to more constructive ways of designing interior climatological spaces.

People experience and imagine their weather always in places, and these experiences and imaginations are infused with cultural meaning. People also seek to bring order to their weather, to construct knowledge of how and why the weather behaves the way it does, to make sense of what otherwise would be too disorderly an atmosphere. The idea of climate is central to such ordering knowledge, even though the idea of climate itself is multifaceted. How people approach the idea of climate is deeply shaped by their cultures, materialities, social relationships, and by their imaginations. Climatic knowledge too is place-based, but it rarely stays put. It travels unevenly around the world through material networks, embodied agents, and encultured discourses. As it does so it brings about unpredictable political and social effects. How climate knowledge comes into public circulation and gains or loses cultural authority has to be studied rather than assumed. In drawing upon historical and cultural geography, the history of science, and science and technology studies, the chapters in this current volume illuminate these realities in fruitful ways.

NOTES

Introduction: Weather, Climate, and the Geographical Imagination

1. Numerous volumes might be referred to here, but as a sampling, see, for example, Hulme, *Weathered*; Strauss and Orlove, *Weather, Climate, Culture*; Jessica Barnes and Michael R. Dove, eds., *Climate Cultures: Anthropological Perspectives on Climate Change* (New Haven, CT: Yale University Press, 2015).

2. Vladimir Janković, *Reading the Skies: A Cultural History of the English Weather, 1650–1820* (Chicago: Chicago University Press, 2000); Anderson, *Predicting the Weather*; Jan Golinski, *British Weather and the Climate of Enlightenment* (Chicago: University of Chicago Press, 2010); Coen, *Climate in Motion*.

3. Stephen Daniels and Georgina H. Endfield, "Narratives of Climate Change: Introduction," *Journal of Historical Geography* 35, no. 2 (April 2009): 215–22; Kathryn Yusoff and Jennifer Gabrys, "Climate Change and the Imagination," *WIREs: Climate Change* 2, no. 4 (2011): 516–34; David N. Livingstone, "Reflections on the Cultural Spaces of Climate," *Climatic Change* 113, no. 1 (2012): 91–93 .

4. Mike Hulme, "Reducing the Future to Climate: A Story of Climate Determinism and Reductionism," *Osiris* 26, no. 1 (2011): 245–66; Livingstone, "Changing Climate."

5. Daniels, "Geographical Imagination," 183.

6. David N. Livingstone, "Landscapes of Knowledge," in *Geographies of Science*, ed. Peter Meusburger, David N. Livingstone, and H. Jons (Heidelberg: Springer-Verlag, 2010), 18.

7. David Bloor, *Knowledge and Social Imagery*, 1st ed. (Chicago: University of Chicago Press, 1976).

8. Steven Shapin and Simon Schaffer, *Leviathan and the Air Pump: Hobbes, Boyle, and the Experimental Life* (Princeton, NJ: Princeton University Press, 1985); Steven Shapin, "The House of Experiment in Seventeenth-Century England," *Isis* 79, no. 3 (January 1988): 373.

9. Richard C. Powell, "Geographies of Science: Histories, Localities, Practices, Futures," *Progress in Human Geography* 31, no. 3 (2007): 309–29.

10. Pickering, *Mangle of Practice*; Bruno Latour and Steve Woolgar, *Laboratory Life: The Construction of Scientific Facts* (Princeton, NJ: Princeton University Press, 1979); Latour, *Science in Action*.

11. Diarmuid A. Finnegan, "The Spatial Turn," 369. See also Simon Schaffer, "The Eighteenth Brumaire of Bruno Latour," *Studies in History and Philosophy of Science* 22, no. 1 (1991): 174–92.

12. Bruno Latour, "Drawing Things Together," in *Representation in Scientific Practice*, ed. M. Lynch and Steve Woolgar (Cambridge, MA: MIT Press, 1990), 19–68; John Law, "On Methods of Long-Distance Control: Vessels, Navigation and the Portuguese Route to India," in *Power, Action & Belief: A New Sociology of Knowledge*, ed. John Law (London: Routledge and Kegan Paul, 1986), 234–63.

13. Steven Shapin, "Placing the View from Nowhere: Historical and Sociological Problems in the Location of Science," *Transactions of the Institute of British Geographers* 23, no. 1 (April 1998): 7; Sheila Jasanoff, "Ordering Knowledge, Ordering Society," in Jasanoff, *States of Knowledge*, 23.

14. Seth, "Putting Knowledge in Its Place," 380.

15. George Basalla, "The Spread of Western Science," *Science* 156, no. 3775 (1967): 611–22. For critiques, see Warwick Anderson and V. Adams, "Pramoedya's Chickens: Postcolonial Studies of Technoscience," in *The Handbook of Science and Technology Studies*, ed. E. J. Hackett, O. Amsterdamska, M. Lynch, and J. Wajcman, 2nd ed. (Cambridge, MA: MIT Press, 2008), 181–204; Seth, "Putting Knowledge in Its Place"; Sandra G. Harding, *Is Science Multicultural? Postcolonialisms, Feminisms, and Epistemologies* (Bloomington: Indiana University Press, 1998); Anne Pollock and Banu Subramaniam, "Resisting Power, Retooling Justice: Promises of Feminist Postcolonial Technosciences," *Science, Technology & Human Values* 41, no. 6 (November 1, 2016): 1–16.

16. Emily O'Gorman, "'Soothsaying' or 'Science'? H. C. Russell, Meteorology, and Environmental Knowledge of Rivers in Colonial Australia," in Beattie, O'Gorman, and Henry, *Climate, Science, and Colonization*, 181.

17. James Beattie, *Empire and Environmental Anxiety: Health, Science, Art and Conservation in South Asia and Australasia, 1800–1920* (Basingstoke: Palgrave, 2011); Claire Fenby, Don Garden, and Joelle Gergis, "'The Usual Weather in New South Wales Is Uncommonly Bright and Clear . . . Equal to the Finest Summer Day in England': Flood and Drought in New South Wales, 1788–1815," in Beattie, O'Gorman, and Henry, *Climate, Science, and Colonization*, 43–59; Chris O'Brien, "Imported Understandings: Calendars, Weather, and Climate in Tropical Australia, 1870s–1940s," in Beattie, O'Gorman, and Henry, *Climate, Science, and Colonization*, 195–212.

18. Peter Holland and Jim Williams, "Pioneer Settlers Recognizing and Responding to the Climatic Challenges of Southern New Zealand," in Beattie, O'Gorman, and Henry, *Climate, Science, and Colonization*, 81–96.

19. Grove, "East India Company."

20. Gregory Allen Barton, *Empire Forestry and the Origins of Environmentalism* (Cambridge: Cambridge University Press, 2002); Sandip Hazareesingh, "Cotton, Climate and Colonialism in Dharwar, Western India, 1840–1880," *Journal of Historical Geography* 38, no. 1 (2012): 1–17.

21. Anderson, *Predicting the Weather*, ch. 6.

22. Mike Davis, *Late Victorian Holocausts: El Niño Famines and the Making of the Third World* (London: Verso, 2001).

23. Coen, *Climate in Motion*.

24. See Grove, "The East Indian Company"; Adamson, this volume; Mahony, "Genie of the Storm."

25. Cushman, "Imperial Politics of Hurricane Prediction."

26. Anderson, "From Subjugated Knowledge."

27. Anna L. Tsing, *Friction: An Ethnography of Global Connection* (Princeton, NJ: Princeton University Press, 2011).

28. Kristine Harper, *Weather by the Numbers* (Cambridge, MA: MIT Press, 2008); Edwards, *Vast Machine*.

29. Anderson, "From Subjugated Knowledge."

30. Miller, "Resisting Empire"; Myanna Lahsen, "Transnational Locals: Brazilian Experiences of the Climate Regime," in Jasanoff and Martello, *Earthly Politics*, 151–72; Mark Carey, M. Jackson, Alessandro Antonello, and Jaclyn Rushing, "Glaciers, Gender, and Science: A Feminist Glaciology Framework for Global Environmental Change Research," *Progress in Human Geography* 40, no. 6 (2016): 770–93; Vetter, *Knowing Global Environments*.

31. Jonathan D. Oldfield, "Imagining Climates Past, Present and Future: Soviet Contributions to the Science of Anthropogenic Climate Change, 1953–1991," *Journal of Historical Geography* 60 (2018): 41–51.

32. Coen, *Climate in Motion*.

33. Anderson's chapter (this volume) bears comparison with Gayatri Spivak's notion of "planetary spaces." See Gayatri C. Spivak, *Death of a Discipline* (New York: Columbia University Press, 2003); Carroll, *Empire of Air and Water*.

34. See, for example, Aubin, Bigg, and Sibum, *Heavens on Earth*.

35. Simon Schaffer, "Astronomers Mark Time: Discipline and the Personal Equation," *Science in Context* 2, no. 1 (1988): 115–45.

36. John Cawood, "The Magnetic Crusade: Science and Politics in Early Victorian Britain," *Isis* 70, no. 4 (1979): 492–518; Peter Collier, "Edward Sabine and the 'Magnetic Crusade,'" in *History of Cartography*, ed. Elri Liebenberg, Peter Collier, and Zsolt Győző Török, 309–23 (Heidelberg: Springer-Verlag, 2013).

37. Simon Schaffer, "Astronomy at the Imperial Meridian: The Colonial Production of Hybrid Spaces," Keynote lecture, International Conference of Historical Geographers, London, July 9, 2015.

38. Robert E. Kohler, *Landscapes and Labscapes: Exploring the Lab-Field Border in Biology* (Chicago: University of Chicago Press, 2002). See also Isla Forsyth, "The More-than-Human Geographies of Field Science," *Geography Compass* 7, no. 8 (2013): 527–39; Kohler and Vetter, "The Field."

39. Latour, *Science in Action*.

40. See for example, Bruce Hevly, "The Heroic Science of Glacier Motion," *Osiris* 11, no. 1996 (1996): 66–86.

41. See Bruno Latour, *Pandora's Hope: Essays on the Reality of Science Studies* (Cambridge, MA: Harvard University Press, 1999), esp. ch. 2.

42. Kohler and Vetter, "The Field."

43. See, for example, David Matless and Laura Cameron, "Experiment in Landscape: The Norfolk Excavations of Marietta Pallis," *Journal of Historical Geography* 32, no. 1 (2006): 96–126.

44. Kohler and Vetter, "The Field," 287.

45. Fleming, "Planetary Scale Fieldwork," 190–211.

46. For a recent exception, see Kelsey Matson, "'The Ozone of Patriotism': Meteorology, Electricity, and the Body in the Nineteenth-Century Yellowstone Region," *History of Meteorology* 8 (2017): 35–53.

47. Georgina Endfield and Samuel Randalls, "Climate and Empire," in *Eco-cultural Networks and the British Empire: New Views on Environmental History*, ed. James Beattie, Edward Melilo, and Emily O'Gorman (London: Bloomsbury, 2014), 25.

48. Michael A. Osborne, "Acclimatizing the World: A History of the Paradigmatic Colonial Science," *Osiris* 15 (2000): 135–51.

49. Livingstone, "Race, Space and Moral Climatology."

50. See Jiat-Hwee Chang and Anthony D. King, "Towards a Genealogy of Tropical Architecture: Historical Fragments of Power-Knowledge, Built Environment and Climate in the British Colonial Territories," *Singapore Journal of Tropical Geography* 32, no. 3 (2011): 283–300; Graham, *Climates*; Daniel Barber, *Climatic Effects: Architecture, Media, and the Globalization of the International Style* (Princeton, NJ: Princeton University Press, 2018).

51. Hulme, *Weathered*; Graham, *Climates*.

52. Daniels, "Geographical Imagination."

53. C. Wright Mills, *The Sociological Imagination* (Oxford: Oxford University Press, 1959).

54. Howie and Lewis, "Geographical Imaginaries."

55. Jasanoff and Kim, *Dreamscapes of Modernity*.

56. The question of whether to refer to "imaginations" or "imaginaries" remains unresolved, and in this book we draw no specific distinction.

57. Radcliffe, "Relating to the Land"; Charles Taylor, *Modern Social Imaginaries* (Durham, NC: Duke University Press, 2003).

58. Annemarie Mol, *The Body Multiple: Ontology in Medical Practice* (Durham, NC: Duke University Press, 2002).

59. Edward W. Said, *Orientalism* (New York: Vintage Books, 1979).

60. Denis Linehan, "Irish Empire: Assembling the Geographical Imagination of Irish Missionaries in Africa," *Cultural Geographies* 21, no. 3 (2014): 429–47; Endfield and Nash, "Good Site for Health"; Diana K. Davis, "Desert 'Wastes' of the Maghreb: Desertification Narratives in French Colonial Environmental History of North Africa," *Cultural Geographies* 11, no. 4 (2004): 359–87.

61. Dee Mack Williams, *Beyond Great Walls: Environment, Identity, and Development on the Chinese Grasslands of Inner Mongolia* (Stanford, CA: Stanford University Press, 2002).

62. Latour, *Science in Action*.

63. Simon Maraud and Sylvain Guyot, "Mobilization of Imaginaries to Build Nordic Indigenous Natures," *Polar Geography* 39, no. 3 (July 2, 2016): 196–216.

64. Bjørn Ingmunn Sletto, "A Swamp and Its Subjects: Conservation Politics, Surveillance and Resistance in Trinidad, the West Indies," *Geoforum* 36, no. 1 (2005): 77–93.

65. Mara J. Goldman, Meaghan Daly, and Eric J. Lovell, "Exploring Multiple Ontologies of Drought in Agro-pastoral Regions of Northern Tanzania: A Topological Approach," *Area* 48, no. 1 (2016): 27–33.

66. Robin Kearns, Gregory O'Brien, Ronan Foley, and Nell Regan, "Four Windows into Geography and Imagination(s)," *New Zealand Geographer* 71, no. 3 (2015): 159–76.

67. Peter Hulme, "Writing on the Land: Cuba's Literary Geography," *Transactions of the Institute of British Geographers* 37, no. 3 (July 2012): 346–58.

68. See also Felix Driver, "Imagining the Tropics: Views and Visions of the Tropical World," *Singapore Journal of Tropical Geography* 25, no. 1 (2004): 1–17.

69. Gordon M. Winder and Michael Schmitt, "Geographical Imaginaries in the *New York Times*' Reports of the Assassinations of Mahatma Gandhi (1948) and Indira Gandhi (1984)," *Journal of Historical Geography* 45 (2014): 106–15.

70. Mark Carey, "Inventing Caribbean Climates: How Science, Medicine, and Tourism Changed Tropical Weather from Deadly to Healthy," *Osiris* 26, no. 26 (2011): 129–41.

71. Edwards, *Vast Machine*. See also Grevsmühl, "Images, Imagination.

72. Hilary Geoghegan, "A New Pattern for Historical Geography: Working with Enthusiast Communities and Public History," *Journal of Historical Geography* 46 (2014): 105–7; H. Hawkins, "Geography and Art. An Expanding Field: Site, the Body and Practice," *Progress in Human Geography* 37, no. 1 (2012): 52–71.

73. Howie and Lewis, "Geographical Imaginaries," 136.

74. For reflection on this point, see Hulme, this volume.

75. Sheila Jasanoff, "Heaven and Earth: The Politics of Environmental Images," in Jasanoff and Martello, *Earthly Politics*, 234.

76. Grevsmühl, "Images, Imagination."

77. Hulme, *Why We Disagree about Climate Change*; Scott Hamilton, "Action, Technology, and the Homogenisation of Place: Why Climate Change Is Antithetical to Political Action," *Globalizations* 13, no. 1 (2016): 62–77; Heymann, "Climate Change Dilemma."

78. Sheila Jasanoff, "A New Climate for Society," *Theory, Culture & Society* 27, no. 2–3 (May 24, 2010): 233–53; Mike Hulme, "Problems with Making and Governing Global Kinds of Knowledge," *Global Environmental Change* 20, no. 4 (2010): 558–64.

79. Martin Mahony and Mike Hulme, "Model Migrations: Mobility and Boundary Crossings in Regional Climate Prediction," *Transactions of the Institute of British Geographers* 37, no. 2 (2012): 197–211; Clark A. Miller, "Globalizing Security: Science and the Transformation of Contemporary Political Imagination," in Jasanoff and Kim, *Dreamscapes of Modernity*, 277–99.

80. Cited in Coen, *Climate in Motion*, 16.

81. Radcliffe, "Relating to the Land."

82. Savage, "Tropicality Imagined and Experienced."

83. See, for example, Sloterdijk, In the *World Interior of Capital*.

84. Coen, *Climate in Motion*.

1. Atmospheric Empire

The research for this chapter was generously funded by a British Academy small research grant and an Arts and Humanities Research Council-funded PhD studentship. Thanks to the archivists who assisted us with our research and in particular to the Royal Society for granting permission to reproduce Wilmot's anemometer sketch. Thanks also to the editors and the other contributors to this volume for their comments on an earlier draft of this chapter.

1. Macdonald, "Making Kew Observatory," 412.

2. John Herschel, "Terrestrial Magnetism," in Herschel, *Essays from the Edinburgh and Quarterly Reviews*, 64; Aubin, Bigg, and Sibum, "Introduction," 7.

3. Schaffer, "Keeping the Books," 131.

4. Taub, "Introduction," 694.

5. Schaffer, "Keeping the Books," 139.

6. Raj, *Relocating Modern Science*, 16.

7. Aubin, Bigg, and Sibum, "Introduction," 19.

8. Herschel, "Address," 653.

9. Edward Sabine, "On What the Colonial Magnetic Observatories Have Accomplished," *Proceedings of the Royal Society of London* 8 (1856–1857): 411.

10. Anonymous, *Report of the President and Council of the Royal Society on the Instructions to be Prepared for the Scientific Expedition to the Antarctic Regions* (London: Richard and John E. Taylor, 1839), 1–2.

11. Herschel, "Address," 652.

12. Macdonald, "Making Kew Observatory," 413.

13. Herschel, "Address," 652. In America, the mathematician Elias Loomis promoted a "grand meteorological crusade" to match Britain's magnetic crusade. Fleming, *Meteorology in America*, 77.

14. Musselman, "Worlds Displaced," 77–78.

15. Edward Sabine, "Report of the Joint Committee of the Royal Society and the British Association for the Advancement of Science for procuring a continuance of the Magnetic and Meteorological Observatories" (1858), 6, RMS RP008(SABI), National Meteorological Library and Archive, Exeter, UK.

16. Edward Sabine, "On What the Colonial Magnetic Observatories Have Accomplished," *Proceedings of the Royal Society of London* 8 (1856–1857): 411.

17. Edward Sabine to Humphrey Lloyd, January 17, 1840, MS/119/I/85, Archives of Royal Society, London.

18. Humphrey Lloyd, *Account of the Magnetical Observatory of Dublin and of the Instruments and Methods of Observation Employed There* (Dublin: University Press, 1842).

19. John Herschel to Francis Beaufort, October 11, 1835, 6, HS/3/343, Archives of the Royal Society, London.

20. Herschel to Beaufort, October 11, 1835, 5, HS/3/343, Archives of the Royal Society, London.

21. James D. Forbes, "Supplementary Report on Meteorology," from the *Report of the British Association for the Advancement of Science for 1840* (London: Richard and John E. Taylor, 1841), 145, 147 (original emphases).

22. Humphrey Lloyd, *Account of the Magnetical Observatory of Dublin and of the Instruments and Methods of Observation Employed There* (Dublin: University Press, 1842), 6. To ensure simultaneous observations were taken, all observatories were required to use Göttingen time.

23. Musselman, "Worlds Displaced," 77.

24. Humphrey Lloyd, "Notes on the Meteorology of Ireland Deduced From the Observations Made in the Year 1851, Under the Direction of the Royal Irish Academy," *Transactions of the Royal Irish Academy* 22 (1854): 411.

25. Frederick Eardly-Wilmot to Lloyd, April 20, 1840, MS/119/II/74, Archives of the Royal Society, London.

26. Wilmot to Herschel, August 12, 1842, HS/7, Herschel Papers, Archives of the Royal Society, London.

27. Josefowicz, "Experience, Pedagogy," 466.

28. Herschel, *Instructions for Making*, 9.

29. Herschel, *Instructions for Making*, 2.

30. Gregory Good, "Between Data, Mathematical Analysis and Physical Theory: Research on Earth's Magnetism in the Nineteenth Century," *Centaurus* 50 (2008): 294.

31. Royal Society, *Report of the Committee of Physics, including Meteorology, on the Objects of Scientific Inquiry in those Sciences* (London: Richard and John E. Taylor, 1840), 58.

32. See, for instance, J. A. Curtis, "Proceedings of the Committee assembled . . . to inspect and report upon such magnetical instruments as may be laid before them by Lieut. Yule, Bengal Engineers," September 21, 1840, MM/11/149, Archives of the Royal Society, London.

33. Charles Riddell to Lloyd, October 8, 1839, MS/119/II/2, Archives of the Royal Society, London (emphasis added).

34. Wilmot to Lloyd, April 20, 1840, MS/119/II/74, Archives of the Royal Society, London.

35. Wilmot to Lloyd, March 31, 1840, MS/119/II/73, Archives of the Royal Society, London.

36. Wilmot to Lloyd, April 20, 1840, MS/119/II/74, Archives of the Royal Society, London (original emphases).

37. Wilmot to Lloyd, April 20, 1840, MS/119/II/74 Archives of the Royal Society, London.

38. John Henry Lefroy to Sabine, March 2, 1840, BJ/3/81/11, The National Archives, Kew, London, UK (TNA).

39. John H. Lefroy, *Autobiography of General Sir J.H. Lefroy published posthumously by his second wife "for private circulation only"* (London, Pardon and Sons, 1895), 35.

40. Lefroy to Sabine, November 4, 1839, BJ/3/81/6, TNA.

41. Henry Yule to James E. Melvill, September 22, 1840, MM/11/148, Archives of the Royal Society, London.

42. Driver, "Yule, Sir Henry."

43. A. F. Yule, "Memoir of Sir Henry Yule," in *The Book of Ser Marco Polo, the Venetian, concerning the kingdoms and marvels of the East*, trans. H. Cordier, 3rd ed. (1903), 1:xxxiv.

44. J. A. Curtis, "Proceedings of the Committee assembled . . . to inspect and report upon such magnetical instruments as may be laid before them by Lieut. Yule, Bengal Engineers," September 21, 1840, MM/11/149, Archives of the Royal Society, London.

45. Yule to Melvill, September 22, 1840, MM/11/148 Archives of the Royal Society, London (original emphasis).

46. Sabine to Herschel, November 21, 1840, HS/15/110, Archives of the Royal Society, London.

47. Sabine to William H. Sykes, November 20, 1840, MM/11/150, Archives of the Royal Society, London.

48. Herschel, *Instructions for Making*, 8.

49. Sabine to Sir Huw Dalrymple Ross, November 10, 1840, BJ/3/27/13 (Part I), TNA; Wilmot to Lloyd, March 31, 1840, MS/119/II/73, Archives of the Royal Society, London, respectively.

50. Musselman, "Worlds Displaced."

51. Wilmot to Lefroy, May 14, 1840, quoted in F. A. Eardley-Wilmot, *Memorials of Fredk. M. Eardley-Wilmot . . . by his Widow* (London: William Clowes and Sons, 1879), 19.

52. Wilmot to Herschel, August 12, 1842, HS/7, Archives of the Royal Society, London.

53. Humphrey Lloyd, "On the Anemometer of Osler," *Proceedings of the Royal Irish Academy* 3 (1844–1847): 270.

54. Wilmot to Lloyd, March 18, 1842, MS/119/II/87, Archives of the Royal Society, London.

55. Wilmot to Herschel, August 12, 1842, HS/7, Archives of the Royal Society, London.

56. Dubow, *A Commonwealth of Knowledge*.

57. Wilmot to Arthur J. Lawrence, Army, 24 12 1840; Sturgis, "Anglicisation," 9.

58. G. Buist, *Provisional Report on the Meteorological Observations Made at Colaba, Bombay, for the Year 1844* (Cupar: G. S. Tuller, 1845); C. Chambers to J. F. Gassiot, Chair of the Kew Committee of the BAAS, BJ1/2, TNA.

59. K. Prior, "Buist, George (1804–1860)," *Oxford Dictionary of National Biography* (Oxford: Oxford University Press, 2004), online edition, January 2013, http://www.oxforddnb.com/view/article/3892/.

60. Edward Sabine, "On Some Points in the Meteorology of Bombay," *Report of the Fifteenth Meeting of the British Association for the Advancement of Science* (London: John Murray, 1846), 75.

61. Company officers and administrators rarely stayed in one posting for very long before promotion, illness, or some other factor necessitated their move to another station. Arnold, *Tropics and the Travelling Gaze*, 15.

62. W. C. Barker, *Report of the Committee of Inquiry on the Colaba Observatory*, 1, Selections from the Records of the Bombay Government, 1865, BJ1/2, India Office Archive, British Library. Further references to this report will be given parenthetically in the text.

63. Sikka, "India Meteorological Department"; J. Sen, *Astronomy in India, 1784–1876* (London: Pickering and Chatto, 2014).

64. Anderson, *Predicting the Weather*; R. Axelby and S. P. Nair, *Science and the Changing Environment in India, 1780–1920* (London: British Library, 2010).

65. Good, "Between Two Empires," 36.

66. Herschel to the earl of Ross, December 28, 1850, quoted in Edward Sabine, "Correspondence Relative to the Continuation of the Toronto Observatory, Ordered by the Council of the Royal Society to Be Printed," 5, T23.A-B RP007 (SABI), Box 54, Archives of the National Meteorological Library and Archive, Exeter, UK.

67. Thomas, "Meteorological Services in Canada," 2.

68. Edward Sabine, "Report on the Meteorology of Toronto in Canada," *Report of the British Association for the Advancement of Science for 1844* (London: John Murray, 1845), 94.

69. Zeller, *Inventing Canada*.

70. George Kingston, "Report of the Director of the Magnetic Observatory for the year ending 30 June 1870," 3 (quote), 4, RG6 A-1 Vol. 10, Library and Archives Canada, Ottawa (LAC).

71. Lt. E. Ashe to the Provincial Secretary of the Legislative Council, November 19, 1856, vol. 82, RG93, LAC.

72. Ian R. Dalton and George D. Garland, "Some Notes on the Toronto Magnetical and Meteorological Observatory" (unpublished manuscript, 1979), University of Toronto Archives, 11.

73. Frederic Stupart, Director, to Major F. Gourdeau, Deputy Minister of Marine and Fisheries, May 17, 1898, vol. 50, RG42, LAC.

74. Hodgins, *Documentary History, Volume XI*, 204; Hodgins, *Documentary History, Volume XII*, 149.

75. Hodgins, *Documentary History, Volume X*, 145.

76. Zeller, *Inventing Canada*, 166.

77. Hodgins, *Documentary History, Volume XII*, 105.

78. Hodgins, *Documentary History, Volume XII*, 118.

79. Edward Sabine, *Observations Made at the Magnetical and Meteorological Observatory, at Toronto in Canada. Vol. III.—1846, 1847, 1848* (London: Longman, Brown, Green, and Longmans, 1857), xv.

80. Edwards, "Meteorology," 230.

81. Edward Sabine, "On Some Points in the Meteorology of Bombay," *Report of the Fifteenth Meeting of the British Association for the Advancement of Science* (London: John Murray, 1846), 82.

2. Imperial Oscillations

1. Hulme, *Weathered*. See also Hulme "Geographical Work"; Brace and Geoghegan, "Human Geographies"; Lahsen, "Social Status"; Mahony, "Climate Change."

2. Broad and Orlove, "Channeling Globality"; Sturken, "Desiring the Weather," 186.

3. Caviedes, "El Niño," 267; Sturken, "Desiring the Weather"; Philander, "El Niño and La Niña."

4. Miller, "Fall of an Angel"; J. Derrida, *De La Grammatologie* (Paris: Editions de Minuit, 1967); Grove and Adamson, *El Niño in World History*.

5. For a discussion of visibility in the context of climate change, see Rudiak-Gould, "We Have Seen It."

6. A. J. Troup, "The Southern Oscillation," *Quarterly Journal of the Royal Meteorological Society* 91 (1965): 490–506.

7. Walker, "Pen Portraits of Presidents."

8. Gilbert T. Walker, "On Boomerangs," *Philosophical Transactions of the Royal Society, Series A* 190 (1897): 23–42; Gilbert T. Walker, "Boomerangs," *Nature* 64 (1901): 338–40; Walker, "Pen Portraits of Presidents."

9. Gilbert T. Walker, "On a Dynamical Top," *Quarterly Journal of Pure and Applied Mathematics* 28 (1896): 175–84; Gilbert T. Walker, *Aberration and Some Other Problems connected with the Electromagnetic Field: One of Two Essays to which the Adams Prize Was Awarded in 1899, in the University of Cambridge* (Cambridge: Cambridge University Press, 1900); Joseph Larmor to Walker, June 8, 1899, MS 2012/13, Science Museum Wroughton Archives (hereafter SMWA).

10. Taylor, "Gilbert Thomas Walker."

11. Cleveland Abbe to Walker, April 1, 1903, MS 2012/43, SMWA; George E. Hale to Walker, October 26, 1903, MS 2012/51, SMWA.

12. Walker, "Pen Portraits of Presidents."

13. Sheppard, "Sir Gilbert Walker," 186.

14. Sikka, "India Meteorological Department," 398.

15. Gilbert T. Walker and Rai Bahadur Hem Raj, "The Cold Weather Storms of Northern India," *Memoirs of the Indian Meteorological Department* 21, no. 7 (1913): 4–12.

16. S. Narayana Lingan to Walker, February 25, 1913, MS 2012/80, SMWA. Walker's subsequent reference was pivotal in the granting of Ramanujan's scholarship to the University of Madras; Kanigel, *Man Who Knew Infinity*.

17. Gilbert T. Walker, "Meteorology and the Non-flapping Flight of Tropical Birds," *Proceedings of the Cambridge Philosophical Society* 21 (1923): 363–75; Gilbert T. Walker, "The Flapping Flight of Birds," *Journal of the Royal Aeronautical Society* 29 (1925): 590–94; Gilbert T. Walker, "The Flapping Flight of Birds, II," *Journal of the Royal Aeronautical Society* 31 (1927): 337–42; Walker, "Pen Portraits of Presidents."

18. Walker, "Pen Portraits of Presidents"; Katz, "Sir Gilbert Walker."

19. Diary of George Simpson, February 5, 1912, MS704/4, Scott Polar Research Institute Archives, Cambridge, UK.

20. Simpson to Walker, May 17, 1912, MS 2012/76, SMWA.

21. "I am glad you like the climate; I fear I should not; but I suppose Simla is very different from what one thinks of and much cooler than most of India." W. McF. Orr to Walker, August 28, 1904, MS 2012/55, SMWA. Here Walker differed from the common experience of British colonists during the nineteenth century. See Livingstone, "Race, Space and Moral Climatology"; Adamson, "Languor of the Hot Weather."

22. Chapman, "Symons Memorial Medal"; Normand, "Sir Gilbert Walker"; Taylor, "Gilbert Thomas Walker."

23. Walker, "Pen Portraits of Presidents"; Katz, "Sir Gilbert Walker."

24. Anderson, *Predicting the Weather*. In the United States, meteorology was not considered a discrete science until the twentieth century. See Willis and Hooke, "Cleveland Abbe and American Meteorology."

25. J. W. Dyson to Walker, June 12, 1899, MS 2012/15, SMWA.

26. Walker, "Pen Portraits of Presidents"; Kelvin to Walker, April 19, 1901, MS 2012/37, SMWA.

27. Gilbert T. Walker, *Outlines of the Theory of Electromagnetism* (Cambridge: Cambridge University Press 1910).

28. Louis Agricola Bauer to Walker, April 26, 1904, MS 2012/54, SMWA.

29. Gustav Hellmann to Walker, April 7, 1904, MS 2012/47, SMWA.

30. Normand, "Monsoon Seasonal Forecasting."

31. J. Eliot, "On the Origin of the Cold Weather Storms of the Year 1893 in India, and the Character of the Air Movement on the Indian Seas and the Equatorial Belt, More Especially during the South-West Monsoon Period (as Shown by the Data of the Indian Monsoon Area Charts for the Year 1893)," *Quarterly Journal of the Royal Meteorological Society* 22, no. 97 (1896): 1–37; Allan, Lindesay, and Parker, *El Niño Southern Oscillation*.

32. George E. Hale to Walker, October 26, 1903, MS 2012/51, SMWA.

33. J. N. Lockyer, "The Meteorology of the Future," *Nature* 7 (1872): 99. See also S. H. Schwabe, "Sonnenbeobachtungen im Jahre 1843," *Astronomische Nachrichten* 21 (1843): 233–36.

34. E. D. Archibald, "Indian Meteorology II," *Nature* 28 (1883): 428–30; Anderson, *Predicting the Weather*.

35. W. J. S. Lockyer, "Barometric Variations of Long Duration over Large Areas," *Proceedings of the Royal Society of London, Series A* 78 (1906): 43. See also J. N. Lockyer, "Simultaneous Solar and Terrestrial Changes," *Science* 18 (1903): 611–23; Lockyer and Lockyer, "Short-Period Atmospheric Pressure Variation"; W. J. S. Lockyer, "Does the Indian Climate Change?" *Nature* 84 (1910): 178.

36. Hildebrandsson, "Quelques Recherches sur les Centres d'Action de l'Atmosphere," *Kunliga Svenska Vetenskapsakademiens Handligar* 29 (1896): 1–33.

37. Walker, "World Weather," 79.

38. William Napier Shaw, "The Pulse of the Atmospheric Circulation," *Nature* 73 (1905): 175–77.

39. Abbe to Walker, April 1, 1903, MS 2012/43, SMWA.

40. "We hope that your new position will not burden you with such heavy executive duties as to interfere with your work in the field of physical research—but, that rather, by suggesting a new class of problems your thoughts may be centered on dynamic meteorology, to the great advantage of this difficult branch of science." Abbe to Walker, February 24, 1902, MS 2012/39, SMWA. See also Abbe, "Long-Range Weather Forecasts"; V. Bjerknes, "Das problem der wettervorhersage, betrachtet vom standpunkte der mechanik und der physic," *Meteorologische Zeitschrift* 21 (1904): 1–7; Lehman, "Whither Climatology?" 49–70.

41. Walker, "Correlation, VIII," 75.

42. A. C. Phillips and Gilbert T. Walker, "The Forms of Stratified Clouds," *Quarterly Journal of the Royal Meteorological Society* 58, no. 243 (1932): 23–30; Gilbert T. Walker, "Clouds and Cells," *Quarterly Journal of the Royal Meteorological Society* 59 (1933): 389–96; Gilbert T. Walker, "Clouds—Natural and Artificial," *Proceedings of the Royal Institution* 28 (1934): 559–71; Gilbert T. Walker, "Clouds in the Sky and in the Laboratory," *Scientific Progress* 29

(1935): 385–94; Gilbert T. Walker, "Some Recent Work on Cloud Forms," *Quarterly Journal of the Royal Meteorological Society* 65, no. 278 (1939): 28–31.

43. Abbe, "Long-Range Weather Forecasts."

44. L. Teisserenc de Bort, "Etude sur l'hiver de 1897–1890 et recherches sur la position de centres de l'action de l'atmosphère dans les hivers anormaux," *Annales, Bureau Central Météorologique de France* 4 (1881): 17–62; Hildebrandsson, "Quelques recherches."

45. Walker, "Correlation, IX"; Walker, "Correlation, VIII."

46. Pincock, "From Sunspots to the Southern Oscillation."

47. Normand, "Monsoon Seasonal Forecasting."

48. A. Schuster, "On the Investigation of Hidden Periodicities with Application to a Supposed 26-Day Period of Meteorological Phenomena," *Terrestrial Magnetism* 3, no. 1 (1898): 13–41; Katz, "Sir Gilbert Walker."

49. Walker, "Correlation III."

50. Gilbert T. Walker, "On the Criterion for the Reality of Relationships or Periodicities," *Memoirs of the Indian Meteorological Department* 21, no. 9 (1914): 13–15.

51. Walker, "Correlation, VIII," 76.

52. Walker, "Correlation, VIII," 87.

53. Pincock, "From Sunspots to the Southern Oscillation."

54. Walker, "Correlation," VIII, 75. See also George E. Hale to Walker, October 26, 1903, MS 2012/51, SMWA; Arthur Schuster to Walker, December 5, 1909, MS 2012/72, SMWA.

55. Normand, "Monsoon Seasonal Forecasting."

56. Walker, "Correlation," VIII.

57. Walker, "Correlation," IX.

58. Eliot, "A Preliminary Discussion of Certain Oscillatory Changes of Pressure of Long Period and Short Period in India," *Indian Meteorological Memoirs* 6 (1895): 89–160.

59. Walker, "Correlation, VIII," 109.

60. Walker, "Correlation, IX," 331. Note that this was an update on his original definition in 1923, which included the Azores, Charleston, Honolulu, Samoa, South America, and the Indian peninsular on one side and Iceland, central Siberia, northwest India, Darwin, Mauritius, and southeast Australia on the other.

61. Walker, "Correlation, IX."

62. Hildebrandsson, "Quelques recherches"; N. Lockyer and WJS Lockyer, "On the Similarity of the Short-Period Pressure Variation over Large Areas," *Proceedings of the Royal Society of London* 71 (1902): 134–35.

63. Pincock, "From Sunspots to the Southern Oscillation." For his good nature, see statements given in his obituaries. "Walker was . . . ever eminently reasonable, liberal minded and very friendly." Normand, "Sir Gilbert Walker." "Withal, Walker was eminently friendly— though somewhat reserved—kind, and a very gentle man." Sheppard, "Sir Gilbert Walker."

64. Walker and Bliss, "World Weather V," 53, 60, 62.

65. Gilbert T. Walker, "On Periodicity in Series of Related Terms," *Proceedings of the Royal Society of London, Series A* 131 (1931): 518–32.

66. Walker, "World Weather," 82.

67. Walker, "Correlation," IX.

68. Walker and Bliss, "World Weather V."

69. Walker, "World Weather," 83; Walker and Bliss, "World Weather V," 64.

70. Walker and Bliss, "World Weather V," 72.

71. Walker, "Pen Portraits of Presidents."

72. Walker, "World Weather."

73. Lehman, "Whither Climatology?"

74. Abbe, "Long-Range Weather Forecasts," 552.

75. Chapman, "Symons Memorial Medal."

76. Robert DeCo Ward to Walker, December 2, 1904, MS 2012/57, SMWA.

77. W. S. Beveridge to A. L. Bowley, September 23, 1922, Beveridge to G. C. Simpson 1922, Simpson to Sir W. Beveridge 1922, all in Beveridge/7/50C, Women's Library Archives, London School of Economics. Using Walker's criterion, however, Beveridge concluded that his thesis was confirmed rather than weakened, and that at least nine of the nineteen periodicities he found are likely to be correct. "On Periodicities," Sir W. H. B.'s comments on Dr Walker's note, Beveridge/7/50C, Women's Library Archives, London School of Economics.

78. M. G. Kendall, "On the Analysis of Oscillatory Time-Series (with Discussion)," *Journal of the Royal Statistical Society* 108 (1945): 137.

79. Harold Jeffreys, *Earthquakes and Mountains* (London: Methuen, 1935), 54; Gilbert T. Walker, "Editorial," *Quarterly Journal of the Royal Meteorological Society* 62, no. 263 (1936): 2.

80. Lockyer and Lockyer, "Short-Period Atmospheric Pressure Variation"; Eliot, "Preliminary Discussion."

81. W. Napier Shaw, "The Pulse of the Atmospheric Circulation," *Nature* 73 (1905): 175–77.

82. Walker, "Correlation, VIII," 109; Walker and Bliss, "World Weather V," 71.

83. A. Schuster, "On the Periodicities of Sunspots," *Philosophical Transactions of the Royal Society Series A* 206 (1906): 69–100.

84. Chapman, "Symons Memorial Medal."

85. "On Periodicities," Beveridge/7/50C, Women's Library Archives, London School of Economics.

86. Schuster to Walker, December 5, 1909, MS 2012/72, SMWA.

87. W. H. Sykes, "Special Report on the Statistics of the Four Collectorates of Dukhun under the British Government," *Report of the Seventh Meeting of the British Association for the Advancement of Science* 6 (1938): 217–336.

88. Anderson, *Predicting the Weather*, 282.

89. See, for example, Kocchar, "Science in British India"; Prakash, *Another Reason*; Kalpagam, "Colonial State"; M. Davis, *Late Victorian Holocausts: El Niño Famines and the Making of the Third World* (London: Verso, 2001); Arnold, *New Cambridge History*; Kalpagam, *Rule by Numbers*; Mahony and Randalls, this volume.

90. Naylor and Goodman (this volume).

91. Wilson, *India Conquered*.

92. Naylor and Goodman (this volume).

93. Kennedy, *Magic Mountains*.

94. This was explained to Walker in a letter from Napier Shaw in 1907, which described the process that led toward the drafting of the convention text so that it should "unify Governments to continue to maintain observatories where they already exist to establish them where they do not exist and to see to their adequate publications of the observations." Shaw to Walker, August 22, 1907, MS 2012/65, SMWA.

95. D. E. Archibald, "Indian Meteorology," *Nature* 28, no. 721 (1883): 405–7.

96. Anderson, *Predicting the Weather*, 236–37.

97. Simpson to Schuster, December 19, 1907, MS 1122, Scott Polar Research Institute Archives, Cambridge, UK.

98. R. A. Fisher and W. A. Mackenzie "The Correlation of Weekly Rainfall (with Discussion)," *Quarterly Journal of the Royal Meteorological Society* 48 (1922): 242. See also Katz, "Sir Gilbert Walker."

99. W. H. Dines, "Review of 'Correlation in Seasonal Variations of Weather, I–VI,' *Memoirs of the Indian Meteorological Department* by G. T. Walker," *Quarterly Journal of the Royal Meteorological Society* 42 (1916): 130. See also Katz, "Sir Gilbert Walker."

100. H. P. Berlage Jr., *Further Researches into the Possibility of Long-Range Forecasting in Netherlands India, by Dr. HP Berlage Jr*, vol. 26 (The Hague: Landsdrukkerij, 1934).

101. Walker, "World Weather," 86.

102. Sheppard, "Sir Gilbert Walker."

103. Walker and Bliss, "World Weather V"; Gilbert T. Walker, "On Periodicity in Series of Related Terms," *Proceedings of the Royal Society of London, Series A* 131 (1931): 525; Katz, "Sir Gilbert Walker."

104. Gilbert T. Walker, "The Atlantic Ocean," *Quarterly Journal of the Royal Meteorological Society* 53 (1927): 112–13.

105. Gilbert T. Walker, "Seasonal Weather and Its Prediction," *Smithsonian Institution Annual Report for 1935* (1936): 136.

106. Cushman, "Choosing between Centers of Action"; Grove and Adamson, *El Niño in World History*.

107. H. P. Berlage Jr. and H. J. de Boer, "On the Extension of the Southern Oscillation throughout the World during the Period July 1, 1949, to July 1, 1957," *Pure and Applied Geophysics* 44 (1959): 287–95; H. P. Berlage Jr. and H. J. de Boer, "On the Southern Oscillation, Its Way of Operation and How It Affects Pressure Patterns in the High Latitudes," *Pure and Applied Geophysics* 46, no. 1 (1960): 329–51; H. P. Berlage Jr., "Variations in the General Atmospheric and Hydrospheric Circulation of Periods of a Few Years Duration Affected by Variations of Solar Activity," *Annals of the New York Academy of Sciences* 95 (1961): 354–67; H. P. Berlage Jr., *The Southern Oscillation and World Weather*, Mededlingen en Verhandelinger No. 88 (Joninklijk Meteorologische Instituut, Staatsdrukkerijs-Gravenhage, Netherlands, 1966).

108. J. Bjerknes, "Atmospheric Teleconnections from the Equatorial Pacific," *Monthly Weather Review* 97, no. 3 (1969): 163–72.

109. For a discussion of the mathematical tradition in Cambridge during the late nineteenth century, see Andrew Warwick, *Masters of Theory: Cambridge and the Rise of Mathematical Physics* (Chicago: University of Chicago Press, 2003).

110. Quoted in F. Fagan, *Floods, Famines and Emperors: El Niño and the Fate of Civilizations* (New York: Basic Books, 1999).

111. Cushman, "Choosing between Centers of Action."

112. Mahony and Endfield, "Climate and Colonialism"; Heymann, "Climate Change Dilemma," doi.org/10.1007/s10113-018-1373-z/.

3. The Weather Ship

I gratefully acknowledge the assistance of the 2015 Ritter Memorial Fellowship of the Scripps Institution of Oceanography, which funded travel to archives, and of the endlessly patient and helpful staff of The National Archives, London (TNA), and the Library and Archives Canada, Ottawa (LAC). I am also grateful to comments from conference participants in 2016, especially the editors of this volume, and to Dror Kahn for permission to use the Kahn image.

1. Petterssen, *Weathering the Storm*, 269.

2. Jones-Imhotep, *Unreliable Nation*.

3. Achbari, "Building Networks"; Hardy, "Every Ship"; Naylor, "Log Books."

4. Anderson, "Marine Meteorology"; Robert Bureau and Michel Coyecque, *Les croisières radiométéorologiques du* Jacques Cartier (Paris: Gauthier Villars, 1926).

5. Anderson, "Marine Meteorology," 222–23.

6. See the summary in US Weather Bureau, *Ocean Station Vessel Meteorological Records Survey Atlantic and Pacific: Key to Meteorological Records Documentation*, No. 23 (Washington, DC: USGPO, 1956), 2–12.

7. Frank Entwhistle, "Atlantic Flight and Its Bearings on Meteorology," *Quarterly Journal of the Royal Meteorological Society* 64 (1938): 389. See also Crewe, "Meteorology and Aerial Navigation."

8. MacKenzie, *ICAO*. On ICAN, see Kranakis, "Good Miracle"; and Kranakis, "European Civil Aviation." On the IMO and the WMO, see Davies, *Forty Years of Progress*.

9. The WMO convention came into force only in March 1950. The subsequent relationship of the ICAO and the WMO, formalized in 1951–1952, was much more cooperative than the relationship of their prewar counterparts, ICAN and IMO. The program and frequency of ship observations followed WMO recommendations, and the two bodies coordinated on technical matters, including such details as the pay scales of meteorologists. *WMO Bulletin* 1 (April 1952): 14; Davies, *Forty Years of Progress*.

10. "Use of LORAN," *Coast Guard Bulletin* 3 (November 1947): 423–24.

11. US Weather Bureau, *Ocean Station Vessel Meteorological Records Survey* (Washington, DC: USGPO, 1956), 2–4; PICAO, *Memorandum on Ocean Weather Observation Stations*, Doc. 1955 C/189, July 25, 1946.

12. PICAO, *Conference of North Atlantic States on Ocean Weather Observation Stations in the North Atlantic Final Act*, Document 2136 C/235, October 1, 1946. For the quotation, see PICAO, *Memorandum on Ocean Weather Observation Stations*, Doc. 1955 C/189, July 25, 1946, 3.

13. On staffing and reconstruction, see Air Ministry/Ministry of Defence, *Ocean Weather Ships: Establishment (1946–1956)*, TNA, AIR 2/12217. For description of the scientific work on weather ships, see J. E. Brown, "The Sea-Going Meteorologist's Aspect of Work in Ocean Weather Ships," *Weather* 3 (July 1948): 216–18; Frankcom, "Ocean Weather Ships"; Downes, "British Ocean Weather Ships."

14. Canadian concerns about its Atlantic obligations are discussed in *Air Services—General—ICAO—North Atlantic Ocean Stations*, Part 1 (January 13, 1940–December 4, 1947), RG 12/2832/602, LAC; for reaction to the US delay, see Andrew Thomson (Controller of Meteorological Services) to Director of Air Services, Department of Transport, April 30, 1947, in the same file.

15. US Senate, *Hearings on S. 1853 and S. 2122*, 9.

16. H. M. Hutchon (liaison meteorologist) to Director of Air Services, April 9, 1949, *Air Services—General—ICAO—North Atlantic Ocean Stations*, Part 2 (December 4, 1947–June 9, 1953), RG12/2832/602, LAC.

17. On seadromes and other ocean installations, see US Senate, *Hearings on S. 1853 and S.*

2122, 17; "Communications Tower in Mid Ocean Proposed," *Civil Aeronautics Administration Journal* 6 (June 15, 1945): 72; Davies, *Fantasies and Fallacies.*

18. US Senate, *Hearings on S. 1853 and S. 2122*, 26.

19. Mr. Clifford P. Burton, the US representative to ICAO in 1949, referred primarily to other nations, but the sentiment applied to US activity as well. "Construction Necessary in Advancing ICAO Plan to Aid World Aviation," *Civil Aeronautics Administration Journal* (June 15, 1949): 62.

20. Dobson, *Peaceful Air Warfare*; Dobson, "The USA, Hegemony."

21. Harold A. Jones, US Representative to ICAO, October 21, 1953, *North Atlantic Ocean Weather Ships Scheme: U.K. policy on follow-up to 1949 agreement*, Appendix C, Item 8A, AIR 2/12199, TNA. The lack of consultation was reported in discussions between Canadian and American meteorologists (the Weather Bureau chief, Reichelderfer was traveling in Europe at the time of the announcement): "Interdepartmental Memorandum—North Atlantic Weather Stations," October 21, 1953, *Air Services—General—ICAO—North Atlantic Ocean Stations*, Part 3 (June 10, 1953–February 29, 1968), RG 12/2832/602, LAC.

22. Member states reactions in 1953 are collected in *North Atlantic Ocean Weather Ships Scheme: U.K. Policy on Follow-Up to 1949 Agreement*, AIR 2/12199, TNA. For the aeronautical research council report, see Cox to Ministry of Transport and Civil Aviation, January 25, 1954, DSIR 23/22319, TNA.

23. "ICAO Working Paper on Ocean Weather Stations," December 28, 1953, *North Atlantic Ocean Weather Ships Scheme: U.K. Policy on Follow-Up to 1949 Agreement*, AIR 2/12199, TNA.

24. "US Withdrawal of Weather Ships," editorial, *Ottawa Citizen*, October 26, 1953. On Canada–United States relations and the radar perimeter, see Herd, "Common Appreciation." On technology and sovereignty in the Cold War, see Jones-Imhotep, *Unreliable Nation.*

25. R. R. G. Watts to D. A. H. Baer, January 12, 1954, *Agreement on North Atlantic Ocean Weather Stations, Paris, 1954*, Item 25, DO 35/4928A, TNA. For the internal Canadian discussions, see *Air Services—General—ICAO—North Atlantic Ocean Stations*, Part 3 (June 10, 1953–February 29, 1968), RG 12/2832/602, LAC.

26. ICAO, *Report on the Third and Fourth ICAO Conferences on North Atlantic Ocean Stations: Brighton, July 1953 and Paris, February 1954*, ICAO Doc. 7510-JS/559.

27. "Ocean Stations at Work," *ICAO Bulletin* 15 (1960): 167.

28. For discussions of alternatives, see ICAO, *Automatic Weather Stations* (December 1949), ICAO Circular 13-AN/11, and "Possible Alternatives to Weather Ships (1978)," *Meteorological Office Administrative Records: Future of North Atlantic Weather Ships*, in BJ5/327, TNA. On automatic buoy programs, see Cushman, "Choosing between Centers of Action."

29. "Ship Weather Stations Solve Secrets of Ocean Blanks," *Popular Mechanics* 70 (November 1938): 669–70.

30. "Position of Canada with Respect to North Atlantic Ocean Station Network," November 2, 1953, *North Atlantic Weather Ships Scheme: UK Policy*, Item 24A, AIR 2/12199, TNA, 5. See also the distinction between routes and regions in charts in Frankcom, "Ocean Weather Ships."

31. Hamblin, "Seeing the Ocean."

32. See Rozwadowski, "Oceans," 450.

33. For a recent account of oceanographic initiatives in the Atlantic in this decade, see Robinson, *Ocean Science and the British Cold War State.*

34. For the survey on the value of NAOS, see *North Atlantic Weather Ships Scheme: UK Policy*, AIR 2/12199, TNA.

35. L. J. Slobinski, "Canadian Ocean Weather Ships: Committee on Oceanography," *Transactions of the Royal Society of Canada* 46 (1952): 1–13.

36. W. C. G. Cribbet to Sir Folliott Standford, November 16, 1953, *Meteorology: Stations and Observatories: North Atlantic Weather Ships Scheme: UK Policy*, AIR 2/12199/31, TNA. BOAC was sensitive because of the problem of metal fatigue in their Comet III aircraft. A popular 1951 film, *No Highway in the Sky*, had dramatized the story of airplane crashes and metal fatigue.

37. Kaplan, *NATO Divided, NATO United*.

38. Andrew Thomson (Controller of Meteorological Service, Canada), "Proposed Change of Position of Ocean Weather Station India: Draft Letter to Undersecretary of State for External Affairs," March 17, 1953, Item no. 239, RG 12/2832/602 (December 4, 1947–June 9, 1953), LAC.

39. Andrew Thomson, "Memorandum [to the] Director of Air Services—Proposed Change of Position of Ocean Weather Station India," March 17, 1953 (December 4, 1947–June 9, 1953) Item no. 240, RG 12/2832/602, LAC.

40. Quoted in MacKenzie, *ICAO*, 5.

41. On watchers and their regulation, see *Committee on Wireless Telegraphy Report and Minutes of Proceedings of the International Conference on Safety of Life at Sea 1913–1914* (London, 1914), 19–25; *International Convention for the Safety of Life at Sea 1929* (London: HMSO, 1932).

42. Rainger, "Edward 'Iceberg' Smith."

43. Station C was located at 52° 45' N, 35° 43' W.

44. "Cutter 'Bibb' Makes Spectacular Rescue of Plane 'Bermuda Sky Queen,'" *Coast Guard Bulletin* 3 (November 1947): 421–23; for the public reception, see "Mid-Ocean Rescue Told by Survivors," *New York Times* (October 20, 1947), 2; for the "lemming" remark, see USCG Report of Assistance (October 14, 1947), http://www.uscg.mil/history/ops/sar/1947_Bermuda Sky Queen/.

45. The pilot was fined two hundred dollars and American International Airways was fined five hundred dollars after the finding of improper flight planning under conditions of excess gross weight. Civil Aeronautics Board Accident Report, "American International Airways, Inc—Bermuda Sky Queen—October 14, 1947," National Transportation Library, Investigation of Aircraft Accidents 1934–1965, dotlibrary.specialcollection.net/, accessed June 12, 2016.

46. Nearly a decade later, another ditching in the Pacific echoed the 1947 event. "Ordeal on Flight 943," *Life* (October 29, 1956); "Supplement 1 to Weekly Report of Activities No. 45–56," *Meteorology: Reports: Ocean Weather Ships: Air/Sea Rescue (1953–1957)*, Item 153B, AIR 2/13348, TNA.

47. "ICAO Working Paper: Views of States on Weather Ships," Appendix A, 3, in *AIR 2/12199/8A Meteorology: Stations and Observatories: North Atlantic Weather Ships Scheme: UK Policy on Follow up to 1949 Agreement*, TNA. The PICAO had emphasized in the original proposal in 1946 that "the knowledge that a station vessel is in the vicinity and that it is available for rescue would have a valuable psychological effect on all concerned." "Memorandum on Ocean Weather Observation Stations," July 25, 1946, PICAO Doc. 1955 C/189, 19.

48. Rube Hornstein, CBC Halifax radio meteorologist, to Andrew Thomson, December 18, 1947 (December 4, 1947–June 9, 1953), RG 12/2832/602, LAC.

49. A. W. Ford, "A Day in an Ocean Weather Ship," *Weather* 4 (February 1949): 61–64.

50. *WMO Bulletin* 3 (April 1954): cover.

51. Jasanoff, "Future Imperfect"; Verchraegen, Vandermoere, Braeckmans, and Segaert, *Imagined Futures*; Sappol, *Body Modern*.

52. For Kahn's career, see Borck, "Communicating the Modern Body"; Sappol, *Body Modern*; Debshitz and Debshitz, *Fritz Kahn*.

53. Fritz Kahn, *Der Mensch Gesund und Krank* (Zurich: A. Muller, 1939); Fritz Kahn, *Man in Structure and Function*, 2 vols. (New York: Knopf, 1943).

54. Fritz Kahn, *Man in Structure and Function*, 2 vols. (New York Alfred Knopf, 1943), 1:472. See also Sappol's discussion of this image and others as Kahn's "universal human." Sappol, *Body Modern*, 95–100.

55. See Emil Du Bois Reymond, "On the Time Required for the Transmission of the Volition and Sensation through the Nerves," *Proceedings of the Royal Institution of Great Britain* 4 (1866): 588.

56. The manuscripts of his Palestine project confirm his interests in the environment, opening with a chapter on weather and climate. *Arthur and Fritz Kahn Collection 1889–1932, Sub-series 4: The Natural History of Palestine*, Internet Archive, https:www.archive.org/.

57. The visual play with comparative scales was a typical feature of a remarkable range of illustration techniques: see Sappol, *Body Modern*, for a full discussion.

58. Pitt, *Pratt: Truant Years*, 88.

59. This convoy SC-42 is well-known to historians as the first instance of the German use of their so-called wolf-pack strategy. Hague, *Allied Convoy System*.

60. Pitt, *Pratt: Master Years*, 375.

61. E. J. Pratt, *Behind the Log* (Toronto: Macmillan, 1947; line references will henceforth be given parenthetically in the text). Pitt notes how the poem highlights "the problem of intelligible communication," whether codes, signals, or different languages, in Pitt, *Pratt: Master Years*, 372.

62. See Pratt's poem of September 1939, "The Radio in the Ivory Tower." On his visit to Signal Hill in 1901, see Pitt, *Pratt: Truant Years*, 59.

63. The negotiation with MacDonald is described in Pitt, *Master Years*, 372.

64. The story appeared in *Windsor Magazine* (England) and *McClure's Magazine* (United States) in November and December 1905 and then in a collection of short stories and poetry, Kipling, *Actions and Reactions* (references will be given parenthetically in the text). On the significance of historical futures, see Verchraegen and Vandermoere, "Introduction," in *Imagined Futures*, 7.

65. On the tradition of pilots as heroes, see Corn, *Winged Gospel*; Van Riper, *Imagining Flight*, 33–60.

66. Kipling writes in the same anthropomorphized manner about ships and oceans in "The Ship that Found Herself" (1895), collected in Kipling, *The Day's Work* (London: Macmillan, 1898).

67. In a sequel, "As Easy as ABC," Kipling gives a more sinister account of the Aerial Board of Control as a repressive governing force. See Harvie, "Sons of Martha."

68. Edwards, *Vast Machine*; Miller, "Climate Science." Eda Kranakis and Nil Disco use the term "cosmopolitan commons" to describe transnational technological and informational resources. Kranakis and Disco, *Cosmopolitan Commons*. The technological and infrastructural sense of globalism here can be usefully compared to ecological ideas: see Kwa, "Romantic and Baroque."

4. Looking for the Leeuwin

1. George Cresswell, "The Leeuwin Current—Observations and Recent Models," *Journal of the Royal Society of Western Australia* 74 (1991): 2; George Cresswell and Terry Golding, "Observations of a South-Flowing Current in the Southeastern Indian Ocean," *Deep-Sea Research*

27A (1980): 449–66; Ken Ridgeway and Scott Condie, "The 5500-Km-Long Boundary Flow off Western and Southern Australia," *Journal of Geophysical Research: Oceans* 109 (2004), doi. org/10.1029/2003JC001921/.

2. Matthew Flinders, *A Voyage to Terra Australis*, vol. 1 (London: W. Bulmer, 1814), 241; George Cresswell, "The Leeuwin Current," *Corrella* 14, no. 4 (1990): 113–18.

3. John Church, George Cresswell, and Stuart Godfrey, "The Leeuwin Current," in *Poleward Flows along Eastern Ocean Boundaries*, ed. Steven Neshyba et al. (New York: American Geophysical Union, 1989), 230–54.

4. Christopher Reason, David Gamble, and Alan Pearce, "The Leeuwin Current in the Parallel Ocean Climate Model and Applications to Regional Meteorology and Fisheries," *Meteorological Applications* (1999): 211–25; Jos Samuel et al., "Influence of Indian Ocean Sea Surface Temperature Variability on Southwest Western Australia Winter Rainfall," *Water Resources Research* 42, no. 8 (2006), doi.org/10.1029/2005WR004672/.

5. Brett Molony et al., "Are Western Australian Waters the Least Productive Waters for Finfish across Two Oceans? A Review with a Focus on Finfish Resources in the Kimberley Region and the North Coast Bioregion," *Journal of the Royal Society of Western Australia* 94 (2011): 323–32; Caputi, Fletcher, Pearce, and Chubb, "Effect of the Leeuwin Current."

6. Nick Caputi, Simon de Lestang, Chris Reid, Alex Hesp, and Jason How, "Maximum Economic Yield of the Western Rock Lobster Fishery of Western Australia after Moving from Effort to Quota Control," *Marine Policy* 51 (2015): 452–64. The western rock lobster became the "official" name for the Western Australian crayfish in 1969, principally for the purposes of its marketing to the United States. For ease of understanding, this chapter will refer to the crayfish and its fishery as "rock lobster" and "western rock lobster." See Gray, *Book 1*, 10–11.

7. Caputi, Fletcher, Pearce, and Chubb, "Effect of the Leeuwin Current."

8. Jo McDonald and Peter Veth, "Rock Art in Arid Landscapes: Pilbara and Western Desert Petroglyths," *Australian Archaeology* 77, no. 1 (2013): 66–81; Jo McDonald, "I Must Go Down to the Seas Again: Or, What Happens When the Sea Comes to You? *Murujuga* Rock Art as an Environmental Indicator for Australia's North-West," *Quaternary International* 385, no. 22 (2015): 124–35.

9. Penn, Caputi, and de Lestang, "Review of Lobster Fishery Management."

10. "Australian Western Rock Lobster," *Marine Stewardship Council* (2018), https://fisheries. msc.org/en/fisheries/australian-western-rock-lobster/about/.

11. Daniels, "Geographical Imagination," 182. For environmental histories of Cold War internationalism, see Hamblin, *Oceanographers and the Cold War*; McNeill and Unger, *Environmental Histories*. For further exploration of the challenge of visibility in studies of the ocean, see Rozwadowski, "Ocean's Depths."

12. Gordon, "Oceanography of the Indonesian Seas"; Pariwono, Ilahude, and Hutomo, "Progress in Oceanography"; Walker and Sobocinska, *Australia's Asia*.

13. See for example, Rozwadowski, *Fathoming the Ocean*; Hamblin, *Oceanographers and the Cold War*; Cushman, *Guano*.

14. Edwards, *Vast Machine*, xiv–xx. See also Rozwadowski, *Fathoming the Ocean*; Anderson and Rozwadowski, *Soundings and Crossings*; Fleming, "Planetary-Scale Fieldwork".

15. Wintersteen, "Fishing for Food and Fodder"; Cushman, *Guano*.

16. Ulanski, *California Current*, 2.

17. This observation echoes O'Gorman, Beattie, and Henry, "Epilogue." For "land-locked"

Australian climate histories, see for example, Tim Flannery, *The Future Eaters: An Ecological History of the Australasian Lands and People* (Chatswood, NSW: Reed Books, 1994); Tim Sherratt, Tom Griffiths, and Libby Robin, eds., *A Change in the Weather: Climate and Culture in Australia* (Canberra: NMA Press, 2005); Don Garden, *Droughts, Floods and Cyclones: El Niños That Shaped Our Colonial Past* (North Melbourne: Australian Scholarly Publishing, 2009); Libby Robin, Robert Heinsohn, and Leo Joseph, eds., *Boom and Bust: Bird Stories for a Dry Country* (Collingwood, Vic.: CSIRO Publishing, 2009); Robyn Ballinger, *An Inch of Rain: A Water History of Northern Victoria* (Melbourne: Australian Scholarly Publishing, 2012); Rebecca Jones, *Slow Catastrophes: Living with Drought in Australia* (Melbourne: Monash University Publishing, 2017).

18. Historians of race, migration, and Australia's relations with Asia have led the way in thinking more critically about the continent's coastal waters as both a resource and a pathway. See for example, Campbell Macknight, *The Voyage to Marege': Macassan Trepangers in Northern Australia* (Carlton, Vic.: Melbourne University Press, 1976); Schnukal, Ramsay, and Nagata, *Navigating Boundaries*; Ganter, *Mixed Relations*; Balint, "Epilogue: The Yellow Sea"; Martinez and Vickers, *Pearl Frontier*.

19. Christensen, "Their Inescapable Portion?"; Christensen, "Unsettled Seas"; Christensen and Jackson, "Shark Bay Snapper"; Christensen, "To the Islands"; Christensen, "Islands as Arks"; Christensen, "Shark Bay"; Tull, "Profits and Lifestyle"; Gaynor, "Shifting Baselines?"; Fowles and Gaynor, "Challenge of Creating."

20. Daniel Pauly, "Anecdotes and the Shifting Baseline Syndrome of Fisheries," *Trends in Ecology and Evolution* 10, no. 10 (1995): 430.

21. Fowles and Gaynor, "Challenge of Creating"; Gaynor, "Shifting Baselines?"; Andrea Gaynor, Amrit Kendrick, and Mark Westera, *An Oral History of Fishing and Diving in the Capes Region of South-West Western Australia* (Perth: South West Catchments Council, 2008).

22. The Oral History of the Australian Fishing Industry Project was an initiative of Malcolm Tull of Murdoch University (Western Australia) for the Australian bicentenary. The interviews were undertaken in each state by Jack Darcey and Howard Smith. See *Oral History of the Australian Fishing Industry* (Perth: Murdoch University and the Fisheries Research & Development Corporation, 1999); Shaw, "Fishing Communities and Climate Change."

23. See, for example, Bolster, "Opportunities in Marine Environmental History"; Rozwadowski, "Ocean's Depths"; Taylor, "Knowing the Black Box."

24. Carroll, *Empire of Air and Water*; Carroll, "Atopia/Non-Place," 159–67.

25. Steinberg, "Of Other Seas"; Steinberg, *Social Construction of the Ocean*; Steinberg, "Navigating to Multiple Horizons"; Peters, "Future Promises"; Steinberg and Peters, "Wet Ontologies."

26. Steinberg, "Of Other Seas," 164 (emphasis in text); Steinberg and Peters, "Wet Ontologies."

27. Ulanski, *Gulf Stream*, 60.

28. Oreskes, "Scaling Up Our Vision."

29. Bashford, "Terraqueous Histories," 261.

30. Joseph Gentilli, "Thermal Anomalies in the Eastern Indian Ocean," *Nature: Physical Science* 238 (1972): 94, 94, 95. For earlier explorations of the eastern Indian Ocean, see Gerald Halligan, "The Ocean Currents around Australia," *Journal of the Royal Society of Western Australia* 55 (1921): 188–95; Gerhard Schott, *Geographie des Indischen und Stillen Ozeans*

(Hamburg: Verlag von C. Boysen, 1935); George Deacon, "The Hydrology of the Southern Ocean," *Discovery Reports* 15, no. 1 (1937): 1–124.

31. Joseph Gentilli, *Atlas of Western Australian Agriculture* (Crawley, WA: UWA Text Books Board, 1941), 49.

32. Godwin, "Fluid Frontier," 116. Thanks to Jodi Frawley for bringing this research to my attention.

33. See, for example, Rozwadowski, *Fathoming the Ocean*; Eva-Maria Stolberg, "From Icy Backwater"; Connery, "Pacific Rim Discourse"; Steinberg, "Mediterranean Metaphors"; Steinberg, *Social Construction of the Ocean*.

34. Carroll, *Empire of Air and Water*, 7.

35. Cresswell, "Leeuwin Current—Observations and Recent Models," 1. See also Joseph Gentilli to Dominic Serventy, November 5, 1941, General Correspondence, K1347, 73, National Archives of Australia, Perth.

36. Rica Erickson, "Serventy, Dominic Louis (1904–1988)," in *Australian Dictionary of Biography* (2012), http://adb.anu.edu.au/biography/serventy-dominic-louis-15496/.

37. "Personal," *West Australian*, October 23, 1943, 2; "W. A. Crayfish," *West Australian*, October 17, 1945, 6.

38. Gray, *Book 1*, 38. For an overview of contemporary techniques, see D. Vaux and A. M. Olsen, "Use of Drift Bottles in Fisheries Research," *Australian Fisheries Newsletter*, Leaflet no. 3 (Sydney: CSIRO, 1961).

39. Dakin, "Percy Sladen Trust Expeditions," 142, 146.

40. William Saville-Kent, *The Naturalist in Australia* (London: Chapman and Hall, 1897), 150–51. For Saville-Kent, see Anthony Harrison, "Saville-Kent, William (1845–1908)," in *Australian Dictionary of Biography* (2005), http://adb.anu.edu.au/biography/saville-kent-william-13185/. John Lort Stokes had also remarked on the tropical features of the Abrolhos while he was on board the HMS *Beagle* (1846), but he did not suggest a cause for the presence of such tropical species. See John Stokes, *Discoveries in Australia* (London: T and W. Boone, 1846), 2:146.

41. Dakin, "Percy Sladen Trust Expeditions," 135, 136. For further exploration of the ways in which marine animals contribute to the structures of marine human communities, see Jones, "Running into Whales." Gray suggests that in the late 1920s, Anglo-Australian fishers resented the industry dominance of fishers from Italian backgrounds. See Gray, *Book 1*, 48–50. See also Godwin, "Fluid Frontier."

42. Judy Dunlop et al., "Recovery of Seabird Colonies on Rat Island (Houtman Abrolhos) following the Eradication of Introduced Predators," *Journal of the Royal Society of Western Australia* 98 (2015): 29–36.

43. Felice Miragliotta, interview with Ronda Jamieson, September 22, 1981, OH431, Battye Library (hereafter BL); Gray, *Book 2*, 42, 49.

For insight into the fortunes of Geraldton and its role in the rock lobster fishery, see Tull, Metcalf, and Gray, "Economic and Social Impacts."

44. Chris Russell, interview with Jack Darcey, November 8, 1989, OH2266/1, BL; Morgan, "Changes in Fishing Practice."

45. Behrman, *Assault on the Largest Unknown*, 9.

46. Revelle, c.1980, cited in Behrman, *Assault on the Largest Unknown*, 19.

47. Rozwadowski, "Ocean's Depths," 522; Rozwadowski, "Arthur C. Clarke."

48. E. Highley, *The International Indian Ocean Expedition: Australia's Contribution* (Mel-

bourne: CSIRO, 1968); David Rochford, "From Estuaries to Oceans," in Mawson, Tranter, and Pearce, *CSIRO at Sea*, 69–76; George Humphrey, "International Collaboration .in Marine Science: Australia's Role," in Mawson, Tranter, and Pearce, *CSIRO at Sea*, 119–24. For example, George Humphrey, the director of the CSIRO oceanography laboratory at Cronulla in New South Wales, was also president of SCOR (1960–1964).

49. Harry Jitts, "Frigate Days," in Mawson, Tranter, and Pearce, *CSIRO at Sea*, 141–42. See also David Rochford, "From Estuaries to Oceans," in Mawson, Tranter, and Pearce, *CSIRO at Sea*, 43.

50. Gilbert Whitley, "Presidential Address: A Survey of Australian Ichthyology," *Proceedings of the Linnaean Society of New South Wales* 89, no. 404 (1964): 30. For Gentilli, see Diana Walker, "100 Years of Marine Science at UWA," *Ocean Institute News* 9 (2013): 7–8.

51. John Apel, "Ocean Science from Space," *EOS* 57, no. 9 (1976): 612–24.

52. George Cresswell, *Wind-Driven Ocean Surface Transport around Australia* (Sydney: CSIRO, 1977); Cresswell and Golding, "Observations of a South-Flowing Current in the Southeastern Indian Ocean," 449–66; Richard Legeckis and George Cresswell, "Satellite Observations of Sea-Surface Temperature Fronts off the Coast of Western and Southern Australia," *Deep-Sea Research* 28A, no. 3 (1981): 297–306.

53. Klaus Wyrtki, "Geopotential Topographies and Associated Circulation in the South-Eastern Indian Ocean," *Australian Journal of Marine and Freshwater Research* 13, no. 3 (1962): 1–17.

54. Cardwell and Thornton, "Fisherly Imagination."

55. Hubbard, "Fisheries Biology," 33.

56. A. J. Fraser, 1953, cited in Gaynor, "Shifting Baselines?" 234.

57. Morgan, "Changes in Fishing Practice," 81.

58. Holm, "World War II"; Butcher, *Closing of the Frontier*, 168–233; Christensen, "Unsettled Seas," 13–40.

59. Gary Morgan, "Initial Allocation of Harvesting Rights in the Rock Lobster Fishery of Western Australia," in *Case Studies on the Allocation of Transferable Quota Rights in Fisheries*, ed. Ross Shotton, FAO Fisheries Technical Paper no. 412 (Rome: FAO, 2001), 136–43.

60. George Bass interview with Jack Darcey, 1989, OH 2266/33, BL; Bernard Bowen interview with Jack Darcey, 1989, OH2266/21, BL.

61. Norman Baxter (Chair), *Report of the Honorary Royal Commission Appointed to Inquire into and Report upon the Fisheries Act 1905–1962 in Its Application to the Crayfishing Industry in Particular* (Perth: Government Printer, 1964), 10.

62. Ray George, "The Species P. Cygnus Identified as an Endemic Species Restricted to Subtropical Western Australia," *Journal of the Royal Society of Western Australia* 45 (1962): 100–110. For an overview of the debates about the classification of the species, see Gray, *Book 1*, 12–17.

63. Klaus Wyrtki, "Geopotential Topographies and Associated Circulation in the South-Eastern Indian Ocean," *Australian Journal of Marine and Freshwater Research* 13, no. 1 (1962): 2.

64. Ray George and Peter Cawthorn, *Investigations on the Phyllosoma Larvae of the Western Australian Crayfish* (Perth: Western Fisheries Research Committee, Fisheries Department, WA, 1962).

65. George Cresswell, *Wind-Driven Ocean Surface Transport around Australia* (Cronulla, NSW: CSIRO Division of Fisheries and Oceanography, 1972), 1.

66. Bruce Phillips, "A Semi-Quantitative Collector of the Puerulus Larvae of the Western Rock Lobster *Panulirus longipes cygnus* George (Decapoda, Palinuridae)," *Crustaceana* 22 (1972): 147–54.

67. Bruce Phillips and Rhys Brown, "The West Australian Rock Lobster Fishery: Research for Management," in *Marine Invertebrate Fisheries: Their Assessment and Management*, ed. John Caddy (New York: Wiley, 1989), 174.

68. Greg Roach interview with Jack Darcey, November 6, 1989, OH2266/44, BL; Morgan, "Changes in Fishing Practice," 82.

69. Irwin (Sonny) Healy interview with Jack Darcey, 1989, OH2266/26, BL; Morgan, "Changes in Fishing Practice," 83.

70. Gary Morgan, Bruce Phillips, and L. M. Joll, "Stock and Recruitment Relationships in *Panulirus cygnus*, the Commercial Rock (Spiny) Lobster of Western Australia," *Fishery Bulletin* 80, no. 3 (1982): 475–86; Gray, *Book 2*, 223.

71. Bruce Phillips, "Prediction of Commercial Catches of the Western Rock Lobster *Panulirus cygnus*," *Canadian Journal of Fisheries and Aquatic Sciences* 43 (1986): 2126–30; Alan Pearce and Bruce Phillips, "ENSO Events, the Leeuwin Current, and Larval Recruitment of the Western Rock Lobster," *Journal du Conseil Permanent International pour l'Exploration de la Mer* 45 (1988): 13–21; David Griffin et al., "Ocean Currents and the Larval Phase of the Australian Western Rock Lobster, *Panulirus cygnus*," *Marine and Freshwater Research* 52 (2001): 1187–99.

72. Rory Thompson, "Observations of the Leeuwin Current off Western Australia," *Journal of Physical Oceanography* 14 (1984): 623–28; Stuart Godfrey and Ken Ridgway, "The Large-Scale Environment of the Poleward-Flowing Leeuwin Current, Western Australia: Longshore Steric Height Gradients, Wind Stresses and Geostrophic Flow," *Journal of Physical Oceanography* 15 (1985): 481–95.

73. Charitha Pattiaratchi and Steve Buchan, "Implications of Long-Term Climate Change for the Leeuwin Current," *Journal of the Royal Society of Western Australia* 74 (1991): 133–40.

74. Stuart Godfrey and Ken Ridgway, "The Large-Scale Environment of the Poleward-Flowing Leeuwin Current, Western Australia: Longshore Steric Height Gradients, Wind Stresses and Geostrophic Flow," *Journal of Physical Oceanography* 15 (1985): 491.

75. Pearce and Phillips, "ENSO Events." See also Rob Allan, "El Niño Southern Oscillation Influences in the Australasian Region," *Progress in Physical Geography* 12, no. 3 (1988): 313–48.

76. Nick Caputi, "Impact of the Leeuwin Current on the Spatial Distribution of the Western Rock Lobster (*Panulirus cygnus*) and Implications for the Fishery of Western Australia," *Fisheries Oceanography* 17 (2008): 149; Morgan, "Changes in Fishing Practice," 83; Rod Lenanton et al., "The Ongoing Influence of the Leeuwin Current on Economically Important Fish and Invertebrates off Temperate Western Australia—Has It Changed?" *Journal of the Royal Society of Western Australia* 92 (2009): 114.

77. Simon de Lestang et al., "What Caused Seven Consecutive Years of Low Puerulus Settlement in the Western Rock Lobster Fishery of Western Australia," *ICES Journal of Marine Science* 72 (2015): 149–158; Tull, Metcalf and Gray, "Economic and Social Impacts."

78. Ming Feng et al., "Decadal Increase in Ningaloo *Niño* since the Late 1990s," *Geophyiscal Research Letters* 42, no. 1 (2015) 104–12; Alan Pearce and Ming Feng, "The Rise and Fall of the 'Marine Heat Wave' off Western Australia during the Summer of 2010/2011," *Journal of Marine Systems* 111–12 (2013): 139–56; Thomas Wernberg et al., "An Extreme Climatic Event Alters Marine Ecosystem Structure in a Global Biodiversity Hotspot," *Nature Climate Change* 3 (2013): 78–82; Alistair Hobday and Gretta Pecl, "Identification of Global Marine Hotspots: Sentinels for Change and Vanguards for Adaptation Action," *Reviews in Fish Biology and Fisheries* 24, no. 2 (2014): 415–25.

79. Scott Evans and Lynda Bellchambers, "Observations on the Effects of High Water Temperatures at the Houtman Abrolhos Islands," in *Fisheries Heatwave Report*, ed. Alan Pearce et al. (Perth: Fisheries Research Division, 2011), 13.

80. Ming Feng et al., "La Niña forces unprecedented Leeuwin Current Warming in 2011," *Scientific Reports* 3 (2013): 1–9.

81. Miller, "Resisting Empire," 83.

82. Martin Mahony and Mike Hulme, "Model Migrations: Mobility and Boundary Cross-ings in Regional Climate Prediction," *Transactions of the Institute of British Geographers* 37 (2012): 197–211; Miller, "Resisting Empire."

83. Wenju Cai et al., "Increased Frequency of Extreme La Niña Events under Greenhouse Warming," *Nature Climate Change* 5 (2015): 132–37.

84. Alan Pearce and Ming Feng, "Observations of Warming on the Western Australian Continental Shelf," *Marine and Freshwater Research* 58 (2007): 914–20; Ming Feng, Evan Weller, and Katy Hill, "The Leeuwin Current," in *Marine Climate Change in Australia: Impacts and Adaptation Responses* (Perth: WA Marine Science Institution, 2009).

85. Nick Caputi, Roy Melville-Smith, Simon de Lestang, Alan Pearce and Ming Feng, "The Effect of Climate Change on the Western Rock Lobster (*Panulirus cygnus*) Fishery of Western Australia," *Canadian Journal of Fisheries and Aquatic Sciences* 67, no. 1 (2010): 85–96.

86. Katherine Mills et al., "Fisheries Management in a Changing Climate: Lessons from the 2012 Ocean Heat Wave in the Northwest Atlantic," *Oceanography* 26, no. 2 (2013): 191–95; Nick Caputi, Mervi Kangas, Ainslie Denham, Ming Feng, Alan Pearce, Yasha Hetzel and Arani Chandrapavan, "Management Adaptation of Invertebrate Fisheries to an Extreme Marine Heat Wave Event at a Global Warming Hot Spot," *Ecology and Evolution* 6, no. 11 (2016): 3583–93.

87. Nick Caputi and Rhys Brown, "The Effect of Environment on Puerulus Settlement of the Western Rock Lobster (*Panulirus cygnus*) in Western Australia," *Fisheries Oceanography* 2, no. 1 (1993): 1–10; Nick Caputi, Chris Chubb, and Alan Pearce, "Environmental Effects on Recruitment of the Western Rock Lobster, *Panulirus Cygnus*," *Marine and Freshwater Research* 52 (2001): 1167–75.

88. Simon de Lestang et al., "What Caused Seven Consecutive Years of Low Puerulus Settlement in the Western Rock Lobster Fishery of Western Australia," *ICES Journal of Marine Science* 72 (2015): 149–i58; Ming Feng, Anya Waite, and Peter Thompson, "Climate Variability and Ocean Production in the Leeuwin Current System off the West Coast of Western Australia," *Journal of the Royal Society of Western Australia* 92 (2009): 67–81; Charathi Pattiaratchi and Steve Buchan, "Implications of Long-Term Climate Change for the Leeuwin Current," *Journal of the Royal Society of Western Australia* 74 (1991): 133–40; Nick Caputi, Ming Feng, Alan Pearce, Jessica Benthuysen, Ainslie Denham, Yasha Hetzel, Richard Matear, *Management Implications of Climate Change Effect on Fisheries in Western Australia, Part 1: Environmental Change and Risk Assessment* (Perth: WA Fisheries Research Division, 2014).

89. William Cheung et al., "Climate-Change Induced Tropicalisation of Marine Communities in Western Australia," *Marine and Freshwater Research* 63 (2012): 415–27.

90. Anon, cited in Jenny Shaw, "Increasing Knowledge in Fishing Communities with Collaboration, Credibility and Co-production," unpublished conference paper at the Seafood Directions Conference: Selling Our Story, Perth, 2015.

91. Jenny Shaw, Laura Stocker, and Leonie Noble, "Climate Change and Social Impacts:

Women's Perspectives from a Fishing Community in Western Australia," *Australian Journal of Maritime and Ocean Affairs* 7, no. 1 (2015): 38–51.

92. Penn, Caputi and de Lestang, "Review of Lobster Fishery Management," 123.

93. Anon., cited in Shaw, "Fishing Communities and Climate Change," 212.

94. Tull, "Profits and Lifestyle."

95. Raleigh Hood et al., "Research Opportunities and Challenges in the Indian Ocean," *EOS*, 89, no. 13 (2008): 125. See also Raleigh Hood et al., "The 2nd International Indian Ocean Expedition (IIOE-2): Motivating New Exploration in a Poor Understood Basin," *Limnology and Oceanography Bulletin* 25, no. 4 (2016): 117.

96. Raleigh Hood et al., eds, *Second International Indian Ocean Expedition: A Basin-Wide Research Program. Science Plan (2015–2020)* (Newark, DE: SCOR, 2015).

97. Sarah Taillier, "Western Rock Lobster Price Falls due to a Drop in China Demand, Cheaper American Exports," *ABC News* April 7, 2017, http://www.abc.net.au/news/2017-04-07/rock-lobster-price-falls-with-demand-lessening-from-china/8426944/.

5. Imagined Geographies of Climate and Race in Anglophone Life Assurance, c. 1840–1930

1. Alborn, *Regulated Lives*; Murphy, *Investing in Life*; McFall, *Devising Consumption*; Bouk, *How Our Days*.

2. Berridge, "Opium Eating and Life Insurance"; Porter, *Trust in Numbers*; Porter, "Life Insurance"; Dupree, "Other than Healing"; Jureidini and White, "Life Insurance"; Kneale and French, "Moderate Drinking."

3. Kneale and Randalls, "Invisible Atmospheric Knowledges."

4. In the terminology of assurance a *life* is not the same thing as a *person* (you can insure your person against other kinds of risk—like accidents—but still not have life assurance). Ultimately life assurance secures *the productive capacity of a person's life*, assuring their life as capital, and compensation is paid when that person dies or reaches an age at which they can no longer work. Francois Ewald, "Insurance and Risk," 204. And while one person might have the same legal rights as another, the cost of a policy depends on the length and quality of their life, so people of different ages pay varying amounts.

5. Herbert Cecil Thiselton, "A Discussion of some points of Life Assurance Administration in respect of which Divergence of Practice exists: A Plea for Uniformity," *JIA* 31, no. 1 (1893).

6. Jureidini and White, "Life Insurance."

7. Jureidini and White, "Life Insurance."

8. Cantlie, "Life Insurance," 110. See also Jolly, "'Foreign Grave' Motif"; Mark Harrison, "Cantlie, Sir James (1851–1926)," *Oxford Dictionary of National Biography*, Oxford University Press, 2004; online edition, October 2008, http://www.oxforddnb.com/view/article/50530/.

9. Minutes of the UK Temperance and General Provident Institution Standing Committee, June 28, 1841, UKP/C8/1840.1, Aviva Archives, Norwich.

10. Gerald Hemmington Ryan, "On a Method for determining the Extra Premiums to be charged in respect of Two-Life Assurances," *JIA* 24, no. 5 (1884): 321.

11. Thomson, "Address"; John Stott, "On the Death-Rate among Assured Lives in the West Indies, being the Experience of the Scottish Amicable Life Assurance Society during Thirty Years, 1846–1876," *JIA* 21, no. 3 (1878); George F. Hardy and Howard J. Rothery, "On the Mortality of Assured Lives in the West Indies (chiefly Barbados)," *JIA* 27, no. 3 (1888).

12. A. Day, in Hardy and Rothery, "Mortality of Assured Lives," *JIA* 27, no. 3 (1888): 189.

13. Harold Edward William Lutt, "On Extra Premiums," *JIA* 41, no. 4 (1907): 474.

14. Thomson, "Address," 179.

15. Arthur F. Burridge, "On the Rates of Mortality in Victoria, and on the Construction of Mortality Tables from Census Returns by the Graphical Method of Graduation," *JIA* 23, no. 5 (1882).

16. Samuel Brown, A. H. Leith, and F. C. Chapman, "On the Rate of Mortality amongst the Natives compared with that of Europeans in India," *JIA* 16, no. 3 (1871): 187. For Marshall and Tulloch, see *Statistical Report on the Sickness, Mortality, and Invaliding among the Troops in the West Indies*, GBPP 1837–1838 XL; Kneale and Randalls, "Invisible Atmospheric Knowledges," 44.

17. Martin, *Influence Of Tropical Climates*, 70.

18. Adolphe Quetelet, *A Treatise on Man and the Development of his Faculties* (Edinburgh: William and Robert Chambers, 1842), 26.

19. Kneale and Randalls, "Invisible Atmospheric Knowledges."

20. Sprague was editor of the *JIA* (1867–1883) and president of the Institute of Actuaries (1882–1886) and of the Faculty of Actuaries in Edinburgh (1894–1896). See George King, "Memoir of Thomas Bond Sprague, MA, LLD," *JIA* 52, no. 3 (1921).

21. Thomas Bond Sprague, "Actuarial Note. On the Rate of Mortality among Europeans in Tropical Africa," *JIA* 25, no. 6 (1886): 437.

22. Sprague, "Review," 65. For more on Felkin and acclimatization, see Livingstone, "Tropical Climate."

23. William Brinton, *On the Medical Selection of Lives for Assurance* (New York: John Hopper, 1863), 29.

24. See, for example, Brockbank, *Life Insurance*.

25. Castellani and Chalmers, *Manual of Tropical Medicine*, 127.

26. Martin, *Influence Of Tropical Climates*, 281; William Brinton, *On the Medical Selection of Lives for Assurance* (New York: John Hopper, 1863), 29.

27. Cantlie, "Life Insurance," 111–12. See also Harrison, "Differences of Degree."

28. Sprague, "Review," 68. See also Lister, *Medical Examination*, 153.

29. Livingstone, "Tropical Hermeneutics."

30. Rupke and Wonders, "Humbolditan Representations."

31. Harrison, "Differences of Degree"; Rupke and Wonders, "Humboldtian Representations"; Rupke, "Adolf Mühry."

32. Rupke and Wonders, "Humboldtian Representations," 167. For Bordier, see Rupke and Wonders, "Humboldtian Representations"; for Berghaus, Camerini, "Heinrich Berghaus's Map of Human Diseases."

33. This was even as Hirsch argued for geographical rather than miasmatic explanations of disease. Barrett, "August Hirsch." For Fuchs, see Vaj, "Medical Geography."

34. Rupke and Wonders "Humboldtian Representations."

35. Anderson, "Geography, Race and Nation."

36. Harrison, "Differences of Degree."

37. See Kneale and Randalls, "Invisible Atmospheric Knowledges."

38. Sprague, "Review," 70.

39. Herbert Cecil Thiselton, "A Discussion of some points of Life Assurance Administration in respect of which Divergence of Practice exists: A Plea for Uniformity," *JIA* 31, no. 1 (1893): 33.

40. Agent's Instructions, Industrial Branch, Prudential Assurance Company, May 1882, Prudential Archive.

41. Francis de Havilland Hall, *The Medical Examination for Life Insurance, with Remarks on the Selection of an Office* (Bristol: John Wright, 1903), 81, 81–82.

42. Sceptre Minute books for 1875–1885, 1885–1894, 1894–1904, and 1904–1915, SC01/05/01/002, 003, 004, 005, London Metropolitan Archives.

43. See Fleming's and Livingstone's chapters (this volume); Anderson, "Geography, Race and Nation."

44. Harrison, "Differences of Degree"; Arnold, *Colonizing the Body*.

45. *Report of the Commissioners appointed to inquire into the Sanitary State of the Army in India* (London: HMSO, 1865), 166.

46. Harrison, "Differences of Degree."

47. Wheeler, "Limited Visions of Africa," 16; Endfield and Nash, "Good Site for Health"; Savage, "Tropicality Imagined."

48. Newbatt in Hardy and Rothery, "Mortality of Assured Lives," *JIA* 27, no. 3 (1888): 190.

49. Hardy and Rothery, "Mortality of Assured Lives," *JIA* 27, no. 3 (1888): 192, 195.

50. Martin, *Influence of Tropical Climates*, 86.

51. Martin, *Influence of Tropical Climates*, 131.

52. Martin, *Influence of Tropical Climates*, 467, 468.

53. Arthur H. Bailey, "On the rates of extra premium for foreign travelling and residence," *JIA* 15 (1869–1870): 94.

54. Thomson, "Address," 181.

55. Sprague, "Review," 66.

56. Brockbank, *Life Insurance*, 277.

57. Brockbank, *Life Insurance*, 273.

58. Castellani and Chalmers, *Manual of Tropical Medicine*, 132, 133.

59. Anderson, "Geography, Race and Nation."

60. Discussed in Anderson, "Geography, Race and Nation"; see Livingstone, "Tropical Hermeneutics" for the response of Brazilian medicine practitioners to European concerns with this line of argument.

61. See Anderson, "Geography, Race and Nation," for a discussion of Barratt's ideas.

62. Livingstone, "Tropical Hermeneutics."

63. Anderson, "Natures of Cultures."

64. Brockbank, *Life Insurance*, 123.

65. Charles Lyman Greene, *The Medical Examination for Life Insurance and Its Associated Clinical Methods, with Chapters on the Insurance of Substandard Lives and Accident Insurance* (London: Rebman, 1905), 365.

66. Lister, *Medical Examination*, 149.

67. Bouk, *How Our Days Became Numbered*.

68. Lister, *Medical Examination*, 148.

69. Lister, *Medical Examination*, 151.

70. Hardy and Rothery, "Mortality of Assured Lives," *JIA* 27, no. 3 (1888); Castellani and Chalmers, *Manual of Tropical Medicine*, 121.

71. Cantlie, "Life Insurance," 115.

72. Martin, *Influence Of Tropical Climates*, 217, 347–48.

73. P. M. Tait, "On the Mortality of Eurasians," *Journal of the Statistical Society of London* 27, no. 3 (1864): 340.

74. Porter, *Trust in Numbers*, 103.

6. The British Women's Emigration Association and the Climate(s) of South Africa

1. Mills, *Gender and Colonial Space*.

2. P. Tinkler, "Introduction," 218. For subaltern women's voices, see Mills, "Gender and Colonial Space"; Rafael, "Colonial Domesticity"; Stern, *Secret History of Gender*; King, *In the Clearing*, https://drum.lib.umd.edu/handle/1903/14525/. For "other" places, see McEwan, "Paradise or Pandemonium?"; McEwan, *Gender, Geography and Empire*; Alison Blunt, *Travel, Gender, and Imperialism*; Blunt and Rose, *Writing Women and Space*; Tinkler, "Introduction"; Riedi, "Teaching Empire," 1346.

3. Burton, *Burdens of History*; Callaway, *Gender, Culture and Empire*; Strobel, *European Women*; Haggis, "Ironies of Emancipation," 108–9; Dagut, "Gender."

4. Haggis, "Heart that Has Felt"; Haggis, "Ironies of Emancipation"; Endfield and Nash "Happy Is the Bride"; Endfield and Nash, "Good Site for Health."

5. Burton "White Woman's Burden."

6. Bush, "Right Sort of Woman"; Chilton, *Agents of Empire*; Pickles, "Forgotten Colonizers"; Pickles, "Link in 'The Great Chain.'"

7. Bush, "Edwardian Ladies," 277.

8. Chilton, *Agents of Empire*; Bush, "Right Sort of Woman."

9. Gowans, "Imperial Geographies," 427.

10. Chilton, *Agents of Empire*, 6. See also Kranidis, *Imperial Objects*; Kranidis, *Victorian Spinster and Colonial Emigration*.

11. For example, MacKenzie, *Propaganda and Empire*.

12. Van Helten and Williams, "Crying Need of South Africa."

13. Levitan, "Redundancy"; William Rathbone Greg, "Why are women Redundant?" *National Review* (April 1862): 434–60.

14. Levitan, "Redundancy," 363.

15. Davin, "Imperialism and Motherhood," 12.

16. Levitan, "Redundancy," 364.

17. Dagut, "Gender," 558.

18. Proponents of imperial migration saw migration as a means of rectifying demographic imbalances, reducing unemployment, and addressing concerns over "deterioration" of the urban proletariat in the domestic sphere, while also ensuring the presence of British populations in the colonies. See Kennedy, "Empire Migration." By the second half of the nineteenth century, however, there were fears that if British population did not increase fast enough in the colonies, then other populations from rival colonial empires would. In a nutshell, "population was power." Davin, "Imperialism and Motherhood," 10.

19. Bush, "Edwardian Ladies," 279.

20. In November 1885, Mrs. Ellen Joyce and Mrs. Adelaide Ross replaced Miss Louisa Hubbard at the head of the organization (GB106 1BWE, Women's Library Archives). Hubbard was a writer and social reformer and set up the Women's Emigration Society in 1880. Joyce and Ross were both widows of Anglican clergymen, while Lefroy was a clergyman's daughter. Joyce was antisuffrage and antifeminist in her speeches. Lefroy's correspondence with

emigrant women was "regularly plundered for good text for the Association's publications." Chilton, *Agents of Empire*, 30.

21. Bush, "Edwardian Ladies."

22. Blakeley, "Society for the Oversea Settlement," 424.

23. Riedi, "Teaching Empire," 1322.

24. Hamilton and Higman "Servants of Empire," 68.

25. Blakeley "Society for the Oversea Settlement," 421; Chilton, *Agents of Empire*, 23.

26. Riedi, "Teaching Empire," 1316–17.

27. Bush, "Right Sort of Woman," 391.

28. File 1, Box 69, D3287, Derbyshire Records Office (DRO).

29. Blakely, "Society for the Overseas Settlement of British Women."

30. Duncan, "Sites of Representation."

31. Ripley, *Races of Europe*.

32. Livingstone, "Race, Space and Moral Climatology," 168.

33. Bell, "Woman's Place."

34. Curtin, *Image of Africa*; Wheeler, "Limited Visions." In 1850 Smyth was the president of the Royal Geographical Society in London. He described Africa as "the least known and least civilised of any" regions of the world. See Barnett, "Impure and Worldly Geography," 242.

35. Curtin, "White Man's Grave."

36. Wheeler, "Limited Visions"; Kupperman, "Fear of Hot Climates," 213.

37. Livingstone, "Tropical Climate and Moral Hygiene," 106. See *Proceedings of the Royal Geographical Society* 13 (1891).

38. Crozier, "Sensationalising Africa," 394.

39. Bell, "Pestilence," 329.

40. McEwan, "Paradise or pandemonium?"; Jennings, "This Mysterious and Intangible Enemy"; Savage, "Tropicality Imagined."

41. Bell, "Pestilence," 329; Livingstone, "Race, Space and Moral Climatology"; Stepan, *Picturing Tropical Nature*.

42. Jennings, "This Mysterious and Intangible Enemy," 68.

43. Bell, "Woman's Place," 144; Chard, "Lassitude and Revival."

44. MacKenzie, *Imperialism and the Natural World*.

45. Avril M. Maddrell, "Empire, Emigration and School Geography," 380.

46. Bush, "Edwardian Ladies," 282–83.

47. *Imperial Colonist* 1, no. 1 (1901): 2, in File 1, Box 69, D3287, DRO.

48. Gell Family of Hopton, Papers relating to Mrs Edith Lyttleton Gell, File 1, Box 76, Relating to the United British Women's Emigration Society, 1900–1904, D3287, DRO (hereafter cited as Gell Family Papers with file and box number).

49. "British Women's Emigration Association Advice to Women and Girls Settling in South Africa," a pamphlet, "priced one penny by Mrs Adele de Ross published by the Offices of the South Africa Expansion Committee, 47 Victoria Street, Westminster, date 1903" (hereafter cited as Ross pamphlet), Gell Family Papers, File 4, Box 76, DRO.

50. Written to Susan, Countess of Malmesbury, *Imperial Colonist* 1, no. 1 (January 1902).

51. Louisa Rhodes to Edith Lyttleton Gell, April 27, 1904, Gell Family Papers, File 3, Box 76, DRO.

52. Ross pamphlet, Gell Family Papers, File 4, Box 76, DRO.

53. Gell FamilyPapers, File 1, Box 75, DRO.

54. SACS Report for 1905, 48–49, DRO; Blakely, "Society for the Overseas Settlement."

55. The Education Sub Committee of the South African Expansion, BWEA, by Mary F. S. Hervey, 12–13, Box 69, D3287, DRO.

56. "The Future of South Africa by Susan Countess of Malmesbury, Re South African Expansion Committee," *Imperial Colonist* 1, no. 1 (January 1902): 10–11, Box 69, D3287, DRO.

57. Gell Family Papers, File 1, Box 76, DRO.

58. Crozier, "Sensationalising Africa," 397. See also Livingstone, "Tropical Climate"; Kennedy, "Empire Migration."

59. Livingstone, "Tropical Climate," 361.

60. Lady Fawsley, *Imperial Colonist* 1, no. 1 (January 1902), Gell FamilyPapers, File 1, Box 69, DRO.

61. Ross pamphlet, Gell Family Papers, File 4, Box 76, DRO.

62. Johnson, "European Cloth," 530.

63. Ross pamphlet, Gell Family Papers, File 4, Box 76, DRO.

64. Ross pamphlet, 9, Gell Family Papers, File 4, Box 76, DRO; also Deacon "Politics of Medical Topography," 279.

65. Ross pamphlet, Gell Family Papers, File 4, Box 76, DRO.

66. Captain MFM Meikeljohn VC, "Natal before and after the War," *Imperial Colonist* 1, no. 4 (April 1902), D3287, DRO.

67. "Openings for Women in South Africa," *Imperial Colonist* (July 1904), in Box 76–77, D3287, DRO. See also *Imperial Colonist* 1, no. 11 (November 1902): 11.

68. *Imperial Colonist* 1, no 12 (December 1902), D3287, DRO.

69. *Imperial Colonist* 1, no 2, 10–11, D3287, DRO.

70. Krogulski, "Turning a Curse into a Blessing," 2.

71. Ross pamphlet, Gell Family Papers, File 4, Box 76, DRO.

72. Livingstone, "Tropical Climate."

73. Chard, "Lassitude and Revival."

74. Gell Family Papers, File 2 cont. 2, Box 76, DRO.

75. *Imperial Colonist* 1, no. 1 (January 1902).

76. *Imperial Colonist* 1, no. 9 (September 1902): 80–82, D3287, DRO.

77. *Imperial Colonist* 1, no. 1 (January 1902), D3287, DRO.

78. *Imperial Colonist* (July 1904): 73–74; Florence Wall, Bulawayo, to Gell, February 13, 1903, Gell Family Papers, Box 76, File 5, DRO.

79. Irene Shene to Gell, April 19, 1905, Gell Family Papers, File 5, Box 75, DRO.

80. Florence Dawson, Hostel at Rosebank, Cape Town, to Miss Tarleton, November 27, 1902, Gell Family Papers, File 3, Box 76, DRO.

81. Gell to Mrs Smith (potential employer), May 28, 1903, Gell Family Papers, Box 76, File 3, DRO.

82. Emily Fox to BWEA, May 24, 1904, Gell Family Papers, File 3, Box 76, DRO.

83. Margaret Grogan, "The Emigrants Return," Gell Family Papers, File 7, Box 76, DRO.

84. Bush, "Edwardian Ladies," 282.

85. Maddrell, "Empire, Emigration and School Geography," 380.

86. Van Helten and Williams, "Crying Need of South Africa."

87. Van Helten and Williams, "Crying Need of South Africa," 36.

7. Race and Rainmaking in Twentieth-Century Southern Africa

I am grateful to James Beattie and the editors of this volume for their helpful comments on earlier versions of this chapter.

1. Beattie, O'Gorman, and Henry, *Climate, Science and Colonization*; Edwards, *Vast Machine*, ch. 3; Fleming, *Historical Perspectives*; Grove, *Green Imperialism*.

2. On British imperial science and experts, see Tilley, *Africa as a Living Laboratory*; Beinart and Hughes, *Environment and Empire*; Drayton, *Nature's Government*; Dubow, *Commonwealth of Knowledge*; Hodge, *Triumph of the Expert*. For works that incorporate the influence of North America, particularly the United States, see Anderson, "Depression, Dust Bowl"; Beinart, "Soil Erosion"; Phillips, "Lessons from the Dust Bowl."

3. Fleming, *Fixing the Sky*. For an early history of popular interest in weather and climate engineering, see Kutzleb, *Rain Follows the Plow*.

4. Dubow, "Earth History," 116.

5. Stephen Legg, "Debating the Climatological Role of Forests in Australia, 1827–1949: A Survey of the Popular Press"; Kristy Douglass, "'For the Sake of a Little Grass': A Comparative History of Settler Science and Environmental Limits in South Australia and the Great Plains"; James Beattie, "Science, Religion, and Drought: Rainmaking Experiments and Prayers in North Otago, 1889–1911," all in Beattie, O'Gorman, and Henry, *Climate, Science and Colonization*.

6. Kutzleb, "Rain Follows the Plow."

7. For a discussion of the role of geomorphology in popular climate knowledge, see Meredith McKittrick, "Theories of 'Reprecipitation.'"

8. The belief in the "drying out" of southern Africa is explored in Georgina Endfield and David Nash, "Missionaries and Morals: Climatic Discourse in Nineteenth-Century Central Southern Africa," *Annals of the Association of American Geographers* 92, no. 4 (2002): 727–42; Meredith McKittrick, "Talking about the Weather: The Language of Environmental Crisis in South Africa, 1915–1945," *Environmental History* 23, no. 1 (2018): 3–27.

9. William Jackson Humphreys, *Rain Making and Other Weather Vagaries* (Baltimore: Williams and Wilkins, 1926), 3–4. Fleming also notes the artificiality of Humphreys's distinction. Fleming, *Fixing the Sky*, 47.

10. Conservative religious whites are not discussed in this paper for lack of space, but their position as antimodern or antiscience also threatened to destabilize racial binaries. Those interested in artificial rainmaking sometimes highlighted the support they received from Afrikaners, who were assumed to be more conservative and religious than their English-speaking counterparts.

11. "Unity," letter to the editor, *Farmer's Weekly*, March 27, 1912, 156.

12. J. Fred Pentz, letter to the editor, *Farmer's Weekly*, May 28, 1913, 1056.

13. White women were almost never topics of these conversations (except as symbols of what needed to be protected), and they received voting rights only in 1930.

14. Charles Stewart, undated report on rainmaking experiments (probably October 1920), 4, file G.1, Director of the Weather Bureau (DWB) 1, National Archives of South Africa (hereafter NASA).

15. The twenty-fifth meridian is southern Africa's equivalent of North America's hundredth meridian, marking the approximate border of rain-fed agriculture. But this is an imperfect boundary in both places; in the case of South Africa, it ignores the distinctive "Mediterranean" climate near the Cape of Good Hope, for example.

16. Peires, *House of Phalo*, 33, 64.

17. D. S. Halacy, *The Weather Changers* (New York: Harper and Row, 1968), 59.

18. On autochthons as rainmakers, see Peires, *House of Phalo*, 24, 65.

19. See, for example, Isaac Schapera, ed., *The Bantu-Speaking Tribes of South Africa: An Ethnographical Survey* (London: Routledge and Kegan Paul, 1937), 234–35, 267.

20. Sanders, *Beyond Bodies*, xiv.

21. Moffat, *Missionary Labors*, 210–11.

22. Moffat, *Missionary Labors*, 213, 214.

23. Shaw, *Story*, 460.

24. One of the most famous reconstructed dialogues is between David Livingstone and a Tswana chief, which Livingstone included in David Livingstone, *Missionary Travels and Researches in South Africa* (London: John Murray, 1857), 23–25. The dialogue and the worldviews it represents are analyzed in Comaroff and Comaroff, *Of Revelation and Revolution*, 206–13, and in Brian Stanley, "The Missionary the Rainmaker: David Livingstone, the Bakwena, and the Nature of Medicine," *Social Sciences and Missions* 27, no. 2–3 (2014): 145–62.

25. Moffat, *Missionary Labors*, 216. See also David Carnegie, *Among the Matabele* (London: Religious Tract Society 1894), 36.

26. Shaw, *Story*, 464.

27. Shaw, *Story*, 464–66.

28. See Meredith McKittrick, *To Dwell Secure: Generation, Christianity and Colonialism in Namibia* (Portsmouth, NH: Heinemann 2002), 28; Moffat, *Missionary Labors*, 218; David Carnegie, *Among the Matabele* (London: Religious Tract Society 1894), 38–39.

29. "Hatfield to Bring 18 Inches of Rain; Then Goes to Africa," *Los Angeles Herald*, September 10, 1905, 6; R. Elliott, letter to the editor, *Farmer's Weekly*, January 22, 1913, 1687; F. Walsh, letter, *Farmer's Weekly*, February 12, 1913, 1982; Charles Stewart to Secretary for Agriculture, January 13, 1926, DWB 1, NASA; "Rain-Making: Hatfield's Scheme Examined: Criticisms by Chief Meteorologist," *Farmer's Weekly*, March 26, 1913, 175; Charles Stewart to Secretary for Agriculture, January 13, 1926, DWB 1, NASA. Legend credited Hatfield with making rain for the Boers during the South African War, but he never actually traveled to South Africa. See A. J. Liebling, *The Honest Rainmaker: The Life and Times of Col. John Stingo* (New York: Doubleday, 1953), 38.

30. "Ptolus," letter to the editor, *Farmer's Weekly*, June 15, 1921, 1457. Ptolus references an earlier letter from "Horlogie" that referred to correspondence with the US Weather Bureau.

31. N. J. Noble to C. K. Hall, August 17, 1921; F. S. Ratcliff & Co. to C. K. Hall, September 3, 1921, both in G.1, DWB 1, NASA.

32. Charles Stewart to B. S. Murray, Swart Ruggens Farmers' Association, April 21, 1926, G.1, DWB 1, NASA.

33. C. H. Hatfield to Valentine Magniac, August 3, 1925; W. T. Clarke to Valentine Magniac, July 22, 1925, both in G.1, DWB 1, NASA.

34. For Balsillie and Bancroft, see Fleming, *Fixing the Sky*, 112–13; C. K. Hall, "Cloud Control: Why Clouds Float; Electrons and Rain," *Farmer's Weekly*, December 8, 1920, 1909.

35. Kanthack to Hall, April 26, 1919, G.1, DWB 1, NASA.

36. Hall to Kanthack, May 3, 1919, G.1, DWB 1, NASA.

37. Stewart to Hall, May 22, 1919, G.1, DWB 1, NASA.

38. Hall to Stewart, June 13, 1919, G.1, DWB 1, NASA.

39. Hall to Stewart, July 31, 1919, G.1, DWB 1, NASA.

40. One of those most well-known was Professor Ernest Schwarz; see Meredith McK-ittrick, "An Empire of Rivers: Climate Anxiety, Imperial Ambition, and the Hydropolitical Imagination in Southern Africa, 1919–1945," *Journal of Southern African Studies* 41, no. 3 (2015): 485–504.

41. Newspaper clipping, *Rand Daily Mail*, January 7, 1927, file 11/162, GG 2291, NASA.

42. Guy Darell to Hawkins, December 27, 1926, file 11/162, GG 2291, NASA.

43. Hargreaves to G. M. Darell, January 8, 1927, file 11/162, GG 2291, NASA.

44. Hargreaves to Darell, January 8, 1927, file 11/162, GG 2291, NASA.

45. James Espy, *The Philosophy of Storms* (Boston: Charles C. Little and James Brown, 1841), xxiv.

46. Fleming, *Fixing the Sky*, 111.

47. "Hat in Seattle: Claims He Made Success in Yukon," *Dawson Daily News*, August 20, 1906, 4.

48. S. H. Boyle, letter, *Farmer's Weekly*, May 4, 1921, 833.

49. Carl Weidner, "The Fallacy of Schwarz's Rain-Making Magic: The Actual Cause of Our Inland Seas Drying Up" (Cape Town: Salesian Press 1925). Weidner reiterated his critique in 1945, after a government expedition was sent to investigate the scheme, expressing his hope that the "rain-making magic" would soon be definitively discarded. C. Weidner, letter, *Farmers' Weekly*, September 19, 1945, 39.

50. Mentor, "Can We Control the Climate?" *Farmer's Weekly*, May 17, 1933, 478–79. The word *Inyanga* actually refers to a traditional healer or diviner, not a rainmaker.

51. *Farmer's Weekly*, July 28, 1926, 1960. While *inyanga* depicted a specialist or professional in many of South Africa's indigenous languages (*inyanga yezulu*, or sky specialist, was one word for rainmaker), white South Africans gave it another, derogatory gloss, one that connoted primitivism.

52. R. T. A. Innes, "Induced Rainfall: A Memorandum upon Mr. C. K. Hall's Proposals for Artificial Rainmaking," *Farmer's Weekly*, February 25, 1920, 3547.

53. Charles Stewart, "Artificial Rain-Making Experiment," undated report (probably October 1920), 4, G.1, DWB 1, NASA.

54. C. K. Hall, "Cloud Control," *Farmer's Weekly*, December 8, 1920, 1911.

55. C. K. Hall, "Cloud Control," *Farmer's Weekly*, December 8, 1920, 1909.

56. R. T. A. Innes, "Induced Rainfall: A Memorandum upon Mr. C. K. Hall's Proposals for Artificial Rainmaking," *Farmer's Weekly*, February 25, 1920, 3547.

57. R. T. A. Innes, "Induced Rainfall: A Memorandum upon Mr. C. K. Hall's Proposals for Artificial Rainmaking," *Farmer's Weekly*, February 25, 1920, 3547.

58. "Desiccated Didymus," letter, *Farmer's Weekly*, January 28, 1920, 2922.

59. George Cox, undated report on rainmaking experiments (probably 1928), DWB 1, G.1, NASA.

60. George Cox, undated report on rainmaking experiments (probably 1928), 9–10, DWB 1, G.1, NASA.

61. Australian Government Memo, "Rainmaking," 4–5, file R2408, Department van Landbou Tengiese Dienste (LTD) 272, NASA.

62. T. Schumann (Weather Bureau) to Secretary of Agriculture, February 4, 1952, R2408, LTD 272, NASA.

63. Inexpensive was key; there was evidence by the 1950s that it was possible to create artificial rainfall—but in very small quantities, at very high cost.

64. Minute No. 783, October 30, 1953, A 462/26, South West Africa Administration (SWAA) 2139, National Archives Namibia (NAN).

65. Director of Works to Secretary for SWA, February 19, 1954, A462/62, SWAA 2139, NAN.

66. Director of Works to Secretary for SWA, February 19, 1954, A462/62, SWAA 2139, NAN.

67. On Bartaune, see Garth Benneyworth, "Bechuanaland's Aerial Pipeline: Intelligence and Counter Intelligence Operations against the South African Liberation Movements, 1960–1965," *South African Historical Journal* 70, no. 1 (2018): 108–23; Neil Parsons, "The Pipeline: Botswana's Reception of Refugees, 1956–1968," *Social Dynamics* 34, no. 1 (2008): 20. As Benneyworth notes, Bartaune was quite likely not a convert to African liberation movements but, rather, a spy for the South African security establishment.

68. A request that arrived the same month was less successful. Another group of farmers, who had formed the Reënvermeerdering Navorsings-Unie (Rainfall Propagation Research Society), also asked for financial support for their rainmaking endeavors. After some debate—and perhaps sensing that there might be no end to such requests—the executive committee announced that it could not offer them money without consulting the South African government Council for Scientific and Industrial Research. Reënvermeerdering Navorsingsunie to Secretary for SWA, February 20, 1954; Secretary for SWA to Reënvermeerdering Navorsingsunie, January 19, 1955, A462/62, SWAA 2139, NAN. The ultimate outcome of this request is unknown; what is notable in this context is that they were still offered the prospect of approval.

69. F. W. London to Secretary for Agriculture, June 6, 1958, R2408, LTD 272, NASA.

70. A. J. H. Stander to Secretary for Agriculture, 16 October 1959, R2408, LTD 272, NASA.

71. "Kunsmatige Reën," notes, July 16–20, 1959, R2408, LTD 272, NASA.

72. F. W. London to Finance Minister, November 4, 1962, R2408, LTD 272, NASA.

73. London to Secretary for Agriculture, October 17, 1962, R2408, LTD 272, NASA.

74. G. J. van de Waal to Prime Minister, South Africa, September 26, 1959, R2408, LTD 272, NASA.

75. H. F. Verwoerd to Water Affairs Minister P. M. K. le Roux, October 19, 1959, R2408, LTD 272, NASA.

76. P. D. Henning to Secretary for Agricultural Technical Services, November 5, 1959, R2408, LTD 272, NASA.

77. Director of Agricultural Services, Réunion, to South African Minister of Agriculture, June 2, 1957, R2408, LTD 272, NASA.

78. New Jansenville Farmers', Wool, and Mohair Growers' Association to Department of Agriculture, May 12, 1960, R2408, LTD 272, NASA.

79. Minister of Agriculture to P. H. Wessels, February 19, 1962; E. J. Mostert to Minister of Agriculture, July 10, 1962; both in R2408, LTD 272, NASA.

80. T. W. Stead to Minister of Agriculture, September 13, 1963; Southern Rhodesian Ministry of Agriculture to South African Department of Agricultural Technical Services, January 30, 1964; both in R2408, LTD 272, NASA.

8. Weather, Climate, and the Colonial Imagination

1. See, for example, Myddelton, *They Meant Well.* On the broader cultural and environmental legacies of the scheme, see Esselborn, "Environment, Memory."

2. Georgina Endfield and Samuel Randalls, "Climate and Empire," in Beattie, Mellilo, and O'Gorman *Eco-cultural Networks and the British Empire*.

3. For a review, see Mahony and Endfield, "Climate and Colonialism."

4. Turnbull, *Masons, Tricksters and Cartographers*.

5. See Bell, "Pestilence That Walketh." See also Endfield, this volume.

6. Hofstra, Haylock, New, Jones, and Frei, "Comparison of Six Methods," D21110.

7. Anderson, *Predicting the Weather*. See also Shapin, "Cordelia's Love"; Heffernan, "A Dream as Frail."

8. Beck, Forsyth, Kohler, Lahsen, and Mahony, "Making of Global Environmental Science," 1072.

9. Jasanoff, "Future Imperfect," 4.

10. See, for example, Storey, "Cecil Rhodes."

11. Hodge, *Triumph of the Expert*, 11 (original emphasis).

12. Tilley, *Africa as a Living Laboratory*. There are important parallels here with Richard Grove's groundbreaking work on earlier relationships between science and the colonial state, which were often conflictual and featured experts in fields like botany emerging as some of most influential critics of empire and as early movers in the development of conservationist thought and practice. See Grove, *Green Imperialism*.

13. Hugh Bunting, interviewed by David Pedgeley, *Distinguished Voices of Meteorology*, Royal Meteorological Society, Reading, England (1997). See http://www.archive.rmets.org/about-us/history-society/distinguished-voices-audio.

14. Wood, *Groundnut Affair*.

15. Quoted in Wood, *Groundnut Affair*, 252.

16. Frank Samuel, quoted in "Groundnut Scheme Director Interviewed," *Tanganyika Standard*, nd., 1947, cutting included with circular letter by C. H. Thornley, May 8, 1947, in MSS Afr. s. 1864, File 1, Box 1, Albert Walter Papers, University of Oxford Special Collections (UOSC).

17. Hogendorn and Scott, "East African Groundnut Scheme," 99.

18. Scott, *Seeing like a State*.

19. Hodge, *Triumph of the Expert*, 211; Wakefield to Samuel, August 1, 1946, UAC 1/15/4/2/1, Unilever Archives, Port Sunlight, UK.

20. "Groundnut Scheme Director Interviewed," *Tanganyika Standard*, nd., 1947 (cutting included with circular letter by C. H. Thornley, May 8, 1947, in MSS Afr. s. 1864, File 1, Box 1, Albert Walter Papers, UOSC).

21. Ministry of Food, Command 7030, "East African Groundnut Scheme," 19, MAF 83/1787, The National Archives, Kew, London (TNA).

22. David Scott-Moncrieff, "More Fats for Us," magazine cutting, 1947, in UAC 1/15/3/3, Unilever Archives.

23. David Martin to W. A. Faure, July 13, 1946, 1, UAC 1/15/4/2/1, Unilever Archives. See also John Rosa to Dawson, March 13, 1947, MAF 83/1775, TNA. For organizing labor, see Rizzo, "What Was Left?" McCloskey observes that the emergence of agricultural insurance in England coincided with the individualization and centralization of farming plots and the move away from spatially extensive collective ownership whereby risk could be spread around. See McCloskey, "Open Fields of England."

24. Martin to Faure, July 13, 1946, UAC 1/15/4/2/1, Unilever Archives.

25. Martin to Faure, June 24, 1946, UAC 1/15/4/2/1, Unilever Archives.

26. For Marenga Makali, see Maddox, "Networks and Frontiers," 446.

27. Ministry of Food, "East African Groundnuts Scheme: Report by Special Section," December 5, 1946, UAC 1/15/2/2/4, Unilever Archives.

28. Sites in northern Rhodesia had also been earmarked for this second wave.

29. Wood, *Groundnut Affair*, 37.

30. Walter to Wakefield, May 19, 1947, UAC 1/15/4/1/4, Unilever Archives.

31. Hugh Bunting, interviewed by David Pedgeley, *Distinguished Voices of Meteorology*, Royal Meteorological Society, Reading, England, 1997. See http://www.archive.rmets.org/about-us/history-society/distinguished-voices-audio.

32. Annexure F.1, UAC 1/15/4/1/4, Unilever Archives. Bunting later recollected that "we had to go by guess and by God and the nature of the vegetation." Hugh Bunting, recorded interview with David Pedgeley, *Distinguished Voices of Meteorology* (Reading: Royal Meteorological Society, 1997), 15.

33. Walter to S. Herbert Frankel, 6 October 1950, MSS Afr. s. 1864, File 4, Box 1, UOSC.

34. Walter to Wakefield, May 5, 1947, MSS Afr. s. 1864, File 1, Box 1, UOSC.

35. See Kenworthy, "Albert Walter, Part I."

36. Walter, *Sugar Industry of Mauritius*, 131.

37. R. H. Hooker, "The Sugar Industry of Mauritius: A Study in Correlation, Including a Scheme of Insurance of the Cane Crop against Damage Caused by Cyclones. By A. Walter, F.R.A.S. London: A. L. Humphreys, 1910. 228 pp. and 22 plates," *Quarterly Journal of the Royal Meteorological Society* 38, no. 164 (August 15, 1912): 327–29; Kenworthy, "Albert Walter, Part I."

38. Walter, "Echoes of a Vanishing Empire," 133, MSS Brit Emp r. 9–10, UOSC.

39. Rothschild, "Recommendations for the Scientific Department," UAC 1/15/4/1/4, Unilever Archives.

40. Walter, "Plan for Climatic Survey of Ground-Nut Project," MSS Brit Emp s. 1864, UOSC.

41. Walter, "Presidential Address," *Proceedings of the Nairobi Scientific and Philosophical Society* 2 (1948): 5–10.

42. Grove, *Green Imperialism*, 192.

43. Stockland, "Policing the Oeconomy of Nature."

44. Hodge, *Triumph of the Expert.*

45. Walter to Wakefield, June 28, 1947, MSS Afr. s. 1864 Box 1, Commonwealth and African Collection, University of Oxford Special Collections (UOSC).

46. Bunting to Walter, 21 October 1947, MSS Afr. s. 1864 Box 1, Commonwealth and African Collection, UOSC.

47. Bunting to Walter, April 14, 1948, MSS Afr. s. 1864 Box 1, Commonwealth and African Collection, UOSC.

48. Bunting to Walter, September 23, 1948, MSS Afr. s. 1864 Box 1, Commonwealth and African Collection, UOSC.

49. David Brunt, holder of the first UK chair in meteorology at Imperial College, was subsequently taken on as an advisor to the Overseas Food Corporation (OFC) in London.

50. Rosa to Dawson, March 13, 1947, MAF 83/1775, TNA.

51. G. W. Raby to OFC Board, September 6, 1950, MAF 83/1988, TNA.

52. See Fleming, *Fixing the Sky*; Harper, *Make It Rain.*

53. Stedman to Bishop, October 10, 1950, MAF 83/1988, TNA.

54. Telegram, MAF 83/1988, TNA.

55. Bunting, "Agricultural Research."

56. Davies, Hepburn, and Samson, "Report on Experiments, 1951," 2.

57. John Wakefield, "Address to the Colonial Office Summer School, Cambridge," August 1948, MSS Afr. S. 353, Wakefield Papers, UOSC.

58. Davies, Hepburn, and Samson, "Report on Experiments, 1951," 7.

59. F. H. Ludlam, "The Production of Showers by the Coalescence of Cloud Droplets," *Quarterly Journal of the Royal Meteorological Society* 77, no. 333 (July 1951): 402–17.

60. Davies, "World Weather Watch: Memoirs of a Welsh Scientist," MSS Eng. C. 4667, UOSC.

61. Davies, Hepburn, and Samson, "Report of Experiments, 1952," 14.

62. Davies, "Experiments on Artificial Stimulation," 256.

63. Anderson, *Predicting the Weather*.

64. Davies, "Experiments on Artifical Stimulation," 256.

65. Quoted in Kenworthy, "Albert Walter, Part II," 48.

66. Wright to Walter, 18 October 1950, MSS Afr. s. 1864, Box 1, UOSC.

67. A. Walter, "Notes on the Utilization of Records from Third Order Climatological Stations for Agricultural Purposes," *Agricultural Meteorology* 4, no. 2 (March 1967): 137–43.

68. Walter, "The Significance of Marginal Climates in the Tropics," MSS Brit Emp s. 391, UOSC.

69. Walter, "Presidential Address," *Proceedings of the Nairobi Scientific and Philosophical Society* 2 (1948): 5–10.

70. Daniels, "Geographical Imagination," 183.

9. Darwinian Hippocratics, Eugenic Enticements, and the Biometeorological Body

1. See, for example, McMichael et al., *Climate Change and Human Health*; Centers for Disease Control and Prevention (CDC), "Climate Change and Health Effects."

2. Rosenberg, "Epilogue," 664. For a useful overview see Hannaway, "Environment and Miasmata."

3. Rosenberg, "Epilogue," 666.

4. US Natural Resources Defense Council, http://www.nrdc.org/health/climate/.

5. McMichael, "Global Climate Change," 8; Haines et al., "Climate Change and Human Health," 585.

6. By way of comparison, see Osborne and Fogarty, "Medical Climatology"; Osborne, "Resurrecting Hippocrates."

7. Petersen, *Man Weather Sun*, 237. In a related arena, Meloni has recently explored the potential political implications of the renewed emphasis (not least in epigenetics) on the "formative influence of the environment on the hereditary material." Meloni, *Political Biology*, 2.

8. "Obituary: William F. Petersen, M.D., 1887–1950," *AMA Archives of Pathology* 51 (1951): 130–32. See also Sargent, *Hippocratic Heritage*, appendix 8, "William Ferdinand Petersen," 524–58. There is a brief sketch of Petersen in Bischof, "Introduction to Integrative Biophysics," 39–40.

9. See, for example, Bouma, "Short History of Human Biometeorology"; Sargent, *Hippocratic Heritage*; Tout, "Biometeorology."

10. For example, Petersen, *Protein Therapy*; Petersen and Levinson, *Skin Reactions, Blood Chemistry*.

11. Petersen, *Patient and the Weather, Vol. 2*, 11.

12. Petersen, *Man Weather Sun*, 237. See also Petersen, *Lincoln–Douglas*, 131–150.

13. Petersen, *Man Weather Sun*, 284, 286.

14. Petersen, *Patient and the Weather*, Vol 3, 327; Petersen, *Patient and the Weather*, Vol. 1, Part 2, 159.

15. Petersen, *Man Weather Sun*, 239.

16. Petersen, *Man Weather Sun*, xii.

17. Petersen, *Man Weather Sun*, 421.

18. Petersen, *Lincoln–Douglas*, cited on dust jacket.

19. Petersen, *Lincoln–Douglas*, 12.

20. Quoted in Sargent, *Hippocratic Heritage*, 546.

21. Petersen, *Hippocratic Wisdom*.

22. Sargent, *Hippocratic Heritage*, 389–90.

23. Tisdale, "Confessions of a Cosmic Resonator"; "Science: Weather as Destiny," *Time Magazine*, August 2, 1943; *Rosen, "Guestwords."*

24. Rosen, *Weathering*.

25. Sargent, "William F. Petersen Foundation," 105.

26. See, for example, Fleming, *Historical Perspectives*; Livingstone, "Changing Climate"; Davis, "Coming Desert." For biographical details, see Martin, *Ellsworth Huntington*.

27. Huntington, *Civilization and Climate*, v, 2, 50.

28. Huntington, "Control of Pneumonia"; Huntington, "Influenza," 471. In discussing climate and upper respiratory infections, Winslow and Herrington insisted that the subject had been "illuminated by the fruitful contributions of the late Ellsworth Huntington" but "confused and misrepresented by other and less scientific writers." Winslow and Herrington, *Temperature and Human Life*, ix. On Winslow, see Viseltear, "C.-E.A. Winslow."

29. Huntington, *Weather and Health*, 9, 10, 120.

30. Huntington, Williams, and van Valkenburg, *Economic and Social Geography*, 118, 120.

31. Meyer claims that Mills was "second only to Huntington as a promoter of determinism." Meyer, *Americans and Their Weather*, 171.

32. See the discussion in Sargent, *Hippocratic Heritage*, 408–21.

33. See, for example, Mills, *Air Pollution*; Mills, *This Air We Breathe*.

34. Mills, *Climate Makes the Man*, 29.

35. Mills, *Climate Makes the Man*, 50, 65–72; Mills, *Living with the Weather*, 8; Mills, *Medical Climatology*, 124–63.

36. Mills, *Climate Makes the Man*, 36–37; Mills, *Medical Climatology*, 34–52.

37. Mills, *Climate Makes the Man*, 59, 62; Mills, *Living with the Weather*, 75, 82.

38. Mills, *Medical Climatology*, 89–175.

39. Huntington, "Review of *Medical Climatology*," 542.

40. Spengler, "Review of *Medical Climatology*," 447.

41. Anonymous, "Review of *Medical Climatology*," Journal of the American Medical Association, 113 (1939): 2263.

42. "Links Fascism with Weather," *Mason City Globe-Gazette*, Thursday, March 27, 1941.

43. Mills, *Climate Makes the Man*, 8, 10.

44. See Mills, *Living with the Weather*, chapter 11, "Weather Stimulation, Human Energy, and Business Cycles."

45. Mills, *Living with the Weather*, 120, 133.

46. Mills, *Living with the Weather*, 3; Mills, *Climate Makes the Man*, 27; Mills, *Medical*

Climatology, 12. On the subject of tropicality, see Arnold, *Problem of Nature*; Livingstone, "Moral Discourse of Climate"; Livingstone, "Race, Space and Moral Climatology."

47. Mills, *Living with the Weather*, 188.

48. Mills, *Medical Climatology*, 14.

49. Mills, *Medical Climatology*, 153; Mills, *Living with the Weather*, 13, 14. Later he confirmed that "From the standpoint of human achievement and basic progress, present day African climates offer little encouragement." Mills, *World Power amid Shifting Climates*, 64.

50. Mills, *Living with the Weather*, 5.

51. Mills, *World Power amid Shifting Climates*, 128.

52. Mills, *Climate Makes the Man*, 8.

53. Mills, *Climate Makes the Man*, 141.

54. Petersen, *Man Weather Sun*, 362.

55. Petersen, *Man Weather Sun*, 365.

56. Petersen, *Man Weather Sun*, 372.

57. Petersen, *Man Weather Sun*, 365, 364.

58. Huntington, *Season of Birth*, v.

59. At times Huntington displayed Neo-Lamarckian tendencies, at others he resorted to the more selectionist vocabulary of the Darwinians. There is reason to think that he moved between these perspectives to suit the topic at hand. See the discussion of his Lamarckism in Campbell and Livingstone, "Neo-Lamarckism."

60. Huntington, *Season of Birth*, 290 (further references to this work will be given parenthetically in the text).

61. Huntington, *Civilization and Climate*, 129. Huntington's studies on temperature, humidity, and industrial production were positively reviewed in Winslow and Herrington, *Temperature and Human Life*, 228–41.

62. Huntington, *Civilization and Climate*, 129, 131.

63. Huntington, *World-Power and Evolution*, especially chapters 5 and 8.

64. Huntington, "Control of Pneumonia," 6, 23.

65. Sargent, *Hippocratic Heritage*, 394–408. See also Dalén, *Season of Birth*; Nonaka, "Effect of Delivery Season"; Miura, "Seasonal Atmospheric Factors"; Norris and Chowning, "Season of Birth and Mental Illness"; Janerich, Porter; and Logrillo, "Season of Birth and Neonatal Mortality."

66. There are numerous biographical sketches, among them, Droessler, "Helmut Landsberg"; Baer, "Helmut E. Landsberg"; Liebowitz, "Landsberg, Helmut Erich"; Baer, Canfield, and Mitchell, *Climate in Human Perspective*. See also Henderson, "Dilemma of Reticence."

67. Landsberg, *Weather and Health*, 102.

68. Landsberg, *Weather and Health*, 39, 64, also 50.

69. According to David Tout, Tromp "probably did more than any other person to establish biometeorology as a distinct discipline." Tout, "Biometeorology," 474. A brief review of the International Society of Biometeorology is available in Weihe, "Review." See also personal historical reflections on the society by Folk, "International Society of Biometeorology."

70. Tromp, *Biometeorology*, chapter 3.

71. See Folk, "International Society of Biometeorology"; Tout, "Biometeorology"; Weihe, "Review."

72. Sargent, *Hippocratic Heritage*, xxxviii.

73. Sargent, "Nature and Scope," 1, 5.

74. Dubos cited from Audy and Dunn, "Health and Disease," 337.

75. Sargent, "Nature and Scope," 18.

76. Galton, *Hereditary Genius*; Galton, *Inquiries into Human Faculty*.

77. Havelock Ellis,"Eugenics and Genius," chapter 13 of *Essays in War-Time: Further Studies in the Task of Social Hygiene* (London: Constable, 1916), 148, 149–50.

78. Among those who considered genius a pathology were Wilhelm Lange-Eichbaum, Arthur Jacobson, and Radoslav A. Tsanoff. See Lange-Eichbaum, *Problem of Genius*; Jacobson, *Genius*; Tsanoff, *Ways of Genius*. See also Lowenberg, "Wilhelm Lange-Eichbaum."

79. Petersen, *Man Weather Sun*, 362.

80. Petersen, *Man Weather Sun*, 373, 363.

81. These are identified in Lavery, "Geography and Eugenics."

82. Huntington, *Season of Birth*, 15, 16 (further references appear parenthetically in the text). In presenting his findings on birth season, Huntington drew on the work of figures like the reproductive biologist Walter Heape FRS, who pioneered embryo transfer experiments; the Finnish philosophical anthropologist Edvard Westermarck, author of *The History of Human Marriage*; the Italian statistician and demographer Corrado Gini, who held the presidency of several eugenics societies; the British travel writer Alleyne Ireland; the Italian physician and criminologist Cesare Lombroso; and Charles Kassel, the Fort Worth writer and founding partner in the law firm of Ledgerwood and Kassel, who wrote articles on the subject of genius, including "Birth Months of Genius."

83. Huntington, Williams, and van Valkenburg, *Economic and Social Geography*, 126.

84. Huntington to Madison Grant, August 27, 1925, Series 3, Box 50, no. 1630 General Correspondence, Ellsworth Huntington Papers, Sterling Memorial Library, Yale University, New Haven, quoted in Lavery, "Geography and Eugenics," 86.

85. Huntington and Whitney, *Builders of America*, 1, 283. This work was reviewed by George S. Schulyer who took the opportunity to attack its racial and class bias and to argue, as Ewa Barbara Luczak puts it, that "eugenics served as a front for a group of privileged Americans to protect their racial and class interests." Luczak, *Breeding and Eugenics*, 174.

86. Huntington, *Tomorrow's Children*, 94–95.

87. Mills, *Medical Climatology*, 139.

88. Mills, *Living with the Weather*, 149–50.

89. Mills, *Living with the Weather*, 171–72.

90. Mills, *Climate Makes the Man*, 114. In the United States these examinations routinely included a blood test to detect syphilis and became increasingly common during the late 1930s. Some of this is discussed in Brandt, *No Magic Bullet*.

91. Mills, *World Power amid Shifting Climates*, 39.

92. Mills, *Medical Climatology*, 34, 35. See also Mills, "Geographic and Time Variation"; Mills, "Climatic Effects on Growth and Development."

93. Mills, *World Power amid Shifting Climates*, 37.

94. Mills, *Living with the Weather*, 22, 38.

95. Mills, *Medical Climatology*, 271–72.

96. Fleming, *Historical Perspectives*, 95.

10. Civilization, Climate, and Ozone

Epigraph: Huntington, *Civilization and Climate*, 294.

1. This sketch of Huntington's life is drawn from Martin, *Ellsworth Huntington*; Chappell, "Huntington and His Critics"; Johnson and Malone, *Dictionary of American Biography*.

2. Champlin, *Raphael Pumpelly.*

3. Huntington, *Pulse of Asia*, 6.

4. Huntington, *Pulse of Asia*, 385, 359.

5. For details, see Fleming, *Historical Perspectives*, chapter 8.

6. Martin, *Ellsworth Huntington*, 73.

7. R. S. Woodward to E. S. Dana, November 16, 1912; E. Brückner to Dana, November 26, 1912; A. Penck to Dana, November 26, 1912; Dana to H. Oertel, December 30, 1912, all cited in Martin, *Ellsworth Huntington*, 85–87.

8. National Research Council, *Geography of Europe*, 87.

9. Martin, *Ellsworth Huntington*, 159–60.

10. Huntington to William Morris Davis, April 5, June 25, 1914, Folder 664, Box 32, Series 3, MS 1, Ellsworth Huntington Papers, Yale University (hereafter Huntington Papers).

11. Harlan Barrows to Isaiah Bowman, June 15, 1921; and Bowman cited in Martin, *Ellsworth Huntington*, 134, 142.

12. William Jackson Humphreys to National Book Buyers Service, February 5, 1923, cited in Martin, *Ellsworth Huntington*, 196.

13. Huntington to A. R. Wallace, mimeo, marked answered December 4, 1913, add. MS 46438, 273–278, Alfred Russell Wallace Papers, Department of Manuscripts, British Library, London.

14. Huntington, *Principles of Human Geography*, 347.

15. Huntington, *Principles of Human Geography*, 347.

16. *New York Times* (January 4, 1939), 20.

17. Kullmer quoted in Martin, *Ellsworth Huntington*, 102–3. Kullmer was the author of a chapter in Huntington, *The Climatic Factor as Illustrated in Arid America*, Carnegie Institution of Washington publication no. 192 (Washington, DC: Carnegie Institution of Washington, 1914). Kullmer also wrote two monographs on storm tracks: Charles J. Kullmer, "The Latitude Shift of the Storm Track in the 11-Year Solar Period; Storm Frequency Maps of the United States, 1883–1930," *Smithsonian Miscellaneous Collections* 89, no. 2 (1933): 34; Charles J. Kullmer, "A Remarkable Reversal in the Distribution of Storm Frequency in the United States in Double Hale Solar Cycles, of Interest in Long-Range Forecasting," *Smithsonian Miscellaneous Collections* 103, no. 10 (1943): 20.

18. Avilés, *Taken by Storm.*

19. Charles A. Peters, "Ozone in the '38 Hurricane," *Science* 90 (November 24, 1939): 491; H. N. Glick, "Hurricane Intelligence," *Science* 91 (May 10, 1940): 450.

20. Lois C. Barail to Huntington, February 11, 1939, Folder 3290, Box 82, Series 3, Huntington Papers.

21. Press release, *New York Times*, September 5, 1939, 22.

22. Louis C. Barail to Huntington, April 26, June 3, 1939; International Congress of Biophysics, Biocosmics and Biocracy, congress announcement and registration form and congress program (with Huntington's annotations), Folder 787, Box 68, Series 3, Huntington Papers.

23. *New York Times*, September 12, 1939, 27.

24. Ellsworth Huntington, "Climatic Pulsations and an Ozone Hypothesis of Libraries and History," *Conservation of Renewable Natural Resources* (Philadelphia: University of Pennsylvania, 1941), 99–147; Huntington, *Mainsprings of Civilization*, 504–34.

25. Typescript entitled "Ozone, Health and Progress," 111 pages, dated September–October 1939, Group 1 V: 49: 362, Huntington Papers.

26. Ellsworth Huntington, "Ozone, Weather, and Health," Folder 595, Box 64, Series 5, Huntington Papers.

27. F. A. R. Russell, "The Atmosphere in Relation to Human Life and Health," *Smithsonian Miscellaneous Collections* 39 (1896): 9–10; Leonard Hill, "The Physiological Influence of Ozone," *Smithsonian Institution Annual Report for 1911* (Washington, DC, 1912), 617–28.

28. *Health Power Beauty*, Elco Electric Health Generators, catalog no. 17 (Chicago: Lindstrom, 1929), 60.

29. Huntington, "Ozone, Health and Progress," 65, Folder 362, Box 49, Series 5, Huntington Papers.

30. C. A. Mills, Professor of Experimental Medicine, University of Cincinnati, to Huntington, 15 October 1941, Folder 411, Box 56, Series 5, Huntington Papers.

31. McGregor, "Huntington's Cyclonic Man."

32. US Department of Education, Office of Educational Research and Improvement, http://www.ed.gov/pubs/parents/Geography/place.html/; *CIA World Factbook, U.S. State Department Travel Facts*, https://www.cia.gov/library/publications/resources/the-world-factbook/attachments/travel/CF-travel-facts.pdf.

33. Report of the Workshop "On Assessing Winners and Losers in the Context of Global Warming," St. Julians, Malta, 18–21 June 1990.

34. Gore, *Earth in the Balance*, 56–80.

35. For Definition 1, see Walter B. Cannon, "Biocracy: Does the Human Body Contain the Secret of Economic Stabilization?" *Technology Review* 35 (1933): 203–6, 227; also Cannon, *The Wisdom of the Body* (New York: Norton, 1932), esp. 287–306; Cannon, "The Body Physiologic and the Body Politic," *Science* 93 (1941): 1–10; Caldwell, "Biocracy and Democracy"; Lee, "Political Homeostasis." For Definition 2, see Robert Jay Lifton, *Nazi Doctors*; Proctor, *Racial Hygiene*. For Definition 3, see Leonardo Boff, "Plenary Speech on the First Day of the Earth Charter Continental Meeting," Mato Grosso, Brazil, December 1998, http://www.earthcharter.org/resources/speeches/boff_en.htm/.

36. Huntington to William Morris Davis, April 5, 1914, series 3, box 32, folder 664, Huntington Papers; Arnold J. Toynbee, foreword to Martin, *Ellsworth Huntington*, ix.

37. Arnold J. Toynbee, foreword to Martin, *Ellsworth Huntington*, ix.

38. Simon, Behmand, and Burke, *Teaching Big History*; Brooke, *Climate Change*.

39. "Ellsworth Huntington," encyclopedia.com, https://www.encyclopedia.com/people/science-and-technology/geography-biographies/ellsworth-huntington; Hackett Fischer, "Climate and History," 248; Dorn, *Geography of Science*, xviii.

40. On climate and society, see, among others, Glantz, *Climate Affairs*; Hulme, *Why We Disagree*; and Fleming and Jankovic, *Klima, Osiris*, 26.

11. The Shaded Modernism of the Global Interior

1. While the concept of thermal comfort was developed in the air-conditioning industry at the beginning of the twentieth century, it was not systematically brought into architecture until the mid 1950s, in large part through reflection on the buildings herein described. See Olgyay and Olgyay, *Solar Control with Shading Devices*. See also Cooper, *Air-Conditioning America*; Murphy, *Sick Building Syndrome*.

2. Sloterdijk, *Spheres II*, 961. See also Sloterdijk, *In the World Interior*.

3. In Dipesh Chakrabarty's well-known text "The Climate of History: Four Theses," he

notes that "the mansion of modern freedom stands on an ever-expanding base of fossil-fuel use"; elegantly conflating the architectural and political aspirations of the last two centuries, and suggesting their inherent contradictions. Chakrabarty, "The Climate of History: Four Theses." See also Mitchell, *Carbon Democracy*.

4. The conditioning of the built interior has in this sense been a primary engine in the Great Acceleration, having, as it does, a massive and destructive reliance on fossil fuels. See McNeil and Engelke, *Great Acceleration*; Bonneiul and Fressoz, *Shock of the Anthropocene*.

5. Coelho da Souza, *Irmaos Robertos Arquitetos*. The youngest brother, Marcelo, didn't join the firm until 1945.

6. Of note in this broad trend: Crinson, *Modern Architecture*; Gyger and Del Real, *Latin American Modern Architectures*; Lu, *Third World Modernism*.

7. This specific thesis relative to Le Corbusier and climate has played out in a number of recent texts, including Kevin Bone, *Lessons from Modernism: Environmental Design Strategies in Architecture, 1925–1970* (New York: Monacelli Press, 2014); Porteous, *New Eco-Architecture*; Requena-Ruiz, "Building Artificial Climates."

8. This is developed, among numerous other places, in Le Corbusier, *Towards an Architecture*. The persistence of the diagram's disciplinary significance can be read in its centrality to the influential article by Somol and Whiting, "Notes around the Doppler Effect."

9. This was the foundation of Le Corbusier's celebrated "Five Points for Architecture," discussed in *Towards an Architecture* and in *L'Esprit Nouveau* in the 1920s. The five points are the use of pilotis to elevate the building off the ground; the principle of the open, or free, plan and the principle of the free design of the façade (both now possible with the use of steel and concrete); the horizontal window; and the roof garden.

10. See Overy, *Light, Air, and Openness*; also Porteous, *The New Eco-Architecture*.

11. Le Corbusier, *Precisions*, 37.

12. Le Corbusier, "Petite historique du brise-soleil extrait de l'oeuvre de Le Corbusier," in *Architecture d'Aujourd'hui* 18 (September 1947): 10–11. The historian Reyner Banham referred, in the 1960s, to the brise-soleil as "one of Le Corbusier's most masterly inventions." Banham, *Well-Tempered Environment*, 158. There are numerous other possible origins, such as Stamo Papadaki in South America, Antonin Raymond in Japan. Papadaki is celebrated in Jeffrey Aronin, Porteous, and others as the first to use the brise-soleil in his proposed Christopher Columbus Memorial Lighthouse of 1928. See Aronin, *Climate and Architecture*. Aronin's text is a collection of climate-related articles from *Progressive Architecture* in the immediate postwar years, a period in which Papadaki was editor of the journal.

13. Passanti, "Modern, the Vernacular," 153; see also Frampton, "Le Corbusier and Oscar Niemeyer"; Frampton, *Le Corbusier*.

14. In Algeria, as in Brazil, the northern façade is the one that requires the most solar protection.

15. On the complexities of architecture and the international universalist premise, see Hitchcock, *International Style*; Ciucci, "Invention of the Modern Movement"; Mumford, *CIAM Discourse*. Light and air were essential aspects of the 1930 CIAM discussion on "Rational Lot Development." See Mumford, *CIAM Discourse*; Gropius, "Houses, Walk-Ups?" (a summary of his presentation at the 1930 meeting).

16. As the Brazilian historian Henrique Mindlin wrote in 1957: "The brise-soleil (in Portuguese *quebra-sol* or 'sun-breaker,' but that the French expression is commonly used indicates its direct derivation from Le Corbusier) has been applied in Brazil in the greatest variety of

ways." Mindlin, *Modern Architecture*, 12. That the French brise-soleil terminology was used in the mid-century period throughout Brazil is suggestive of the importance of this strategy as a symbol of regional interpretation of a seemingly European modernism. Mindlin's text was introduced by the prominent modernist architectural historian and critic Sigfried Geideon, who had also been the guest editor for the *Architecture d'Aujourd'hui* issue on Brazil in 1947; in numerous places he cites Brazil and Sweden as centers of the regional development of modernist ideals.

17. Costa was removed from his ENBA post after a little over a year, because of strong counter reformist figures who were still holding sway at the school. Williams, *Culture Wars*, 54.

18. Le Corbusier drew a low-slung longer building that he thought more appropriate to the region—it is not unlike, in its general approach, the Museu de Arte Moderna do Rio de Janeiro that Affonso Reidy later designed for the same site. See Patricio del Real, "Paternity Rights: The Brise-Soleil and the Sources of Modernity in the Ministry of Education and Health in Rio de Janeiro, Brazil," *ACSA Proceedings*, Portland, Oregon, 2002. As the title indicates, del Real is concerned with understanding the precise role of Le Corbusier's influence, read through American interest in the building and in the *Brazil Builds* exhibition at MoMA in 1943. Although much North American research often suggested that Le Corbusier had an important role, Brazilian historians have more recently downplayed the specific influence of Le Corbusier on the MES, while still acknowledging his importance to the culture of architecture in Rio more generally. See, for example, Segawa, *Architecture of Brazil*, 84–87.

19. Rocha-Peixoto, "Prefacio," 15.

20. Barry Bergdoll, *Latin America in Construction*, 17. These "lessons from the 'underdeveloped' world," Bergdoll continues, are "useful even for the 'developed' world to contemplate."

21. Schneider, *"Order and Progress."*

22. The primary figure here, in Rio, is that of Oswaldo Cruz, who was essential to the spread of vaccines in Brazil and to public health in general. See Azevedo and Ferreira, "Dilemmas of a Scientific Tradition." On media and Cruz's practice, see Villela, de Mello, and Lowes de Lacerda, "Images of Public Health."

23. Brazil had also just attempted, in 1926, to become a part of the permanent council of the League of Nations. That this prospect was—in the end, and unexpectedly—rejected is seen by some historians as an important push toward the coup that brought Vargas to power. Williams, *Culture Wars*, 54.

24. Williams, *Culture Wars*, 54.

25. Numerous maps were drawn on these general terms of institutional racism and climatic determinism. Most of Brazil is here classified in a "Low" or "Very Low" condition of "human health and energy on the basis of climate," although Rio and São Paulo, along with the provinces that surround them, show the more promising "Medium" condition. S. F. Markham, *Climate and the Energy of Nations* (London: Oxford University Press, 1942), 21, 27, 31. See also Chang, "Thermal Comfort and Climatic Design."

26. Le Corbusier, *Precisions*, 66.

27. Richard Neutra, *Architecture of Social Concern*.

28. Richard Neutra, "A Planetary Test" (c. 1942) from the Richard and Dion Neutra Archives, University of California Los Angeles (unpublished).

29. See Escobar, *Encountering Development*.

30. On tropical architecture and other related developments, see Jackson, *Edwin Maxwell*

Fry and Jane Drew; Jiat-Hwee Chang, *A Genealogy of Tropical Architecture: Colonial Networks, Nature and Technoscience* (London: Routledge, 2016); Crinson, *Modern Architecture.*

31. Although the Olgyays and von Neumann were compatriots, I have found no evidence (in the Olgyay archive) of substantive interaction between their research projects or even of a casual meeting. Perhaps the disciplinary gap between the hard sciences and architecture (then closely affiliated, on Princeton's campus, with the arts) was too wide. The Olgyays arrived in Princeton in 1952; von Neumann died in 1957

32. James Marston Fitch, *American Building and the Forces that Shape It* (Boston: Houghton-Mifflin, 1948). This was revised and republished in 1972 as *American Building: The Environmental Forces that Shape It.* Fitch was a pioneer in historical preservation, establishing a department at Columbia. See Barber, "Modernism and Microclimatology." Fitch would start the Preservation Program at the Columbia University Graduate School of Architecture, Planning, and Preservation.

33. Fleming, *Inventing Atmospheric Science*, 193ff; Geiger, *Climate near the Ground*; Turner, "Weathering Heights," 11–14.

34. James Marston Fitch, "Microclimatology" in *Architectural Forum* 36, no. 2 (February 1947): 21. See also G. Manley, "Microclimatology: Local Variations of Climate Likely to Affect the Design and Siting of Buildings," *RIBA Journal* (May 1949): 317–23; W. E. Graham, "The Influence of Micro-climate on Planning," *Planning Outlook* (March 1949): 40–52; Helmut Landsberg, "Microclimatology," *Architectural Forum* 36, no. 2 (March 1947): 114–20.

35. Although the analyses included "some well known facts" such as relative sun angle in summer and winter, Siple insisted that "these have in many cases been consistently ignored in residential design." Paul Siple, "Climatic Criteria for Building Construction," *Proceedings of the Research Correlation Conference on Weather and the Building Industry* (Washington, DC: National Academy of Sciences, 1950): 7. The series of analyses developed by the American Institute of Architects (AIA) were published in the *Bulletin of the American Institute of Architects* every two months from September 1949 to January 1952.

36. Siple, "Climatic Criteria," 14.

37. One could perhaps see this as a genealogy of the so-called Smart City. See Halpern, Mitchell, and Geoghegan, "Smartness Mandate."

38. Ibrahim Carone, "Qual a localidade mais sálubre do Rio," in *Arquitetura e Urbanismo* (May–June 1936): 61.

39. Paola Sá, "Ventilacao e indices de conforto," *Arquitetura e Urbanismo* (November–December 1936): 194–95.

40. See Porteous, *New Eco-architecture.* See also Bowler and Brimblecombe, "Environmental Pressures"; Maldonado, "Idea of Comfort."

41. Requena-Ruiz, "Building Artificial Climates," 1–3. See also André Missenard, *L'Homme et le climat* (Paris: Plan, 1936).

42. Paola Sá, "Ventilacao e indices de conforto," *Arquitetura e Urbanismo* (November–December 1936): 195.

43. See Mitchell, "Economentality"; and more generally, Foucault, *Security, Territory, Population*, lecture 4, "The Problem of Government," 87–114.

44. This Distrito Federal would be moved to Brasilia in the early 1960s, the new city constructed according to Costa's plan.

45. Mindlin, *Modern Architecture*, 194.

46. Mindlin, *Modern Architecture*, 202. See also the large collection of images in "IRB Building, Rio de Janeiro, Brazil," *Architectural Forum* 18 (August 1944): 66–77; "IRB Building at Rio de Janeiro," *Architectural Record* (January 1948).

47. The IRB still operates today, in this same capacity, with funding from the government and from global reinsurance giants such as Swiss Re and Lloyds of London.

48. See Olgyay and Olgyay, *Solar Control and Shading Devices*.

49. The drawings were published in most of the European and North American press on the buildings, including in *Architectural Forum* (January 1944) and in "Bâtiment d'Administration 'I.R.B.'" in *Architecture d'Aujourd'hui* 22 (August 1952).

50. Further annotations, not published, identified some additional complexities. A pair of diagrams that could have replaced the third one identified above indicates that one band of glass at eye level had the disadvantages of poorly lighting the room (without the band at the top to reflect light off the ceiling), the potential for excessive glare, "not to mention the excessive cost of glass in Brazil." The final drawing, identical to the third above, is described differently: "In this way, the whole room is well-lighted. The window proper has a human scale—and in cold weather, when air conditioning is not used, the upper part, which is moveable, supplies the ventilation required, while the lower part is used exclusively to provide light and view." Item #26758, MMM Roberto Papers, Research and Documentation Center, Universidade Federal do Rio de Janeiro.

51. See, for example, "Vocational School SENAI [1946]," in *Architectural Forum* (November 1947); "School for Industrial Apprentices [1953]," in Mindlin, *Modern Architecture*, 140; *Architecture d'Aujourd'hui* 22 (August 1952).

52. In addition to the "Architectural Examples" section of the Olgyays' *Solar Control and Shading Devices*, already mentioned, numerous contemporaneous publications surveyed and promoted Brazilian modernism, often with some attention to shading and climate. See, for example, Mindlin, *Modern Architecture*; Goodwin, *Brazil Builds*. More contemporary works tend to reduce the importance of shading to these developments. See, for example, Zilah Quezado Deckker, *Brazil Built*; Segawa, *Architecture of Brazil*. Generally speaking, this period of innovation in Brazil is not discussed relative to experiments in shading, but relative to reflection of the lessons of European modernism.

53. Ewald, "Insurance and Risk," in Graham Burchell, Colin Gordon, and Peter Miller, eds., *The Foucault Effect: Studies in Governmentality with Two Lectures by and an Interview with Michel Foucault* (Chicago: Chicago University Press, 1991), 197–210. This volume was the first publication of Foucault's governmentality lectures in English, and Ewald's was one of a number of entries concerned with insurance— the volume also included relevant material from Daniel Defert, Ian Hacking, and Robert Castel.

54. Krüger, Daston, and Heidelberger, *Probabilistic Revolution: Volume 1*; Krüger, Gigerenzer, and Morgan, *Probabilistic Revolution: Volume 2*; also Daston and Galison, *Obectivity*. This is not, I should note, the *Risk Society* of Ulrich Beck, well-known to many scholars, that casts forward toward a new phase of advanced modernity.

55. Mindlin, *Modern Architecture*, 274. The photograph of the brothers adjusting the shading device is in Mindlin and is from the archive. Images or information about the building were not widely published.

56. Mindlin, *Modern Architecture*, 194; "Colonie de Vacances a Gavea," *Architecture d'Aujourd'hui* 18 (September 1947): 60–61. One of the few color photographs in the issue is of the hillside elevation of the main building; it was also published as "Holiday Hostel at Rio de

Janeiro," *Architectural Forum* (December 1947): 185–88, and "Small Resort Hotel in Tijuca," in *Architectural Review* (November 1947).

57. Carvalho, "Mapping the Urbanized Beaches."

58. See, for example, Gosseye and Heynen, "Architecture for Leisure." The classic text Gosseye and Heynen draw on is Johan Huizinga, *Homo Ludens: A Study of the Play-Element in Culture* (Boston: Beacon Press, 1955).

59. All of these images come from the extensive "Architectural Examples" in Olgyay and Olgyay, *Solar Control and Shading Devices*, 1957.

60. Mindlin, *Modern Architecture*, 194. See also "Flats in Rio de Janeiro" in *Architecture d'Aujourd'hui* 18 (September 1947): 60–61. It is also included in a review of the Roberto brothers' extensive residential work around Rio, titled "Edifícios de Apartmentos, Copacabana, Rio," *Habitat* 6 (1956): 6–17.

61. Ghogh, *Great Derangement*.

62. Chun, "On Hypo-Real Models." See also Chun, "Crisis, Crisis, Crisis"; Stengers, *In Catastrophic Times*.

63. See the blog entry from Patrimônio, the Brazilian cultural heritage organization, entitled "Pedregulho e a complexidade do Restauro do Patimônio Moderno," http://patrimoniohistori co.prefeitura.sp.gov.br/pedregulho-e-a-complexidade-do-restauro-do-patrimonio-moderno/ accessed April 17, 2017; Britto, *Pedregulho*.

Afterword: Historiographies and Geographies of Climate

1. Kagan, *Three Cultures*, 266.

2. R. G. Barry and R. J. Chorley, *Atmosphere, Weather and Climate* (London: Methuen, 1968).

3. J. Houghton, *Global Warming: The Complete Briefing* (Cambridge: Cambridge University Press, 1994).

4. Randalls, "Contributions and Perspectives," DOI: 10.1002/wcc.466/.

5. Weart, *Discovery of Global Warming*.

6. Fleming, *Fixing the Sky*. See also Boettcher and Schäfer, "Geoengineering Research."

7. Howe, *Behind the Curve*.

8. Mike Hulme, "Problems with Making and Governing Global Kinds of Knowledge," *Global Environmental Change* 20, no. 4 (2010): 558–64.

9. Grove and Adamson, *El Niño in World History*.

10. *Clement* and *DiNezio*, "Tropical Pacific Ocean."

11. Latour, *Never Been Modern*.

12. Beck and Mahony, "IPCC and the Politics of Anticipation."

13. Orlove, Lazrus, Hovelsrud, and Giannini, "Recognitions and Responsibilities."

14. Mahony and Hulme, "Modelling and the Nation."

15. Pickering, *Mangle of Practice*.

16. Jasanoff, *States of Knowledge*.

17. Turnhout, Dewulf, and Hulme, "Policy-Relevant Global Environmental Knowledge"; Rudiak-Gould, "We Have Seen It."

18. Strengers and Maller, "Adapting to 'Extreme' Weather."

19. Fleming, *Fixing the Sky*.

SELECTED BIBLIOGRAPHY

List of Archives

Archives of the Royal Society, London, UK
Commonwealth and African Collection, University of Oxford Special Collections, UK
Derbyshire Records Office (DRO), UK
Ellsworth Huntington Archives, Yale University, USA
India Office Archive, British Library, London, UK
Library and Archives Canada, Ottawa (LAC)
The National Archives, Kew, London, UK (TNA)
National Archives of Australia, Perth
National Archives Namibia (NAN)
National Archives of South Africa, Pretoria (NASA)
National Meteorological Library and Archive, Exeter, UK
Science Museum Archives, Wroughton, UK
Scott Polar Research Institute Archives, Cambridge, UK
Unilever Archives, Port Sunlight, UK
University of Toronto Archives, Canada.
Women's Library Archives, London School of Economics, UK

Works Cited

Abbe, Cleveland. "The Physical Basis of Long-Range Weather Forecasts." *Monthly Weather Review* 29 (1901): 551–61.

Achbari, Azedah. "Building Networks for Science: Conflict and Cooperation in Nineteenth Century Global Marine Science." *Isis* 106 (2015): 257–82.

Adamson, George C. D. "'The Languor of the Hot Weather': Everyday Perspectives on Weather and Climate in Colonial Bombay, 1819–1828." *Journal of Historical Geography* 38 (2012): 143–54.

Alborn, Timothy. *Regulated Lives: Life Assurance and British Society, 1800–1914*. Toronto: University of Toronto Press, 2009.

Allan R., J. Lindesay, and D. Parker. *El Niño Southern Oscillation and Climate Variability*. Collingwood: CSIRO Publishing, 1996.

Anderson, David. "Depression, Dust Bowl, Demography and Drought: The Colonial State and Soil Conservation in East Africa during the 1930s." *African Affairs* 83, no. 332 (July 1984): 321–43.

Anderson, Katharine. "Marine Meteorology: Observing Regimes and Global Visions." In Anderson and Rozwadowski, *Soundings and Crossings*, 213–44.

Anderson, Katharine. *Predicting the Weather: Victorians and the Science of Meteorology*. Chicago: University of Chicago Press, 2005.

Anderson, Katharine, and Helen Rozwadowski, eds. *Soundings and Crossings: Doing Science at Sea, 1800–1970*. Sagamore Beach, MA: Science History Publications, 2016.

Anderson, Warwick. "From Subjugated Knowledge to Conjugated Subjects: Science and Globalisation, or Postcolonial Studies of Science?" *Postcolonial Studies* 12, no. 4 (2009): 389–400.

Anderson, Warwick. "Geography, Race and Nation: Remapping 'Tropical' Australia, 1890–1930." *Medical History* 44, no. S20 (2000): 146–59.

Anderson, Warwick. "The Natures of Cultures: Environment and Race in the Colonial Tropics." In *Nature and Culture in the Global South: Environmental Projects in South and Southeast Asia*, edited by Paul Greenough and Anna L. Tsing, 29–46. Durham, NC: Duke University Press, 2003.

Anderson, Warwick, and V. Adams. "Pramoedya's Chickens: Postcolonial Studies of Technoscience." In *The Handbook of Science and Technology Studies*, edited by E. J. Hackett, O. Amsterdamska, M. Lynch, and J. Wajcman, 181–204. 2nd ed. Cambridge, MA: MIT Press, 2008.

Arnold, David. *Colonizing the Body: State, Medicine and Epidemic Disease in Nineteenth-Century India*. Los Angeles: University of California Press, 2003.

Arnold, David. *The New Cambridge History of India III: Science, Technology and Medicine in Colonial India*. Cambridge: Cambridge University Press, 2004.

Arnold, David. *The Problem of Nature: Environment, Culture and European Expansion*. Oxford: Blackwell, 1996.

Arnold, David. *The Tropics and the Traveling Gaze: India, Landscape, and Science, 1800–1856*. Seattle: University of Washington Press, 2006.

Aronin, Jeffrey. *Climate and Architecture*. New York: Reinhold, 1953.

Aubin, David, Charlotte Bigg, and H. Otto Sibum. *The Heavens on Earth*. Durham, NC: Duke University Press, 2010.

Aubin, David, Charlotte Bigg, and H. Otto Sibum. "Introduction: Observatory Techniques in Nineteenth-Century Science and Society." In Aubin, Bigg, and Sibum, *The Heavens on Earth*, 1–32.

Audy, J. Ralph, and Frederick L. Dunn. "Health and Disease." In *Human Ecology*, edited by Frederick Sargent, 325–43. Amsterdam: North-Holland, 1974.

Avilés, Lourdes B. *Taken by Storm, 1938: A Social and Meteorological History of the Great New England Hurricane*. Boston: AMS Books, 2013.

Azevedo, Nara, and Luiz O. Ferreira. "The Dilemmas of a Scientific Tradition: Higher Education, Science, and Public Health at the Instituto Oswaldo Cruz, 1908–1953." *Historia, Ciencies, Suade—Manguinhos* 19, no. 2 (2012): 581–610.

Baer, Ferdinand. "Helmut E. Landsberg, 1906–1985." *Bulletin of the American Meteorological Society* 67, no. 12 (1986): 1522–23.

Baer, Ferdinand, Norman L. Canfield, and J. Murray Mitchell, eds. *Climate in Human Perspective: A Tribute to Helmut E. Landsberg*. Dordrecht, Netherlands: Kluwer Academic Publishers, 1991.

Balint, Ruth. "Epilogue: The Yellow Sea." In Walker and Sobocinska, *Australia's Asia*, 345–65.

Banham, Reyner. *The Architecture of the Well-Tempered Environment*. Chicago: University of Chicago Press, 1968.

Barber, Daniel A. *Climatic Effects: Architecture, Media, and the Globalization of the International Style*. Princeton, NJ: Princeton University Press, 2018.

Barber, Daniel A. "Modernism and Microclimatology." In *Energy Accounts: Architectural Representations of Energy, Climate, and the Future*, edited by William Braham, Dan Willis, and Daniel A. Barber, 216–31. New York: Routledge, 2016.

Barnes, Jessica, and Michael R. Dove, eds. *Climate Cultures: Anthropological Perspectives on Climate Change*. New Haven, CT: Yale University Press, 2015.

Barnett, Clive. "Impure and Worldly Geography: (1) The Africanist Discourse of the Royal Geographical Society, 1831–1873." *Transactions of the Institute of British Geographers* 23, no. 2 (1998): 239–51.

Barrett, Frank A. "August Hirsch: As Critic of, and Contributor to, Geographical Medicine and Medical Geography." *Medical History* 44, no. S20 (2000): 98–117.

Barton, Gregory Allen. *Empire Forestry and the Origins of Environmentalism*. Cambridge: Cambridge University Press, 2002.

Basalla, George. "The Spread of Western Science." *Science* 156, no. 3775 (1967): 611–22.

Bashford, Alison. "Terraqueous Histories." *Historical Journal* 60, no. 2 (2017): 253–72.

Beattie, James. *Empire and Environmental Anxiety: Health, Science, Art and Conservation in South Asia and Australasia, 1800–1920*. Basingstoke: Palgrave, 2011.

Beattie, James, Edward Mellilo, and Emily O'Gorman, eds. *Eco-Cultural Networks and the British Empire: New Views on Environmental History*. London: Bloomsbury Academic, 2015

Beattie, James, Emily O'Gorman, and Matthew Henry, eds. *Climate, Science and Colonization: Histories from Australia and New Zealand*. London: Palgrave 2014.

Beck, Silke, Tim Forsyth, Pia M. Kohler, Myanna Lahsen, and Martin Mahony. "The Making of Global Environmental Science and Politics." In *The Handbook of Science and Technology Studies*, edited by Ulrike Felt, Rayvon Fouché, Clark A. Miller, and Laurel Smith-Doerr, 1059–86. 4th ed. Cambridge: MIT Press, 2016.

Beck, Silke, and Martin Mahony. "The IPCC and the Politics of Anticipation." *Nature Climate Change* 7, no. 5 (2017): 311–13.

Beck, Ulrich. *Risk Society: Towards a New Modernity*. London: Sage Publications, 1992.

Behrman, Daniel. *Assault on the Largest Unknown: The International Indian Ocean Expedition, 1959–1965*. Paris: UNESCO Press, 1981.

Beinart, William. "Soil Erosion, Conservationism, and Ideas about Development: A Southern African Exploration, 1900–1960." *Journal of Southern African Studies* 11, no. 1 (1984): 52–83.

Beinart, William, and Lotte Hughes. *Environment and Empire*. Oxford: Oxford University Press, 2007.

Bell, Morag. "'The Pestilence That Walketh in Darkness': Imperial Health, Gender and Images of South Africa c. 1880–1910." *Transaction of the Institute of British Geographers* 18, no. 3 (1993): 327–41.

Bell, Morag. "A Woman's Place in 'a White Man's Country': Rights, Duties and Citizenship for the 'New' South Africa, c. 1902." *Ecumene* 2, no. 2 (1995): 129–48.

Bergdoll, Barry, C. E. Comas, J. F. Liemur, and P. Del Real. *Latin America in Construction, 1955–1980*. New York: Museum of Modern Art, 2015.

Berridge, Virginia. "Opium Eating and Life Insurance." *British Journal of Addiction* 72 (1977): 371–77.

Bischof, Marco. "Introduction to Integrative Biophysics." In *Integrative Biophysics, Biophotonics*, edited by Fritz-Albert Popp and Lev Belonssov, 1–112. Dordrecht: Kluwer, 2003.

Blakeley, Brian L. "The Society for the Oversea Settlement of British Women and the Problems of Empire Settlement, 1917–1936." *Albion* 20, no. 3 (1988): 421–44.

Blakely, Brian L. "Women and Imperialism: The Colonial Office and Female Emigration to South Africa, 1901–1910." *Albion* 13, no. 2 (1981): 131–49.

Bloor, David. *Knowledge and Social Imagery*. 1st ed. Chicago: University of Chicago Press, 1976.

Blunt, Alison. "Imperial Geographies of Home: British Domesticity in India, 1886–1925." *Transactions of the Institute of British Geographers* 24, no. 4 (1999): 421–40.

Blunt, Alison. *Travel, Gender, and Imperialism: Mary Kingsley and West Africa*. New York: Guilford Press, 1994.

Blunt, Alison, and Gillian Rose, eds. *Writing Women and Space: Colonial and Postcolonial Geographies*. New York: Guilford Press, 1994.

Boettcher, Miranda, and Stefan Schäfer. "Reflecting upon 10 Years of Geoengineering Research: Introduction to the Crutzen + 10 Special Issue." *Earth's Future* 5 (2017): 266–77.

Bolster, Jeffrey. "Opportunities in Marine Environmental History." *Environmental History* 11, no. 3 (2006): 567–97.

Bonneuil, Christophe, and Jean-Baptiste Fressoz. *The Shock of the Anthropocene*. New York: Verso, 2016.

Borck, Cornelius. "Communicating the Modern Body: Fritz Kahn's Popular Images of Human Physiology as an Industrialized World." *Canadian Journal of Communication* 32 (2007): 495–520.

Bouk, Dan. *How Our Days Became Numbered: Risk and the Rise of the Statistical Individual*. Chicago: University of Chicago Press, 2015.

Bouma, J. J. S. H. J. W. "A Short History of Human Biometeorology." *Experientia* 43 (1987): 1–6.

Bowler, Catherine, and Peter Brimblecombe. "Environmental Pressures on Building Design and Manchester's John Rylands Library." *Journal of Design History* 13, no. 3 (2000): 175–91.

Brace, Catherine, and Hilary Geoghegan. "Human Geographies of Climate Change: Landscape, Temporality, and Lay Knowledges." *Progress in Human Geography* 35, no. 3 (2011): 284–302.

Brandt, Allan. *No Magic Bullet: A Social History of Venereal Disease in the United States since 1880*. New York: Oxford University Press, 1985.

Britto, Alfredo. *Pedregulho: O sonho pioneiro da habitação popular no Brasil*. Rio: Edições de Janiero, 2015.

Broad, Kenneth, and Ben Orlove. "Channeling Globality: The 1997–1998 El Niño Climate Event in Peru." *American Ethnologist* 34, no. 2 (2007): 285–302.

Brockbank, E. M. *Life Insurance and General Practice*. London: Henry Frowde, Hodder and Stoughton, 1908.

Brooke, John L. *Climate Change and the Course of Global History: A Rough Journey*. Cambridge: Cambridge University Press, 2014.

Bunting, A. H. "Agricultural Research in the Groundnut Scheme, 1947–1951." *Nature* 168, no. 4280 (1951): 804–6.

Burton, Antoinette M. *Burdens of History: British Feminists, Indian Women, and Imperial Culture, 1865–1915*. Chapel Hill: University of North Carolina Press, 1994.

Burton, Antoinette M. "The White Woman's Burden: British Feminists and the Indian Woman, 1865–1915." *Women's Studies International Forum* 13, no. 4 (1990): 295–308.

Bush, Julia. "Edwardian Ladies and the 'Race': Dimensions of British Imperialism." *Women's Studies International Forum* 21, no. 3 (1998): 277–89.

Bush, Julia. "'The Right Sort of Woman': Female Emigrators and Emigration to the British Empire, 1890–1910." *Women's History Review* 3, no. 3 (1994): 385–409.

Butcher, John G. *The Closing of the Frontier: A History of the Marine Fisheries of Southeast Asia, c. 1850–2000*. Singapore: Institute of Southeast Asian Studies, 2004.

Caldwell, Lynton Keith. "Biocracy and Democracy: Science, Ethics, and the Law." *Politics and the Life Sciences* 3 (1985): 137–49.

Callaway, Helen. *Gender, Culture and Empire: European Women in Colonial Nigeria*. Dordrecht: Springer, 1986.

Camerini, Jane R. "Heinrich Berghaus's Map of Human Diseases." *Medical History* 44, no. S20 (2000): 186–208.

Campbell, John A., and David N. Livingstone. "Neo-Lamarckism and the Development of Geography in the United States and Great Britain." *Transactions of the Institute of British Geographers* ns 8 (1983): 267–94.

Cantlie, James. "Life Insurance in the Tropics." *Transactions of the Royal Society of Tropical Medicine and Hygiene* 15, no. 4 (1921): 109–16.

Caputi, Nick, Warrick Fletcher, Alan Pearce, and Chris Chubb. "Effect of the Leeuwin Current on the Recruitment of Fish and Invertebrates along the Western Australian Coast." *Marine and Freshwater Research* 47 (1996): 147–55.

Cardwell, Emma, and Thomas Thornton. "The Fisherly Imagination: The Promise of Geographical Approaches to Marine Management." *Geoforum* 64 (2015): 157–67.

Carey, Mark. "Inventing Caribbean Climates: How Science, Medicine, and Tourism Changed Tropical Weather from Deadly to Healthy." *Osiris* 26, no. 26 (2011): 129–41.

Carey, Mark, M. Jackson, Alessandro Antonello, and Jaclyn Rushing. "Glaciers, Gender, and Science: A Feminist Glaciology Framework for Global Environmental Change Research." *Progress in Human Geography* 40, no. 6 (2016): 770–93.

Carroll, Siobhan. "Atopia /Non-Place." In *The Routledge Handbook of Literature and Space*, edited by Robert Tally Jr., 159–67. New York: Routledge, 2017.

Carroll, Siobhan. *An Empire of Air and Water: Uncolonizable Space in the British Imagination, 1750–1850*. Philadelphia: University of Pennsylvania Press, 2015.

Carvalho, Bruno. "Mapping the Urbanized Beaches of Rio de Janeiro: Modernization, Modernity and Everyday Life." *Travesia: Journal of Latin American Cultural Studies* 16, no. 3 (2007): 325–39.

Castellani, Aldo, and Albert J. Chalmers. *Manual of Tropical Medicine*. London: Baillière, Tindall and Cox, 1919.

Caviedes, Cesar N. "El Niño, 1982–1983." *Geographical Review* 74, no. 3 (1984): 267–90.

Cawood, John. "The Magnetic Crusade: Science and Politics in Early Victorian Britain." *Isis* 70, no. 4 (1979): 492–518.

Chakrabarty, Dipesh. "The Climate of History: Four Theses." *Critical Inquiry* 35, no. 2 (2009): 197–222.

Champlin, Peggy. *Raphael Pumpelly: Gentleman Geologist of the Gilded Age*. Tuscaloosa: University of Alabama Press, 1994.

Chang, Jiat-Hwee. *A Genealogy of Tropical Architecture: Colonial Networks, Nature and Technoscience*. New York: Routledge, 2016.

Chang, Jiat-Hwee. "Thermal Comfort and Climatic Design in the Tropics: A Historical Critique." *Journal of Architecture* 21, no. 8 (2016): 1171–202.

Chang, Jiat-Hwee, and Anthony D. King. "Towards a Genealogy of Tropical Architecture: Historical Fragments of Power-Knowledge, Built Environment and Climate in the British Colonial Territories." *Singapore Journal of Tropical Geography* 32, no. 3 (2011): 283–300.

Chapman, S. "Symons Memorial Medal, 1934." *Quarterly Journal of the Royal Meteorological Society* 60, no. 254 (1943) 184–85.

Chappell, John E. Jr. "Huntington and His Critics: The Influence of Climate on Civilization." PhD dissertation, University of Kansas, Lawrence, 1968.

Chard, Chloe. "Lassitude and Revival in the Warm South: Relaxing and Exciting Travel, 1750–1830." *Clio Medica/The Wellcome Series in the History of Medicine* 56, no. 1 (2000): 179–205.

Chilton, Lisa. *Agents of Empire: British Female Migration to Canada and Australia, 1860s–1930.* Toronto: University of Toronto Press, 2007.

Chilton, Lisa. "A New Class of Women for the Colonies: The Imperial Colonist and the Construction of Empire." *Journal of Imperial and Commonwealth History* 31, no. 2 (2003): 36–56.

Christensen, Joseph. "Islands as Arks: Nature Protection and the Preservation Ethic, 1898–1918." *Studies in Western Australian History* 27 (2011): 13–30.

Christensen, Joseph. "Shark Bay, 1616–1991: The Spread of Science and the Emergence of Ecology in a World Heritage Area." PhD dissertation, University of Western Australia, 2008.

Christensen, Joseph. "Their Inescapable Portion? Cyclones, Disaster Relief and the Political Economy of Australian Pearlshelling, 1865–1935." In *Bordering on Danger: Natural Disasters and People in the Indian Ocean World*, edited by Greg Bankoff and Joseph Christensen, 283–411. New York: Palgrave Macmillan, 2016.

Christensen, Joseph. "To the Islands: Ecological Imperialism on the North-West Australian Coast." In *Fluid Frontiers: New Currents in Marine Environmental History*, edited by John R. Gillis and Franziska Torma, 63–75. Cambridge, UK: White Horse Press, 2015.

Christensen, Joseph. "Unsettled Seas: Towards a History of Marine Animal Populations in the Central Indo-Pacific." In *Historical Perspectives on Fisheries Exploitation in the Indo-Pacific*, edited by Joseph Christensen and Malcolm Tull, 13–39. Dordrecht: Springer, 2014.

Christensen, Joseph, and Gary Jackson. "Shark Bay Snapper: Science, Policy and the Decline and Recovery of a Marine Recreational Fishery." In *Historical Perspectives on Fisheries Exploitation in the Indo-Pacific*, edited by Joseph Christensen and Malcolm Tull, 251–68. Dordrecht: Springer, 2014.

Chun, Wendy Hui Kyong. "Crisis, Crisis, Crisis: Sovereignty and Networks." In *Theory, Culture and Society* 28, no. 6 (2011): 91–112.

Chun, Wendy Hui Kyong. "On Hypo-real Models or Global Climate Change: A Challenge for the Humanities." *Critical Inquiry* 41 (Spring 2015): 675–703.

Ciucci, Giorgio. "The Invention of the Modern Movement." *Oppositions* 24 (1981): 68–91.

Clement, Amy, and Pedro DiNezio. "The Tropical Pacific Ocean—Back in the Driver's Seat?" *Science* 343, no. 6174 (2014): 976–78.

Coelho de Souza, Luiz Felipe Machado. *Irmãos Roberto, Arquitetos.* Rio de Janeiro: Rio Books, 2013.

Coen, Deborah R. *Climate in Motion: Science, Empire, and the Problem of Scale.* Chicago: University of Chicago Press, 2018.

Collier, Peter. "Edward Sabine and the 'Magnetic Crusade.'" In *History of Cartography*, edited by Elri Liebenberg, Peter Collier, and Zsolt Gyözö Török, 309–23. Heidelberg: Springer-Verlag, 2013.

Collingham, E. M. *Imperial Bodies: The Physical Experience of the Raj, c. 1800–1947*. Cambridge, UK: Polity Press, 2001.

Comaroff, Jean, and John Comaroff. *Of Revelation and Revolution: Christianity, Colonialism, and Consciousness in South Africa*. Vol. 1. Chicago: University of Chicago Press, 1991.

Connery, Chris. "Pacific Rim Discourse: The US Global Imaginary in the Late Cold War Years." *Boundary* 21, no. 1 (1994): 30–56.

Cooper, Gail. *Air-Conditioning America: Engineers and the Controlled Environment*. Baltimore, MD: Johns Hopkins University Press, 2002.

Corn, Joseph. *The Winged Gospel: America's Romance with Aviation, 1900–1950*. New York: Oxford University Press, 1983.

Crewe, Maurice E. "Meteorology and Aerial Navigation." *Royal Meteorological Society Occasional Papers on Meteorological History* 4 (September 2002).

Crinson, Mark. *Modern Architecture at the End of Empire*. London: Ashgate, 2003.

Crozier, Anna. "Sensationalising Africa: British Medical Impressions of Sub-Saharan Africa 1890–1939." *Journal of Imperial and Commonwealth History* 35, no. 3 (2007): 393–415.

Curtin, Philip D. *The Image of Africa: British Ideas and Action, 1780–1850*. Madison: University of Wisconsin Press, 1964.

Curtin, Philip D. "'The White Man's Grave': Image and Reality, 1780–1850." *Journal of British Studies* 1, no. 1 (1961): 94–110.

Cushman, Gregory T. "Choosing between Centers of Action: Instrument Buoys, El Niño, and Scientific Internationalism in the Pacific, 1957–1982." In *The Machine in Neptune's Garden: Historical Perspectives on Technology and the Marine Environment*, edited by Helen M. Rozwadowski and David K. van Keuren, 133–82. Sagamore Beach, MA: Science History Publications, 2004.

Cushman, Gregory T. *Guano and the Opening of the Pacific World: A Global Ecological History*. Cambridge: Cambridge University Press, 2013.

Cushman, Gregory T. "The Imperial Politics of Hurricane Prediction: From Calcutta and Havana to Manila and Galveston, 1839–1900." In *Nation-States and the Global Environment*, edited by Mark Lawrence, Erika Bsumek, and David Kinkela, 137–62. Oxford: Oxford University Press, 2013.

Dagut, Simon. "Gender, Colonial 'Women's History' and the Construction of Social Distance: Middle-Class British Women in Later Nineteenth-Century South Africa." *Journal of Southern African Studies* 26, no. 3 (2000): 555–72.

Dakin, William. "The Percy Sladen Trust Expeditions to the Albrolhos Islands (Indian Ocean): Report 1—Introduction, General Description of the Coral Islands forming the Houtman Abrolhos Group, the Formation of the Islands." *Journal of the Linnaean Society—Zoology* 34 (1919): 127–80.

Dalén, Per. *Season of Birth: A Study of Schizophrenia and Other Mental Disorders*. New York: Elsevier, 1975.

Daniels, Stephen. "Geographical Imagination." *Transactions of the Institute of British Geographers* 36, no. 2 (2011): 182–87.

Daniels, Stephen, and Georgina H. Endfield. "Narratives of Climate Change: Introduction." *Journal of Historical Geography* 35, no. 2 (April 2009): 215–22.

Daston, Lorraine, and Peter Galison. *Objectivity*. New York: Zone Books, 2007.

Davies, Arthur. *Forty Years of Progress and Achievement*. Geneva: WMO, 1990.

Davies, David A. "Experiments on Artificial Stimulation of Rain in East Africa." *Nature* 174, no. 4423 (1954): 256–57.

Davies, David A., D. Hepburn, and H. W. Samson. "Report of Experiments at Kongwa on Artificial Control of Rainfall, January–April 1952." East African Meteorological Department Memoirs. Nairobi, 1952.

Davies, David A., D. Hepburn, and H. W. Samson. "Report on Experiments at Kongwa on Artificial Stimulation of Rain, January–April 1951." *East African Meteorological Department Memoirs*. Vol. 9. East African Meteorological Department Memoirs. Nairobi, 1951.

Davies, R. E. G. *Fantasies and Fallacies of Air Transport History*. McLean, VA: Paladwr Press, 1994.

Davin, Anna. "Imperialism and Motherhood." *History Workshop* 5 (1978): 9–65.

Davis, Diana K. "Desert 'Wastes' of the Maghreb: Desertification Narratives in French Colonial Environmental History of North Africa." *Cultural Geographies* 11, no. 4 (2004): 359–87.

Davis, Mike. "The Coming Desert: Kropotkin, Mars and the Pulse of Asia." *New Left Review* 97 (2016): 23–43.

Davis, Mike. *Late Victorian Holocausts: El Niño Famines and the Making of the Third World*. London: Verso, 2001.

Deacon, Harriet. "The Politics of Medical Topography: Seeking Healthiness at the Cape during the Nineteenth Century." *Clio Medica/The Wellcome Series in the History of Medicine* 56, no. 1 (2000): 279–97.

Deckker, Zilah Quezado. *Brazil Built: The Architecture of the Modern Movement in Brazil*. London: Spon Press, 2001.

del Real, Patricio. "Paternity Rights: The Brise-Soleil and the Sources of Modernism in the Ministry of Education and Health in Rio de Janeiro, Brazil." *ACSA Proceedings* (2002): 199–207.

Diamond, Marion. *Emigration and Empire: The Life of Maria S. Rye*. Abingdon: Routledge, 2013.

Disco, Nil, and Eda Kranakis, eds. *Cosmopolitan Commons: Sharing Resources and Risks across Borders*. Cambridge, MA: MIT Press, 2013.

Dobson, Alan. *Peaceful Air Warfare: The United States, Britain, and the Politics of International Aviation*. Oxford: Clarendon Press, 1991.

Dobson, Alan. "The USA, Hegemony, and Airline Market Access to Britain and Western Europe, 1945–1996." *Diplomacy & Statecraft* 9 (1998): 129–59.

Dorn, Harold. *The Geography of Science*. Baltimore, MD: Johns Hopkins University Press, 1991.

Downes, C. R. "History of the British Ocean Weather Ships." *Marine Observer* 67 (1977): 179–86.

Drayton, Richard. *Nature's Government: Science, Imperial Britain, and the "Improvement" of the World*. New Haven, CT: Yale University Press, 2000.

Driver, Felix. "Imagining the Tropics: Views and Visions of the Tropical World." *Singapore Journal of Tropical Geography* 25, no. 1 (2004): 1–17.

Driver, Felix. "Yule, Sir Henry." Oxford Dictionary of National Biography. Oxford: Oxford University Press, 2004. Online edition, January 2013.

Droessler, E. G. "Helmut Landsberg, 1906–1985." *Eos* 67, no. 19 (1986): 457.

Dubow, Saul. *A Commonwealth of Knowledge: Science, Sensibility and White South Africa 1820–2000*. Oxford: Oxford University Press, 2006.

Dubow, Saul. "Earth History, Natural History, and Prehistory at the Cape, 1860–1875." *Comparative Studies in Society and History* 46, no. 1 (2004): 107–33.

Duncan, James S. "Sites of Representation: Place, Time and the Discourse of the Other."

In *Place, Culture and Representation*, edited by James S. Duncan and David Ley, 39–56. London: Routledge, 1993.

Dupree, Marguerite. "Other than Healing." *Social History of Medicine* 10 (1997): 79–103.

Edwards, Paul N. "Meteorology as Infrastructural Globalism." *Osiris* 21 (2006): 229–50.

Edwards, Paul N. *A Vast Machine: Computer Models, Climate Data and the Politics of Global Warming*. Cambridge, MA: MIT Press, 2010.

Eliot, J. "A Preliminary Discussion of Certain Oscillatory Changes of Pressure of Long Period and Short Period in India." *Indian Meteorological Memoirs* 6 (1895): 89–160.

Endfield, Georgina H., and David J. Nash. "'A Good Site for Health': Missionaries and the Pathological Geography of Central Southern Africa." *Singapore Journal of Tropical Geography* 28 (2007): 142–57.

Endfield, Georgina H., and David J. Nash. "'Happy Is the Bride the Rain Falls On': Climate, Health and 'The Woman Question' in Nineteenth-Century Missionary Documentation." *Transactions of the Institute of British Geographers* 30, no. 3 (2005): 368–86.

Endfield, Georgina, and David Nash. "Missionaries and Morals: Climatic Discourse in Nineteenth-Century Central Southern Africa." *Annals of the Association of American Geographers* 92, no. 4 (2002): 727–42.

Endfield, Georgina H., and Samuel Randalls. "Climate and Empire." In Beattie, Mellilo, and O'Gorman, *Eco-Cultural Networks*, 21–43.

Escobar, Arturo. *Encountering Development: The Making and Unmaking of the Third World*. Princeton, NJ: Princeton University Press, 2011.

Esselborn, Stefan. "Environment, Memory, and the Groundnut Scheme: Britain's Largest Colonial Agricultural Development Project and Its Global Legacy." *Global Environment* 11 (2013): 58–93.

Ewald, Francois. "Insurance and Risk." In *The Foucault Effect: Studies in Governmentality with Two Lectures by and an Interview with Michel Foucault*, edited by Graham Burchell, Colin Gordon, and Peter Miller, 197–210. Chicago: Chicago University Press, 1991.

Fagan, F. *Floods, Famines and Emperors: El Niño and the Fate of Civilizations*. New York: Basic Books, 1999.

Finnegan, Diarmuid A. "The Spatial Turn: Geographical Approaches in the History of Science." *Journal of the History of Biology* 41, no. 2 (2008): 369–88.

Fitch, James Marston. *American Building: The Forces that Shape It*. Boston: Houghton Mifflin, 1947.

Fitch, James Marston. *American Building and the Environmental Forces that Shape It*. New York: Schocken Books, 1972.

Fleming, James R. *Fixing the Sky: The Checkered History of Weather and Climate Control*. New York: Columbia University Press, 2010.

Fleming, James R. *Historical Perspectives on Climate Change*. New York: Oxford University Press, 1998.

Fleming, James R. *Inventing Atmospheric Science: Bjerkness, Rossby, Wexler and the Foundations of Modern Meteorology*. Cambridge, MA: MIT Press, 2015.

Fleming, James R. *Meteorology in America, 1800–1870*. Baltimore, MD: Johns Hopkins University Press, 1990.

Fleming, James R. "Planetary-Scale Fieldwork: Harry Wexler on the Possibilities of Ozone Depletion and Climate Control." In *Knowing Global Environments: New Historical*

Perspectives on the Field Sciences, edited by Jeremy Vetter, 190–211. New Brunswick, NJ: Rutgers University Press, 2010.

Fleming, James R., and Vladimir Jankovic, eds. *Klima, Osiris 26*. Chicago: University of Chicago Press, 2011.

Folk, G. Edgar. "International Society of Biometeorology: A Fifty Year History." https://uwm.edu/biometeorology/wp-content/uploads/sites/439/2017/06/ISB_50YearHistoryof Biometeorology.pdf.

Forsyth, Isla. "The More-than-Human Geographies of Field Science." *Geography Compass* 7, no. 8 (2013): 527–39.

Foucault, Michel. *Security, Territory, Population: Lectures at the College de France, 1977–1978*. London: Picador, 2009.

Fowles, Brooke, and Andrea Gaynor. "The Challenge of Creating a Scientifically-Robust Historical Description of Changing Finfish Populations in the Ningaloo Marine Park." *Studies in Western Australian History* 27 (2011): 99–124.

Frampton, Kenneth. *Le Corbusier*. New York: Thames and Hudson, 2001.

Frampton, Kenneth. "Le Corbusier and Oscar Niemeyer: Influence and Counterinfluence, 1929–1965." In *Latin American Architecture, 1929–1960: Contemporary Reflections*, edited by Carlos Brillembourg, 34-49. New York: Monacelli Press, 2004.

Frankcom, C. E. N. "Ocean Weather Ships: Some Navigational and Oceanographical Aspects." *International Hydrographic Review* 40 (1963): 141–53.

Galton, Francis. *Hereditary Genius*. London: Macmillan, 1869.

Galton, Francis. *Inquiries into Human Faculty and Its Development*. London: Macmillan 1883.

Ganter, Regina. *Mixed Relations: Asian-Aboriginal Contact in North Australia*. Perth: University of Western Australia Press, 2006.

Gaynor, Andrea. "Shifting Baselines or Shifting Currents? An Environmental History of Fish and Fishing in the South-West Capes Region of Western Australia." In *Historical Perspectives on Fisheries Exploitation in the Indo-Pacific*, edited by Joseph Christensen and Malcolm Tull, 231–50. Dordrecht: Springer, 2014.

Geiger, Rudolf. *The Climate near the Ground*. Cambridge, MA: Harvard University Press, 1965.

Geoghegan, Hilary. "A New Pattern for Historical Geography: Working with Enthusiast Communities and Public History." *Journal of Historical Geography* 46 (2014): 105–7.

Ghogh, Amitav. *The Great Derangement: Climate Change and the Unthinkable*. Chicago: University of Chicago Press, 2016.

Glantz, Michael H. *Climate Affairs: A Primer*. Washington, DC: Island Press, 2003.

Glantz, Michael H. "Science, Politics and Economics of the Peruvian Anchoveta Fishery." *Marine Policy* 3, no. 3 (1979): 201–10.

Godwin, Luke. "The Fluid Frontier: Central Queensland, 1845–1863." In *Colonial Frontiers: Indigenous-European Encounters in Settler Societies*, edited by Lynette Russell, 101–18. Manchester: Manchester University Press, 2001.

Goldman, Mara J., Meaghan Daly, and Eric J. Lovell. "Exploring Multiple Ontologies of Drought in Agro-pastoral Regions of Northern Tanzania: A Topological Approach." *Area* 48, no. 1 (2016): 27–33.

Golinski, Jan. *British Weather and the Climate of Enlightenment*. Chicago: University of

Chicago Press, 2010.

Good, Gregory. "Between Two Empires: The Toronto Magnetic Observatory and American Science before Confederation." *Scientia Canadensis* 10 (1986): 34–52.

Goodwin, Philip L. *Brazil Builds: Architecture New and Old, 1652–1942*. New York: Museum of Modern Art, 1943.

Gordon, Arnold. "Oceanography of the Indonesian Seas and Their Throughflow." *Oceanography* 18, no. 4 (2005): 14–27.

Gore, Al. *Earth in the Balance*. Boston: Houghton Mifflin, 1992.

Gosseye, Janina, and Hilde Heynen. "Architecture for Leisure in Post-war Europe: Between Experimentation, Liberation, and Patronisation." *Journal of Architecture* 18, no. 5 (November 2013): 623–31.

Gowans, Georgina. "Imperial Geographies of Home: Memsahibs and Miss-Sahibs in India and Britain, 1915–1947." *Cultural Geographies* 10, no. 4 (2003): 424–41.

Graham, James, ed. *Climates: Architecture and the Planetary Imaginary*. Zurich: Lars Müller, 2016.

Gray, Howard. *The Western Rock Lobster, Panulirus cygnus: Book 1, A Natural History*. Geraldton, WA: Westralian Books, 1992.

Gray, Howard. *The Western Rock Lobster, Panulirus cygnus: Book 2, A History of the Fishery*. Geraldton, WA: Westralian Books, 1999.

Grevsmühl, Sebastian V. "Images, Imagination and the Global Environment: Towards an Interdisciplinary Research Agenda on Global Environmental Images." *Geo: Geography and Environment* 3, no. 2 (2016): e00020.

Gropius, Walter. "Houses, Walk-Ups, or High Rise Apartment Blocks?" In *The Scope of Total Architecture*, edited by Walter Gropius, 119—135. New York: Harper and Row, 1955.

Grove, Richard H. "The East India Company, the Raj and the El Nino: The Critical Role Played by Colonial Scientists in Establishing the Mechanisms of Global Climate Teleconnections, 1770–1930." In *Nature and the Orient*, edited by Richard H. Grove, Vinita Damodaran, and Satpal Sangwan, 301–23. Oxford: Oxford University Press, 1998.

Grove, Richard H. *Green Imperialism: Colonial Expansion, Tropical Island Edens and the Origins of Environmentalism, 1600–1860*. Cambridge: Cambridge University Press, 1995.

Grove, Richard H., and George C. D. Adamson. *El Niño in World History*. London: Palgrave Macmillan, 2018.

Gyger, Helen, and Patricio Del Real, eds. *Latin American Modern Architectures: Ambiguous Territories*. New York: Routledge, 2012.

Hackett Fischer, David. "Climate and History: Priorities for Research." In *Climate and History: Studies in Interdisciplinary History*, edited by R. Rotbert and T. Rabb. Princeton, NJ: Princeton University Press, 1981.

Haggis, J. "Ironies of Emancipation: Changing Configurations of 'Women's Work' in the 'Mission of Sisterhood' to Indian Women." *Feminist Review* 65, no. 1 (2000): 108–26.

Haggis, Jane. "Gendering Colonialism or Colonising Gender? Recent Women's Studies Approaches to White Women and the History of British Colonialism." *Women's Studies International Forum* 13, no. 1 (1990): 105–15.

Haggis, Jane. "'A Heart that Has Felt the Love of God and Longs for Others to Know It': Conventions of Gender, Tensions of Self and Constructions of Difference in Offering to Be a Lady Missionary." *Women's History Review* 7, no. 2 (1998): 171–93.

Hague, Arnold. *Allied Convoy System, 1939–1945: Its Organization, Defence and Operation*. St. Catharine's, Ontario: Vanwell, 2000.

Haines, A., R. S. Kovats, D. Campbell-Lendrum, and C. Corvalan. "Climate Change and Human Health: Impacts, Vulnerability and Public Health." *Public Health* 120 (2006): 585–96.

Halacy, D. C. *The Weather Changers*. New York: Harper and Row, 1968.

Halpern, Orit, Robert Mitchell, and Bernard Dionysius Geoghegan. "The Smartness Mandate: Notes towards a Critique." *Grey Room* 68 (Summer 2017): 106–29.

Hamblin, Jacob Darwin. *Oceanographers and the Cold War: Disciplines of Marine Science*. Seattle: University of Washington Press, 2005.

Hamblin, Jacob Darwin. "Seeing the Ocean in the Shadow of Bergen Values." *Isis* 105 (2014): 352–63.

Hamilton, Paula, and B. W. Higman. "Servants of Empire: The British Training of Domestics for Australia, 1926–1931." *Social History* 28, no. 1 (2003):67–82.

Hamilton, Scott. "Action, Technology, and the Homogenisation of Place: Why Climate Change Is Antithetical to Political Action." *Globalizations* 13, no. 1 (2016): 62–77.

Hannaway, Caroline. "Environment and Miasmata." In *Companion Encyclopedia of the History of Medicine*, edited by W. F. Bynum and Roy Porter, 292–308. London: Routledge, 1993.

Harding, Sandra G. *Is Science Multicultural? Postcolonialisms, Feminisms, and Epistemologies*. Bloomington: Indiana University Press, 1998.

Hardy, Penelope K. "Every Ship a Floating Observatory: Matthew Fontaine Maury and the Acquisition of Knowledge at Sea." In Anderson and Rozwadowski, *Soundings and Crossing*, 17–48.

Harper, Kristine. *Make It Rain: State Control of the Atmosphere in Twentieth-Century America*. Chicago: University of Chicago Press, 2017.

Harper, Kristine. *Weather by the Numbers*. Cambridge, MA: MIT Press, 2008.

Harrison, Mark. "Differences of Degree: Representations of India in British Medical Topography, 1820–c.1870." *Medical History* 44, no. S20 (2000): 51–69.

Harrison, Mark. "'The Tender Frame of Man': Disease, Climate and Racial Difference in India and the West Indies, 1760–1860." *Bulletin of the History of Medicine* 70, no. 1 (1996): 68–93.

Harvie, Christopher. "'The Sons of Martha': Technology, Transport and Rudyard Kipling." *Victorian Studies* 20 (1977): 269–82.

Hawkins, Harriet. "Geography and Art: An Expanding Field: Site, the Body and Practice." *Progress in Human Geography* 37, no. 1 (2012): 52–71.

Hazareesingh, Sandip. "Cotton, Climate and Colonialism in Dharwar, Western India, 1840–1880." *Journal of Historical Geography* 38, no. 1 (2012): 1–17.

Heffernan, Michael. "'A Dream as Frail as Those of Ancient Time': The In-Credible Geographies of Timbuctoo." *Environment and Planning D: Society and Space* 19, no. 2 (2001): 203–25.

Henderson, Gabriel. "The Dilemma of Reticence: Helmut Landsberg, Stephen Schneider, and Public Communication of Climate Risk, 1971–1976." *History of Meteorology* 6 (2014): 53–78.

Herd, Alexander W. G. "A 'Common Appreciation': Eisenhower, Canada, and Continental Air Defense, 1953–1954." *Journal of Cold War Studies* 13 (2011): 4–26.

Herschel, John. "An Address to the British Association for the Advancement of Science at the Opening of their Meeting at Cambridge, June 19th, 1845." In Herschel, *Essays from the Edinburgh and Quarterly Reviews*, 653.

Herschel, John. *Essays from the Edinburgh and Quarterly Reviews.* 1840. London: Longman, Brown, Green, Longmans & Roberts, 1857.

Herschel, John. *Instructions for Making and Registering Meteorological Observations in Southern Africa, and Other Countries in the South Seas, and also at Sea.* London: Bradbury and Evans, 1835.

Hevly, Bruce. "The Heroic Science of Glacier Motion." *Osiris* 11, no. 1996 (1996): 66–86.

Heymann, Matthias. "The Climate Change Dilemma: Big Science, the Globalizing of Climate and the Loss of the Human Scale." *Regional Environmental Change* 19 (2019): 1549–60.

Hildebrandsson, H. H. "Quelques recherches sur les centres d'action de l'atmosphère." *Kunliga Svenska vetenskapsakademiens handligar* 29 (1897): 33.

Hitchcock, Henry Russel. *The International Style.* New York: Norton, 1932.

Hodge, Joseph Morgan. *Triumph of the Expert: Agrarian Doctrines of Development and the Legacies of British Colonialism.* Athens: Ohio University Press, 2007.

Hodgins, J. George. *Documentary History of Education in Upper Canada. Volume X, 1851–1852; Volume XI, 1853–1855; Volume XII, 1855–1856.* Toronto: L. K. Cameron, 1903, 1903, 1904.

Hofstra, Nynke, Malcolm Haylock, Mark New, Phil Jones, and Christoph Frei. "Comparison of Six Methods for the Interpolation of Daily, European Climate Data." *Journal of Geophysical Research Atmospheres* 113, no. 21 (2008): D21110.

Hogendorn, Jan S., and K. M. Scott. "The East African Groundnut Scheme: Lessons of a Large-Scale Agricultural Failure." *African Economic History* 10, no. 10 (1981): 81–115.

Holm, Paul. "World War II and the 'Great Acceleration' of North Atlantic Fisheries." *Global Environment* 10 (2012): 66–91.

Howe, Joshua P. *Behind the Curve: Science and the Politics of Global Warming.* Seattle: University of Washington Press, 2014.

Howie, Bill, and Nick Lewis. "Geographical Imaginaries: Articulating the Values of Geography." *New Zealand Geographer* 70, no. 2 (August 2014): 131–39.

Hubbard, Jennifer. "Fisheries Biology and the Dismal Science: Economists and the Rational Exploitation of Fisheries for Social Progress." In *Fisheries, Quota Management and Quota Transfer: Realization through Bio-Economics,* edited by Gordon Winder, 31–61. Dordrecht: Springer, 2018.

Hubbard, Jennifer. "Mediating the North Atlantic Environment: Fisheries Biologists, Technology, and Marine Spaces." *Environmental History* 18 (2013): 88–100.

Huizinga, Johan. *Homo Ludens: A Study of the Play-Element in Culture.* Boston: Beacon Press, 1955.

Hulme, Mike. "Geographical Work at the Boundaries of Climate Change." *Transactions of the Institute of British Geographers* 35 (2008): 5–11.

Hulme, Mike. "Problems with Making and Governing Global Kinds of Knowledge." *Global Environmental Change* 20, no. 4 (2010): 558–64.

Hulme, Mike "Reducing the Future to Climate: A Story of Climate Determinism and Reductionism." *Osiris* 26, no. 1 (2011): 245–66.

Hulme, Mike. *Weathered: Cultures of Climate.* London: Sage, 2016.

Hulme, Mike. *Why We Disagree about Climate Change: Understanding Controversy, Inaction and Opportunity.* Cambridge: Cambridge University Press, 2009.

Hulme, Peter. "Writing on the Land: Cuba's Literary Geography." *Transactions of the Institute of British Geographers* 37, no. 3 (July 2012): 346–58.

Humphreys, William Jackson. *Rain Making and Other Weather Vagaries.* Baltimore, MD: Williams and Wilkins, 1926.

Huntington, Ellsworth. *Civilization and Climate.* New Haven, CT: Yale University Press, 1915.

Huntington, Ellsworth. "The Control of Pneumonia and Influenza by the Weather." *Ecology* 1 (1920): 6–23.

Huntington, Ellsworth. *The Human Habitat.* New York: D. Van Nostrand, 1927.

Huntington, Ellsworth. *Mainsprings of Civilization.* New York: John Wiley, 1945.

Huntington, Ellsworth. *Principles of Human Geography.* 2nd ed. New York: John Wiley, 1922.

Huntington, Ellsworth. *The Pulse of Asia: A Journey in Central Asia Illustrating the Geographic Basis of History.* Boston: Houghton Mifflin, 1907.

Huntington, Ellsworth, "Review of Medical Climatology by Clarence A. Mills." *Science* 90, no. 2345 (1939): 540–42.

Huntington, Ellsworth. *Season of Birth: Its Relation to Human Abilities.* New York: John Wiley, 1938.

Huntington, Ellsworth. *Tomorrow's Children: The Goal of Eugenics.* New York: John Wiley, 1935.

Huntington, Ellsworth. *Weather and Health: A Study of Daily Mortality in New York City.* Bulletin of the National Research Council, no. 75 (1930): 1–161. Washington, DC.

Huntington, Ellsworth. *World-Power and Evolution.* New Haven, CT: Yale University Press, 1919.

Huntington, Ellsworth, and Leon F. Whitney. *Builders of America.* New York: Morrow, 1927.

Huntington, Ellsworth, Frank E. Williams, and Samuel van Valkenburg. *Economic and Social Geography.* New York: Wiley, 1933.

Jackson, Iain. *The Architecture of Edwin Maxwell Fry and Jane Drew: Twentieth Century Architecture, Pioneer Modernism and the Tropics.* London: Routledge, 2014.

Jacobson, Arthur C. *Genius.* New York: Greenberg, 1926.

Janerich, Dwight T., Ian H. Porter, and Vito Logrillo. "Season of Birth and Neonatal Mortality." *American Journal of Public Health* 61 (1971): 1119–25.

Janković, Vladimir. *Reading the Skies: A Cultural History of the English Weather, 1650–1820.* Chicago: University of Chicago Press, 2000.

Jasanoff, Sheila. "Future Imperfect: Science, Technology, and the Imaginations of Modernity." In Jasanoff and Kim, *Dreamscapes of Modernity*, 1–33.

Jasanoff, Sheila, ed. *States of Knowledge: The Co-production of Science and the Social Order.* London: Routledge, 2004.

Jasanoff, Sheila. "A New Climate for Society." *Theory, Culture & Society* 27, no. 2–3 (May 24, 2010): 233–53.

Jasanoff, Sheila, and Sang-Hyun Kim, eds. *Dreamscapes of Modernity: Sociotechnical Imaginaries and the Fabrication of Power.* Chicago: University of Chicago Press, 2015.

Jasanoff, Sheila, and Marybeth Long Martello, eds. *Earthly Politics: Local and Global in Environmental Governance.* Cambridge, MA: MIT Press, 2004.

Jennings, Mike. "'This Mysterious and Intangible Enemy': Health and Disease amongst the Early UMCA Missionaries, 1860–1918." *Social History of Medicine* 15, no. 1 (2002): 65–87.

Johnson, Allen, and Dumas Malone, eds. *Dictionary of American Biography.* 10 vols. New York: Scribner, 1964.

Johnson, Ryan. "European Cloth and 'Tropical' Skin: Clothing Material and British Ideas of Health and Hygiene in Tropical Climates." *Bulletin of the History of Medicine* 83, no. 3 (2009): 530–60.

Jolly, Roslyn. "The 'Foreign Grave' Motif in Victorian Medicine and Literature." In *A Cul-*

tural History of Climate Change, edited by Tom Bristow and Thomas H. Ford, 138–56. Abingdon: Routledge, 2016.

Jones, Ryan Tucker. "Running into Whales: The History of the North Pacific from below the Whales." *American Historical Review* 118 (2013): 349–77.

Jones-Imhotep, Edward. "Maintaining Humans." In *Cold War Science: Knowledge Production, Liberal Democracy and Human Nature*, edited by Mark Solovey and Hamilton Cravens, 225–43. New York: Palgrave Macmillan, 2012.

Jones- Imhotep, Edward. *The Unreliable Nation: Hostile Nature and Technological Failure in the Cold War*. Cambridge, MA: MIT Press, 2017.

Josefowicz, Diane G. "Experience, Pedagogy, and the Study of Terrestrial Magnetism." *Perspectives on Science* 13 (2005): 452–94.

Jureidini, Ray, and Kevin White. "Life Insurance, the Medical Examination and Cultural Values." *Journal of Historical Sociology* 13 (2000): 190–214.

Kagan, Jerome. *The Three Cultures: Natural Sciences, Social Sciences and the Humanities in the Twenty-First Century*. Cambridge: Cambridge University Press, 2009.

Kalpagam, U. "The Colonial State and Statistical Knowledge." *History of the Human Sciences* 13, no. 2 (2000): 37–55.

Kalpagam, U. *Rule by Numbers: Governmentality in Colonial India*. New Delhi: Orient Black Swan, 2014.

Kanigel, Robert. *The Man Who Knew Infinity: A Life of the Genius Ramanujan*. New York: Washington Square Press, 1991.

Kaplan, Lawrence S. *NATO Divided, NATO United: The Evolution of an Alliance*. Westport, CT: Praeger, 2004.

Katz, Richard W. "Sir Gilbert Walker and a Connection between El Niño and Statistics." *Statistical Science* 17, no. 1 (2002): 97–112.

Kearns, Robin, Gregory O'Brien, Ronan Foley, and Nell Regan. "Four Windows into Geography and Imagination(s)." *New Zealand Geographer* 71, no. 3 (2015): 159–76.

Kennedy, Dane. "Empire Migration in Post-war Reconstruction: The Role of the Oversea Settlement Committee, 1919–1922." *Albion* 20, no. 3 (1988): 403–19.

Kennedy, Dane. *The Magic Mountains: Hill Stations and the British Raj*. Berkeley: University of California Press, 1996.

Kenworthy, Joan M. "Albert Walter, O.B.E (1877–1972) Meteorologist in the Colonial Service. Part I: His Early Life and Work in Mauritius." Vol. 44. Occasional Papers on Meteorological History. Royal Meteorological Society, Reading, 2013.

Kenworthy, Joan M. "Albert Walter, O.B.E (1877–1972): Meteorologist in the Colonial Service. Part II." Vol. 44. Occasional Papers on Meteorological History. Royal Meteorological Society, Reading, 2014.

King, T. J. *In the Clearing: Black Female Bodies, Space and Settler Colonial Landscapes*. University of Maryland dissertation, 2013. https://drum.lib.umd.edu/handle/1903/14525/.

Kipling, Rudyard. *Actions and Reactions*. London: Macmillan, 1909.

Kneale, James, and Shaun French. "Moderate Drinking before the Unit: Medicine and Life Assurance in Britain and the US, c.1860–1930." *Drugs: Education, Prevention and Policy* 22, no. 2 (2015): 111–17.

Kneale, James, and Samuel Randalls. "Invisible Atmospheric Knowledges in British Insurance Companies, 1830–1914." *History of Meteorology* 6 (2014): 35–52.

Kocchar, R. K. "Science in British India. I. Colonial Tool." *Current Science* 63, no. 11 (1992): 689–94.

Kohler, Robert E. *Landscapes and Labscapes: Exploring the Lab-Field Border in Biology.* Chicago: University of Chicago Press, 2002.

Kohler, Robert E., and Jeremy Vetter. "The Field." In *A Companion to the History of Science*, edited by Bernard Lightman, 282–95. Oxford: Wiley, 2016.

Kranakis, Eda. "European Civil Aviation in an Era of Hegemonic Nationalism: Infrastructure, Air Mobility, and European Identity Formation, 1919–1933." In *Materializing Europe Transnational Infrastructures and the Project of Europe*, edited by Alexander Badenoch and Andreas Fickers, 290–326. New York: Palgrave Macmillan, 2010.

Kranakis, Eda. "The 'Good Miracle': Building a European Airspace Commons, 1919–1939." In *Cosmopolitan Commons: Sharing Risks and Resources across Borders*, edited by Eda Kranakis and Nils Disco, 57–96. Cambridge, MA: MIT Press, 2013.

Kranidis, Rita S., ed. *Imperial Objects: Essays on Victorian Women's Emigration and the Unauthorized Imperial Experience.* New York: Twayne Publishers, 1998.

Kranidis, Rita S. *The Victorian Spinster and Colonial Emigration: Contested Subjects.* London: Macmillan, 1999.

Krogulski, S. "Turning a Curse into a Blessing: Propaganda and the Emigration of British Single Women." *Concept* 33 (2010): 2. https://concept.journals.villanova.edu/article/view/328/291/.

Krüger, Lorenz, Lorraine Daston, and Michael Heidelberger, eds. *The Probabilistic Revolution, Volume 1: Ideas in History.* Cambridge, MA: MIT Press, 1987.

Krüger, Lorenz, Gerd Gigerenzer, and Mary S. Morgan, eds. *The Probabilistic Revolution, Volume 2: Ideas in the Sciences.* Cambridge, MA: MIT Press, 1987.

Kullmer, Charles J. "The Latitude Shift of the Storm Track in the 11-Year Solar Period; Storm Frequency Maps of the United States, 1883–1930." *Smithsonian Miscellaneous Collections* 89, no. 2 (1933): 1–34.

Kupperman, Karen Ordahl. "Fear of Hot Climates in the Anglo-American Colonial Experience." *William and Mary Quarterly: A Magazine of Early American History* (1984): 213–40.

Kutzleb, Charles Robert. *Rain Follows the Plow: The History of an Idea.* PhD dissertation, University of Colorado, Boulder, CO, 1968.

Kwa, Chunglin. "Romantic and Baroque Conceptions of Complex Wholes in the Sciences." In *Complexities: Social Studies of Knowledge Practices*, edited by John Law and Annemarie Mol, 23–52. Durham, NC: Duke University Press, 2002.

Lahsen, Myanna. "The Social Status of Climate Change Knowledge: An Editorial Essay." *WIREs Climate Change* 1 (2010): 162–71.

Lambert, David, Luciana Martins, and Miles Ogborn. "Currents, Visions and Voyages: Historical Geographies of the Sea." *Journal of Historical Geography* 32 (2006): 479–93.

Landsberg, Helmut E. "Microclimatology: Facts for Architects, Realtors and City Planners on Climatic Conditions at the Breathing Line." *Architectural Forum* 86, no. 3 (March 1947): 114–18.

Landsberg, Helmut E. *Weather and Health: An Introduction to Biometeorology.* New York: Doubleday, 1969.

Lange-Eighbaum, Wilhelm. *The Problem of Genius.* London: K. Paul, Trench, Trübner, 1931.

Lavery, Colm. "Geography and Eugenics in Britain and the United States, 1900–1950." PhD dissertation, Queen's University, Belfast, 2015.

Law, John. "On Methods of Long-Distance Control: Vessels, Navigation and the Portuguese Route to India." In *Power, Action & Belief: A New Sociology of Knowledge*, edited by John Law, 234–63. London: Routledge and Kegan Paul, 1986.

Latour, Bruno. "Drawing Things Together." In *Representation in Scientific Practice*, edited by M. Lynch and Steve Woolgar, 19–68. Cambridge, MA: MIT Press, 1990.

Latour, Bruno. *Pandora's Hope: Essays on the Reality of Science Studies.* Cambridge, MA: Harvard University Press, 1999.

Latour, Bruno. *Science in Action: How to Follow Scientists and Engineers through Society.* Cambridge, MA: Harvard University Press, 1987.

Latour, Bruno. *We Have Never Been Modern.* Translated by C. Porter. New York: Harvester/ Wheatsheaf, 1993.

Latour, Bruno, and Steve Woolgar. *Laboratory Life: The Construction of Scientific Facts.* Princeton, NJ: Princeton University Press, 1979.

Le Corbusier. *Precisions on the Present State of Architecture and City Planning.* Cambridge, MA: MIT Press, 1986.

Le Corbusier. *Towards an Architecture.* Los Angeles: Getty Research Institute, 2004.

Lee, Young Kuk. "Political Homeostasis: Walter B. Cannon's Organic Analogy and Depression America." B.A. honors thesis, history of science, Harvard University, Cambridge, MA., 1999.

Lehmann, Philipp. "Whither Climatology? Brückner's *Climate Oscillations*, Data Debates and Dynamic Climatology." *History of Meteorology* 7 (2015): 49–70.

Levitan, Kathrin. "Redundancy, the 'Surplus Woman' Problem, and the British Census, 1851–1861." *Women's History Review* 17, no. 3 (2008): 359–76.

Liebowitz, Ruth Prelowski. "Landsberg, Helmut Erich." *Complete Dictionary of Scientific Biography* (2008). https://www.encyclopedia.com/science/dictionaries-thesauruses-pictures -and-press-releases/landsberg-helmut-erich/.

Lifton, Robert Jay. *The Nazi Doctors: Medical Killing and the Psychology of Genocide.* New York: Basic Books, 1986.

Linehan, Denis. "Irish Empire: Assembling the Geographical Imagination of Irish Missionaries in Africa." *Cultural Geographies* 21, no. 3 (2014): 429–47.

Lister, Thomas David. *Medical Examination for Life Insurance.* London: Edward Arnold, 1921.

Livingstone, David N. "Changing Climate, Human Evolution and the Revival of Environmental Determinism." *Bulletin of the History of Medicine* 86, no. 4 (2012): 564–95.

Livingstone, David N. "Landscapes of Knowledge." In *Geographies of Science*, edited by Peter Meusburger, David N. Livingstone, and Heike Jöns, 3–22. Heidelberg: Springer-Verlag, 2010.

Livingstone, David N. "The Moral Discourse of Climate: Historical Considerations on Race, Place, and Virtue." *Journal of Historical Geography* 17 (1991): 413–34.

Livingstone, David N. "Race, Space and Moral Climatology: Notes toward a Genealogy." *Journal of Historical Geography* 28, no. 2 (2002): 159–80.

Livingstone, David N. "Reflections on the Cultural Spaces of Climate." *Climatic Change* 113, no. 1 (2012): 91–93 .

Livingstone, David N. "Tropical Climate and Moral Hygiene: The Anatomy of a Victorian Debate." *British Journal for the History of Science* 32, no. 1 (1999): 93–110.

Livingstone, David N. "Tropical Hermeneutics and the Climatic Imagination." *Geographische Zeitschrift* 90 (2002): 65–88.

Lockyer, J. N. , and W. J. S. Lockyer. "The Behaviour of the Short-Period Atmospheric Pressure Variation over the Earth's Surface." *Proceedings of the Royal Society of London* 73 (1904): 457–70.

Lu, Duanfang, ed. *Third World Modernism*. New York: Routledge, 2010.

Luczak, Ewa Barbara. *Breeding and Eugenics in the American Literary Imagination: Heredity Rules in the Twentieth Century*. London: Palgrave, 2015.

Macdonald, Lee T. "Making Kew Observatory: The Royal Society, the British Association and the Politics of Early Victorian Science." *British Journal for the History of Science* 48, no. 3 (2015): 409–33.

MacKenzie, David. *ICAO: A History of the International Civil Aviation Organization*. Toronto: University of Toronto Press, 2010.

MacKenzie, John M., ed. *Imperialism and the Natural World*. Manchester: Manchester University Press, 1990.

MacKenzie, John M. *Propaganda and Empire: The Manipulation of British Public Opinion (1880–1960)*. Manchester: Manchester University Press, 1984.

Maddox, Gregory H. "Networks and Frontiers in Colonial Tanzania." *Environmental History* 3, no. 4 (1998): 436–59.

Maddrell, Avril M. "Empire, Emigration and School Geography: Changing Discourses of Imperial Citizenship, 1880–1925." *Journal of Historical Geography* 22, no. 4 (1996): 373–87.

Mahony, Martin. "Climate Change and the Geographies of Objectivity: The Case of the IPCC's Burning Embers Diagram." *Transactions of the Institute of British Geographers* 40, no. 2 (2015): 153–67.

Mahony, Martin. "The 'Genie of the Storm': Cyclonic Reasoning and the Spaces of Weather Observation in the Southern Indian Ocean, 1851–1925." *British Journal for the History of Science* 51, no. 4 (2018): 607–33.

Mahony, Martin, and Georgina H. Endfield. "Climate and Colonialism." *WIREs Climate Change* 9, no. 2 (2018): e510.

Mahony, Martin, and Mike Hulme. "Modelling and the Nation: Institutionalizing Climate Prediction in the UK, 1988–1992." *Minerva* 54 (2016): 445–70.

Mahony, Martin, and Mike Hulme. "Model Migrations: Mobility and Boundary Crossings in Regional Climate Prediction." *Transactions of the Institute of British Geographers* 37 (2012): 197–211.

Maldonado, Tomas. "The Idea of Comfort." *Design Issues* 8, no. 1 (1991): 35–43.

Maraud, Simon, and Sylvain Guyot. "Mobilization of Imaginaries to Build Nordic Indigenous Natures." *Polar Geography* 39, no. 3 (July 2, 2016): 196–216.

Markham, S. F. *Climate and the Energy of Nations*. London: Oxford University Press, 1942.

Martin, Geoffrey J. *Ellsworth Huntington: His Life and Thought*. Hamden, CT: Archon Books, 1973.

Martin, James Ranald. *The Influence Of Tropical Climates on European Constitutions, Including Practical Observations On The Nature And Treatment Of The Diseases Of Europeans On Their Return From Tropical Climates*. London: John Churchill, 1856.

Martinez, Julia, and Adrian Vickers. *The Pearl Frontier: Indonesian Labor and Indigenous Encounters in Australia's Northern Trading Network*. Honolulu: University of Hawai'i Press, 2015.

Matless, David, and Laura Cameron. "Experiment in Landscape: The Norfolk Excavations of Marietta Pallis." *Journal of Historical Geography* 32, no. 1 (2006): 96–126.

Matson, Kelsey. "'The Ozone of Patriotism': Meteorology, Electricity, and the Body in the Nineteenth-Century Yellowstone Region." *History of Meteorology* 8 (2017): 35–53.

Mawson, Vivienne, David Tranter, and Alan Pearce, eds. *CSIRO at Sea: 50 Years of Marine Science.* Hobart: CSIRO Marine Laboratories, 1988.

McCloskey, Donald. "The Open Fields of England." In *Markets in History*, edited by David Galenson, 5–51. Cambridge: Cambridge University Press, 1989.

McEwan, Cheryl. *Gender, Geography and Empire: Victorian Women Travellers in West Africa.* London: Ashgate, 2000.

McEwan, Cheryl. "Paradise or Pandemonium? West African Landscapes in the Travel Accounts of Victorian Women." *Journal of Historical Geography* 22, no. 1 (1996): 68–83.

McFall, Liz. *Devising Consumption: Cultural Economies of Insurance, Credit and Spending.* London: Routledge, 2014.

McGregor, Kent. "Huntington's Cyclonic Man Theory and Ozone." *History of Geography Newsletter* 5 (1986): 14–17.

McKittrick, Meredith. "An Empire of Rivers: Climate Anxiety, Imperial Ambition, and the Hydropolitical Imagination in Southern Africa, 1919–1945." *Journal of Southern African Studies* 41, no. 3 (2015): 485–504.

McKittrick, Meredith. "Talking about the Weather: The Language of Environmental Crisis in South Africa, 1915–1945." *Environmental History* 23, no. 1 (2018): 3–27.

McKittrick, Meredith. "Theories of 'Reprecipitation' and Climate Change in the Settler Colonial World." *History of Meteorology* 8 (2017): 74–94.

McMichael, A. J. "Global Climate Change and Health: An Old Story Writ Large." In McMichael, Campbell-Lendrum, Crvalán, Ebi, Githeko, Scheraga, and Woodward, *Climate Change and Human Health*, 1–17.

McMichael, A. J., D. H. Campbell-Lendrum, C. F. Crvalán, K. L. Ebi, A. K. Githeko, J. D. Scheraga, and A. Woodward, eds. *Climate Change and Human Health.* Geneva: World Health Organization, 2003.

McNeil, John R., and Peter Engelke. *The Great Acceleration: An Environmental History of the Anthropocene since 1945.* Cambridge, MA: Belknap Press of Harvard University Press, 2014.

McNeill, John, and Corinna Unger, eds. *Environmental Histories of the Cold War.* New York: Cambridge University Press, 2010.

Meloni, Maurizio. *Political Biology: Science and Social Values in Human Heredity from Eugenics to Epigenetics.* London: Palgrave, 2016.

Meyer, William B. *Americans and Their Weather.* Oxford: Oxford University Press, 2000.

Miller, Clark. "Climate Science and the Making of a Global Political Order." In Jasanoff, *States of Knowledge*, 46–66.

Miller, Clark A. "Resisting Empire: Globalism, Relocalization, and the Politics of Knowledge." In Jasanoff and Martello, *Earthly Politics*, 81–102.

Miller, Julia. "The Fall of an Angel: Gendering and Demonizing El Niño." *World History Connected* 4, no. 3 (2007).

Mills, Clarence A. *Air Pollution and Community Health.* Boston: Christopher Publishing House, 1954.

Mills, Clarence A. *Climate Makes the Man.* London: Victor Gollancz, 1946.

Mills, Clarence A. "Climatic Effects on Growth and Development." *American Anthropologist* 44 (1942): 1–13.

Mills, Clarence A. *Living with the Weather.* Cincinnati: Caxton Press, 1934.

Mills, Clarence A. *Medical Climatology: Climatic and Weather Influences in Health and Disease.* London: Baillière, Tindall and Cox, 1939.

Mills, Clarence A. *This Air We Breathe.* Boston: Christopher Publishing House, 1962.

Mills, Clarence A. *World Power amid Shifting Climates.* Boston: Christopher Publishing House, 1963.

Mills, Sara. "Gender and Colonial Space." *Gender, Place and Culture: A Journal of Feminist Geography* 3, no. 2 (1996): 125–48.

Mills, Sara. *Gender and Colonial Space.* Manchester: Manchester University Press, 2009.

Mindlin, Henrique. *Modern Architecture in Brazil.* New York: Reinhold, 1957.

Missenard, André. *L'Homme et le climat.* Paris: Librairie Plon, 1936.

Mitchell, Timothy. *Carbon Democracy: Political Power in the Age of Oil.* New York: Verso, 2011.

Mitchell, Timothy. "Econ<i>mentality: How the Future Entered Government." *Critical Inquiry* 40, no. 4 (Summer 2014): 479–507.

Miura, T. "The Influence of Seasonal Atmospheric Factors on Human Reproduction." *Experientia* 43 (1987): 48–54.

Moffat, Robert. *Missionary Labors and Scenes in Southern Africa.* 9th ed. New York: Robert Carter, 1846.

Mol, Annemarie. *The Body Multiple: Ontology in Medical Practice.* Durham, NC: Duke University Press, 2002.

Morgan, Gary. "Changes in Fishing Practice, Fleet Capacity and Ownership of Harvesting Rights in the Rock Lobster Fishery of Western Australia." In *Case Studies on the Effects of Transferable Fishing Rights on Fleet Capacity and Concentration of Quota Ownership*, edited by Ross Shotton, 80–88. Rome: FAO, 2001.

Mumford, Eric. *The CIAM Discourse on Urbanism, 1928–1960.* Cambridge, MA: MIT Press, 2000.

Murphy, Michelle. *Sick Building Syndrome and the Problem of Uncertainty: Environmental Politics, Technoscience, and Women Workers.* Durham, NC: Duke University Press, 2006.

Murphy, Sharon Ann. *Investing in Life: Insurance in Antebellum America.* Baltimore, MD: Johns Hopkins University Press, 2010.

Musselman, Elizabeth G. "Worlds Displaced: Projecting the Celestial Environment from the Cape Colony." *Kronos: Journal of Cape History* 29 (2003): 64–85.

Myddelton, David R. *They Meant Well: Government Project Disasters.* London: Institute of Economic Affairs, 2007.

National Research Council. *The Geography of Europe.* Edited by Ellsworth Huntington and Herbert E. Gregory. New Haven, CT: Yale University Press, 1918.

Naylor, Simon. "Log Books and the Law of Storms: Maritime Meteorology and the British Admiralty in the Nineteenth Century." *Isis* 106 (2015): 771–97.

Neutra, Richard. *Architecture of Social Concern in Regions of Mild Climate.* São Paulo: Gerth Todtmann, 1948.

Nonaka, K. "Effect of Delivery Season on Subsequent Birth Interval in the Early Twentieth Century in Japan." *International Journal of Biometeorology* 33 (1989): 238–45.

Normand, C. "Monsoon Seasonal Forecasting." *Quarterly Journal of the Royal Meteorological Society* 79, no. 342 (1953): 463–73.

Normand, C. "Sir Gilbert Walker, CSI, FRS." *Nature* 182 (1958): 1706.

O'Gorman, Emily, James Beattie, and Matthew Henry. "Epilogue: Future Research Directions." In Beattie, O'Gorman, and Henry, *Climate, Science and Colonization*, 251–52.

Oldfield, Jonathan D. "Imagining Climates Past, Present and Future: Soviet Contributions to the Science of Anthropogenic Climate Change, 1953–1991." *Journal of Historical Geography* 60 (2018): 41–51.

Olgyay, Aladar, and Victor Olgyay. *Solar Control and Shading Devices*. New York: Reinhold, 1957.

Oreskes, Naomi. "Scaling Up Our Vision." *Isis* 105, no. 2 (2014): 379–91.

Orlove, Ben, Heather Lazrus, Grete K. Hovelsrud, and Alessandra Giannini. "Recognitions and Responsibilities: On the Origins and Consequences of the Uneven Attention to Climate Change around the World." *Current Anthropology* 55, no. 3 (2014): 249–75.

Osborne, Michael A. "Acclimatizing the World: A History of the Paradigmatic Colonial Science." *Osiris* 15 (2000): 135–51.

Osborne, Michael A. "Resurrecting Hippocrates: Hygienic Sciences and the French Scientific Expeditions to Egypt, Morea and Algeria." In *Warm Climates and Western Medicine: The Emergence of Topical Medicine, 1500–1900*, edited by David Arnold, 80–98. Amsterdam: Rodopi, 1996.

Osborne, Michael A., and Richard S. Fogarty. "Medical Climatology in France: The Persistence of Neo-Hippocratic Ideas in the First Half of the Twentieth Century." *Bulletin of the History of Medicine* 86 (2012): 543–63.

Overy, Paul. *Light, Air, and Openness: Modern Architecture between the Wars*. London: Thames and Hudson, 2004.

Pariwono, John, Abdul Gani Ilahude, and Malikusworo Hutomo. "Progress in Oceanography of Indonesian Seas: A Historical Perspective." *Oceanography* 18, no. 4 (2005): 42–49.

Passanti, Francesco. "The Modern, the Vernacular, Le Corbusier." In *Vernacular Modernism: Heimat, Globalization, and the Built Environment*, edited by Maiken Umbach and Bernd Huppauf, 141–56. Stanford, CA: Stanford University Press, 2005.

Pearce, Alan, and Bruce Phillips. "ENSO Events, the Leeuwin Current, and Larval Recruitment of the Western Rock Lobster." *Journal du Conseil International pour l'Exploration de la Mer* 45, no. 1 (1988): 13–21.

Peires, Jeffrey B. *The House of Phalo: A History of the Xhosa People in the Days of Their Independence*. Berkeley: University of California Press, 1982.

Penn, Jim, Nick Caputi, and Simon de Lestang. "A Review of Lobster Fishery Management: The Western Australian Fishery for *Panuliris cygnus*, A Case Study in the Development and Implementation of Input and Output-Based Management Systems." *ICES Journal of Marine Science* 72 (2015): 122–134.

Peters, Kimberley. "Future Promises for Contemporary Social and Cultural Geographies of the Sea." *Geography Compass* 4, no. 9 (2010): 1260–72.

Petersen, William F. *Hippocratic Wisdom, for Him Who Wishes to Pursue Properly the Science of Medicine: A Modern Appreciation of Ancient Scientific Achievement*. Springfield, IL: Charles C. Thomas, 1946.

Petersen, William F. *Lincoln–Douglas: The Weather as Destiny*. Springfield, IL: Charles C. Thomas, 1943.

Petersen, William F. *Man Weather Sun*. 1878. Springfield, IL: Charles C. Thomas, 1947.

Petersen, William F. *The Patient and the Weather. Vol. 1, Part 2: Autonomic Integration*. Ann Arbor: Edward Bros., 1936.

Petersen, William F. *The Patient and the Weather. Vol. 2: Autonomic Dysintegration.* Ann Arbor: Edward Bros., 1934.

Petersen, William F. *The Patient and the Weather. Vol. 3: Mental and Nervous Diseases.* Ann Arbor: Edward Bros., 1937.

Petersen, William F. *Protein Therapy and Non-specific Reactions.* New York: Macmillan, 1922.

Petersen, William F., and Samuel A. Levinson. *The Skin Reactions, Blood Chemistry and Physical Status of "Normal" Men and Clinical Patients.* Chicago: American Medical Association, 1930.

Petterssen, Sverre. *Weathering the Storm: Sverre Petterssen, the D-Day Forecast and the Rise of Modern Meteorology.* Edited by James Rodger Fleming. Boston, MA: American Meteorological Society, 2001.

Philander, S. G. H. "El Niño and La Niña." *Journal of the Atmospheric Sciences* 42, no. 23 (1985): 2652–62.

Phillips, Sarah. "Lessons from the Dust Bowl: Dryland Agriculture and Soil Erosion in the United States and South Africa, 1900–1950." *Environmental History* 4, no. 2 (1999): 245–66.

Pickering, Andrew. *The Mangle of Practice: Time, Agency and Science.* Chicago: University of Chicago Press, 1995.

Pickles, Katie. "Forgotten Colonizers: The Imperial Order Daughters of the Empire (IODE) and the Canadian North." *Canadian Geographer/Le Géographe Canadien* 42, no. 2 (1998): 193–204.

Pickles, Katie. "A Link in 'The Great Chain of Empire Friendship': The Victoria League in New Zealand." *Journal of Imperial and Commonwealth History* 33, no. 10 (2005): 29–50.

Pincock Christopher. "From Sunspots to the Southern Oscillation: Confirming Models of Large-Scale Phenomena in Meteorology." *Studies in the History and Philosophy of Science* 40 (2009): 45–56.

Pitt, David. *E. J. Pratt: The Master Years, 1927–1964.* Toronto: University of Toronto Press, 1984.

Pitt, David. *E. J. Pratt: The Truant Years, 1882–1927.* Toronto: University of Toronto Press, 1984.

Pollock, Anne, and Banu Subramaniam. "Resisting Power, Retooling Justice: Promises of Feminist Postcolonial Technosciences." *Science, Technology and Human Values* 41, no. 6 (November 1, 2016): 1–16.

Porteous, Colin. *The New Eco-architecture: Alternatives from the Modern Movement.* London: Taylor and Francis, 2002.

Porter, Theodore M. "Life Insurance, Medical Testing, and the Management of Mortality." In *Biographies of Scientific Objects*, edited by Lorraine Daston, 226–46. Chicago: University of Chicago Press, 2000.

Porter, Theodore M. *Trust in Numbers: The Pursuit of Objectivity in Science and Public Life.* Princeton, NJ: Princeton University Press, 1995.

Powell, Richard C. "Geographies of Science: Histories, Localities, Practices, Futures." *Progress in Human Geography* 31, no. 3 (2007): 309–29.

Prakash Gyan. *Another Reason: Science and the Imagination of Modern India.* Princeton, NJ: Princeton University Press, 1999.

Pratt, E. J. *Behind the Log.* Toronto: Macmillan, 1947.

Prior, K. "Buist, George (1804–1860)." *Oxford Dictionary of National Biography.* Oxford: Oxford University Press, 2004. Online edition, January 2013. https://doi.org/10.1093/ref:odnb/3892/.

Proctor, Robert. *Racial Hygiene: Medicine under the Nazis.* Cambridge, MA: Harvard University Press, 1988.

Radcliffe, Sarah A. "Relating to the Land: Multiple Geographical Imaginations and Lived-In Landscapes." *Transactions of the Institute of British Geographers* 37, no. 3 (2012): 359–64.

Rafael, Vicente L. "Colonial Domesticity: White Women and United States Rule in the Philippines." *American Literature* 67, no. 4 (1995): 639–66.

Rainger, Ronald. "Edward 'Iceberg': Smith and the Changing Character of American Arctic Oceanography." In *Extremes: Oceanography's Adventures at the Poles*, edited by Keith Benson and Helen Rozwadowski, 133–72. Sagamore Beach, MA: Science History Publications, 2007.

Raj, Kapil. *Relocating Modern Science: Circulation and the Construction of Knowledge in South Asia and Europe, 1650–1900*. Basingstoke: Palgrave Macmillan, 2007.

Randalls, Samuel. "Contributions and Perspectives from Geography to the Study of Climate." *WIREs Climate Change* 8, no. 4 (2017): e466.

Requena-Ruiz, Ignacio. "Building Artificial Climates: Thermal Control and Comfort in Modern Architecture." *Ambiances: Environment Sensible, Architecture et Espace Urbain* (2016): 1–22.

Riedi, Eliza. "Teaching Empire: British and Dominions Women Teachers in the South African War Concentration Camps." *English Historical Review* 120, no. 489 (2005): 1316–47.

Riedi, Eliza. "Women, Gender, and the Promotion of Empire: The Victoria League, 1901–1914." *Historical Journal* 45, no. 3 (2002): 569–99.

Ripley, William Z. *The Races of Europe: A Sociological Study*. New York: D. Appleton, 1899.

Ritter, David. "The Rejection of *Terra Nullius* in *Mabo*: A Critical Analysis." *Sydney Law Review* 18, no. 5 (1996): 5–33.

Rizzo, Matteo. "What Was Left of the Groundnut Scheme? Development Disaster and Labour Market in Southern Tanganyika, 1946–1952." *Journal of Agrarian Change* 6, no. 2 (2006): 205–38.

Robinson, Samuel A. *Ocean Science and the British Cold War State*. London: Palgrave Macmillan, 2018.

Rocha-Peixoto, Gustavo. "Prefacio: Rigor Arejado e Solar." In Luis Felipe Machado Coelho de Souza, *Irmaos Roberto, Arquitetos*, 15–25. Rio de Janeiro: Rio Books, 2013.

Rosen, Stephen. "Guestwords: Spring Fever." *East Hampton Star*, April 17, 2013. https://easthamptonstar.com/archive/guestwords-spring-fever/.

Rosen, Stephen. *Weathering: How the Atmosphere Conditions Your Body, Your Mind, Your Moods—and Your Health*. New York: M. Evans, 1975.

Rosenberg, Charles E. "Epilogue: *Airs, Waters, Places*. A Status Report." *Bulletin of the History of Medicine* 86 (2012): 661–70.

Rozwadowski, Helen. "Arthur C. Clarke and the Limitations of the Ocean as a Frontier." *Environmental History* 17, no. 3 (2012): 1–25.

Rozwadowski, Helen. *Fathoming the Ocean: The Discovery and Exploration of the Deep Sea*. Cambridge, MA: Harvard University Press, 2005.

Rozwadowski, Helen. "Oceans: Fusing the History of Science and Technology with Environmental History." In *A Companion to American Environmental History*, edited by D.C. Sackman, 442–61. Malden, MA: Wiley-Blackwell, 2010.

Rozwadowski, Helen. "Ocean's Depths." *Environmental History* 15, no. 3 (2010): 520–25.

Rozwadowski, Helen. *The Sea Knows No Boundaries: A Century of Marine Science under ICES*. Seattle: University of Washington Press, 2002.

Rudiak-Gould, Peter. "'We Have Seen It with Our Own Eyes': Why We Disagree about Climate Change Visibility." *Weather, Climate and Society* 5, no. 2 (2013): 120–32.

Rupke, Nicolaas A. "Adolf Mühry (1810–1888): Göttingen's Humboldtian Medical Geographer." *Medical History* 44, no. S20 (2000): 86–97.

Rupke, Nicolaas A., and Karen E. Wonders. "Humboldtian Representations in Medical Cartography." *Medical History* 44, no. S20 (2000): 163–75.

Said, Edward W. *Orientalism*. New York: Vintage Books, 1979.

Sanders, Todd. *Beyond Bodies: Rain-Making and Sense-Making in Tanzania*. Toronto: University of Toronto Press, 2008.

Sappol, Michael. *Body Modern: Fritz Kahn, Scientific Illustration and the Homuncular Subject*. Minneapolis: University of Minneapolis Press, 2017.

Sargent, Frederick II. *Hippocratic Heritage: A History of Ideas about Weather and Human Health*. New York: Pergamon Press, 1982.

Sargent, Frederick II. "Nature and Scope of Human Ecology." In *Human Ecology*, edited by Frederick Sargent II, 1–25. Amsterdam: North-Holland Publishing, 1974.

Savage Victor R. "Tropicality Imagined and Experienced. A Commentary on Felix Driver's 'Imagining the Tropics': Views and Visions of the Tropical World." *Singapore Journal of Tropical Geography* 25, no. 1 (2004): 26–31.

Schaffer, Simon. "Astronomers Mark Time: Discipline and the Personal Equation." *Science in Context* 2, no. 1 (1988): 115–45.

Schaffer, Simon. "Astronomy at the Imperial Meridian: The Colonial Production of Hybrid Spaces." Keynote lecture, International Conference of Historical Geographers, London, July 9, 2015.

Schaffer, Simon. "The Eighteenth Brumaire of Bruno Latour." *Studies in History and Philosophy of Science* 22, no. 1 (1991): 174–92.

Schaffer, Simon. "Keeping the Books at Paramatta Observatory." In Aubin, Bigg, and Sibum, *Heavens on Earth*, 118–47.

Schneider, Ronald M. *"Order and Progress": A Political History of Brazil*. San Francisco: Westview Press, 1991.

Schnukal, Anna, Guy Ramsay, and Yuriko Nagata, eds. *Navigating Boundaries: The Asian Diaspora in Torres Strait*. Canberra: Pandanus Books, 2004.

"Science: Weather as Destiny." *Time Magazine*, August 2, 1943.

Scott, James C. *Seeing like a State: How Certain Schemes to Improve the Human Condition Have Failed*. New Haven, CT: Yale University Press, 1998.

Segawa, Hugo. *Architecture of Brazil, 1900–1990*. New York: Springer, 2013.

Sen, Joydeep. *Astronomy in India, 1784–1876*. London: Pickering and Chatto, 2014.

Seth, Suman. "Putting Knowledge in Its Place: Science, Colonialism, and the Postcolonial." *Postcolonial Studies* 12, no. 4 (2009): 380.

Shapin, Steven. "Cordelia's Love: Credibility and the Social Studies of Science." *Perspectives on Science* 3, no. 3 (1995): 255–75.

Shapin, Steven. "The House of Experiment in Seventeenth-Century England." *Isis* 79, no. 3 (January 1988): 373–404.

Shapin, Steven. "Placing the View from Nowhere: Historical and Sociological Problems in the Location of Science." *Transactions of the Institute of British Geographers* 23, no. 1 (April 1998): 5–12.

Shapin, Steven, and Simon Schaffer. *Leviathan and the Air Pump: Hobbes, Boyle, and the Experimental Life*. Princeton, NJ: Princeton University Press, 1985.

Shaw, Jenny. "Fishing Communities and Climate Change: Impacts and the Co-production of Knowledge." PhD dissertation, Curtin University, Bentley and Perth, Western Australia, 2016.

Shaw, Jenny, Laura Stocker, and Leonie Noble. "Climate Change and Social Impacts: Women's Perspectives from a Fishing Community in Western Australia." *Australian Journal of Maritime and Ocean Affairs* 7, no. 1 (2015): 38–51.

Shaw, William. *The Story of My Mission in South-Eastern Africa*. London: Hamilton, Adams, 1860.

Sheppard, P. A. "Sir Gilbert Walker, CSI FRS." *Quarterly Journal of the Royal Meteorological Society* 85, no. 364 (1959): 186.

Sikka, D. R. "The Role of the India Meteorological Department, 1875–1947." In *Science and Modern India: An Institutional History, c. 1784–1947*, edited by Uma Das Gupta, 381–427. Delhi: Pearson, 2011.

Simon, Richard B., Mojan Behmand, and Thomas Burke, eds. *Teaching Big History*. Oakland: University of California Press, 2015.

Sletto, Bjørn Ingmunn. "A Swamp and Its Subjects: Conservation Politics, Surveillance and Resistance in Trinidad, the West Indies." *Geoforum* 36, no. 1 (2005): 77–93.

Sloterdijk, Peter. *In the World Interior of Capital: Towards a Philosophical Theory of Globalization*. Cambridge, UK: Polity Press, 2013.

Sloterdijk, Peter. *Spheres II: Globes*. Cambridge, MA: MIT Press, 2014.

Somol, Robert, and Sárah Whiting. "Notes around the Doppler Effect and Other Moods of Modernism." *Perspecta* 33 (2002): 72–77.

Spate, O. H. K. "Ellsworth Huntington." *International Encyclopedia of the Social Sciences* 7 (1968): 26.

Spivak, Gayatri C. *Death of a Discipline*. New York: Columbia University Press, 2003.

Sprague, Alfred. "Review: Distribution of Diseases in Africa." *JIA* 32, no. 1 (1895): 65–70.

Stanley, Brian. "The Missionary the Rainmaker: David Livingstone, the Bakwena, and the Nature of Medicine." *Social Sciences and Missions* 27, no. 2–3 (2014): 145–62.

Steinberg, Philip. "Mediterranean Metaphors: Travel, Translation and Oceanic Imaginaries in the 'New Mediterraneans' of the Arctic Ocean, the Gulf of Mexico and the Caribbean." In *Water Worlds: Human Geographies of the Ocean*, edited by Jon Anderson and Kimberley Peters, 23–47. New York: Ashgate Publishing, 2014.

Steinberg, Philip. "Navigating to Multiple Horizons: Towards a Geography of Ocean Space." *Professional Geographer* 51, no. 3 (1999): 366–75.

Steinberg, Philip. "Of Other Seas: Metaphors and Materialities in Maritime Regions." *Atlantic Studies* 10, no. 2 (2013): 156–69.

Steinberg, Philip. *The Social Construction of the Ocean*. Cambridge: Cambridge University Press, 2001.

Steinberg, Philip, and Kimberley Peters. "Wet Ontologies, Fluid Spaces: Giving Depth to Volume through Oceanic Thinking." *Environment and Planning D: Society and Space* 33 (2015): 247–64.

Stengers, Isabelle. *In Catastrophic Times: Resisting the Coming Barbarism*. Paris: Open Humanities Press, 2015.

Stepan, Nancy. *Picturing Tropical Nature*. Ithaca, NY: Cornell University Press, 2001.

Stern, Steve J. *The Secret History of Gender: Women, Men, and Power in Late Colonial Mexico.* Chapel Hill: University of North Carolina Press, 1997.

Stockland, Etienne. "Policing the Oeconomy of Nature: The Oiseau Martin as an Instrument of Oeconomic Management in the Eighteenth-Century French Maritime World." *History and Technology* 30, no. 3 (2014): 207–31.

Stolberg, Eva-Maria. "'From Icy Backwater to Nuclear Waste Ground': The Russian Arctic Ocean in the Twentieth Century." In *Sea Narratives: Cultural Responses to the Sea, 1600–Present*, edited by Charlotte Mathieson, 111–38. London: Palgrave Macmillan, 2016.

Storey, William K. "Cecil Rhodes and the Making of a Sociotechnical Imaginary for South Africa." In *Dreamscapes of Modernity: Sociotechnical Imaginaries and the Fabrication of Power*, edited by Sheila Jasanoff and Sang-Hyun Kim, 34–55. Chicago: University of Chicago Press, 2015.

Strengers, Yolande, and Cecily Maller. "Adapting to 'Extreme' Weather: Mobile Practice Memories of Keeping Warm and Cool as a Climate Change Adaptation Strategy." *Environment & Planning: A* 49, no. 6 (2017): 1432–50.

Strobel, M. *European Women and the Second British Empire.* Bloomington: Indiana University Press, 1991.

Sturgis, James. "Anglicisation at the Cape of Good Hope in the Early Nineteenth Century." *Journal of Imperial and Commonwealth History* 11, no. 1 (1982): 5–32.

Sturken, Marita. "Desiring the Weather: El Niño, the Media, and California Identity." *Public Culture* 13, no. 2 (2001): 161–89.

Strauss, Sarah, and Ben S. Orlove, eds. *Weather, Climate, Culture.* Oxford: Berg Publishers, 2003.

Swaisland, C. *Servants and Gentlewomen to the Golden Land: The Emigration of Single Women from Britain to Southern Africa, 1820–1939.* London: Bloomsbury, 1993.

Taub, Liba. "Introduction: Reengaging with Instruments." *Isis* 102, no. 4 (2011): 689–96.

Taylor, Charles. *Modern Social Imaginaries.* Durham, NC: Duke University Press, 2003.

Taylor, G. I. "Gilbert Thomas Walker, 1868–1958." *Biographical Memoirs of Fellows of the Royal Society* 8 (1962): 166–74.

Taylor, Joseph. "Knowing the Black Box: Methodological Challenges in Marine Environmental History." *Environmental History* 18, no. 1 (2013): 60–75.

Thomas, Morley K. "A Brief History of Meteorological Services in Canada, Part I: 1839–1930." *Atmosphere* 9, no. 1 (1971): 1–8.

Thomson, Spencer C. "Address to the Members of the Actuarial Society of Edinburgh, 1st November 1877." *JIA* 21, no. 3 (1878).

Tilley, Helen. *Africa as a Living Laboratory: Empire, Development, and the Problem of Scientific Knowledge, 1870–1950.* Chicago: University of Chicago Press, 2011.

Tinkler, P. "Introduction to Special Issue: Women, Imperialism and Identity." In *Women's Studies International Forum* 21 (May 1998): 217–22.

Tisdale, Sallie. "Confessions of a Cosmic Resonator." *Outside Magazine*, May 2, 2004. http://www.outsideonline.com/1837011/confessions-cosmic-resonator/.

Tout, David G. "Biometeorology." *Progress in Physical Geography* 11 (1987): 473–86.

Tromp, S. W. *Biometeorology: The Impact of the Weather and Climate on Humans and Their Environment.* London: Heyden, 1980.

Tsanoff, Radoslav A. *The Ways of Genius.* New York: Harper, 1949.

Tsing, Anna L. *Friction: An Ethnography of Global Connection.* Princeton, NJ: Princeton University Press, 2011.

Tull, Malcolm. "Profits and Lifestyle: Western Australia's Fishers." *Studies in Western Australian History* 8 (1992): 92–111.

Tull, Malcolm, Sarah Metcalf, and Howard Gray. "The Economic and Social Impacts of Environmental Change on Fishing Towns and Coastal Communities: A Historical Case Study of Geraldton, Western Australia." *ICES Journal of Marine Science* 73, no.5 (2016): 1437–46.

Turnbull, David. *Masons, Tricksters and Cartographers: Comparative Studies in the Sociology of Scientific and Indigenous Knowledge.* London: Routledge, 2000.

Turner, Roger. "'Weathering Heights': The Emergence of Aeronautical Climatology as an Infrastructural Science." PhD dissertation, University of Pennsylvania, Philadelphia, 2010.

Turnhout, Esther, Art Dewulf, and Mike Hulme. "What Does Policy-Relevant Global Environmental Knowledge Do? The Cases of Climate and Biodiversity." *Current Opinions in Environmental Sustainability* 18 (2016): 65–72.

Ulanski, Stan. *The California Current: A Pacific Ecosystem and Its Fliers, Divers and Swimmers.* Chapel Hill: University of North Carolina Press, 2016.

Ulanski, Stan. *The Gulf Stream: Tiny Plankton, Giant Bluefin, and the Amazing Story of the Power River in the Atlantic.* Chapel Hill: University of North Carolina Press, 2008.

US Centers for Disease Control and Prevention (CDC). "Climate Change and Health Effects." http://www.cdc.gov/climateandhealth/effects/.

US Senate, Subcommittee of the Committee on Interstate and Foreign Commerce. *Hearings on S. 1853, A Bill to Authorize the Coast Guard to Establish, Maintain and Operate Aids to Navigation, and S. 2122, A Bill to Authorize the Coast Guard to Operate and Maintain Ocean Stations.* 80th Congress, 2nd sess., March 29, 1948.

Vaj, Daniela. "Medical Geography and Phthisic Immunity in the High Altitudes: The Origins of a Therapeutic Hypothesis." *Journal of Alpine Research* 93, no. 1 (2005): 34–42.

Van Helten, Jean Jacques, and Keith Williams. "'The Crying Need of South Africa': The Emigration of Single British Women to the Transvaal, 1901–1910." *Journal of Southern African Studies* 10, no. 1 (1983): 17–38.

Van Riper, Bowdoin. *Imagining Flight: Aviation and Popular Culture.* College Station: Texas A&M Press, 2004.

Verchraegen, Gert, Frédéric Vandermoere, Luc Braeckmans, and Barbara Segaert, eds. *Imagined Futures in Science, Technology and Society.* New York: Routledge, 2017.

Vetter, Jeremy, ed. *Knowing Global Environments: New Historical Perspectives on the Field Sciences.* New Brunswick, NJ: Rutgers University Press, 2011.

Villela, Maria Teresa, Bandeira de Mello, and Aline Lowes de Lacerda. "Images of Public Health: The Institutionalization of the Institute Oswaldo Cruz in Brazil." *Dynamis* 25 (2005): 179–98.

Viseltear, Arthur J. "C.-E.A. Winslow and the Early Years of Public Health at Yale, 1915–1925." *Yale Journal of Biology and Medicine* 55 (1982): 137–51.

von Debshitz, Uta, and Thilo von Debshitz. *Fritz Kahn: Infographics Pioneer.* Taschen: Bibliotheca Universalis, 2017.

Walker, David, and Agnieszka Sobocinska, eds. *Australia's Asia: From Yellow Peril to Asian Century.* Perth, WA: University of Western Australia Publishing, 2012.

Walker, Gilbert T. "World Weather." *Quarterly Journal of the Royal Meteorological Society* 54, no. 226 (1928): 79–87.

Walker, Gilbert T., and E. W. Bliss. "World Weather V." *Memoirs of the Royal Meteorological Society* 4, no. 36 (1932): 53–84.

Walker, James M. "Pen Portraits of Presidents—Sir Gilbert Walker, CSI, ScD, MA, FRS." *Weather* 52, no. 7 (1997): 217–20.

Walter, Albert. *The Sugar Industry of Mauritius: A Study in Correlation; Including a Scheme of Insurance of the Cane Crop against Damage Caused by Cyclones.* London: A. L. Humphreys, 1910.

Weart, Spencer. *The Discovery of Global Warming.* 1st ed. Cambridge, MA: Harvard University Press, 2003.

Weihe, Wolf H. "Review on the History of the International Society of Biometeorology." *International Journal of Biometeorology* 40 (1997): 9–15.

Wheeler, Roxann. "Limited Visions of Africa: Geographies of Savagery and Civility in Early Eighteenth-Century Narratives." In *Writes of Passage: Reading Travel Writing*, edited by James S. Duncan and Derek Gregory, 14–48. London: Routledge, 1999.

White, Richard. "'Are You an Environmentalist or Do You Work for a Living?' Work and Nature." In *Uncommon Ground: Rethinking the Human Place in Nature*, edited by William Cronon, 171–85. New York: W. W. Norton, 1996.

Williams, Daryle. *Culture Wars in Brazil: The First Vargas Regime, 1930–1945.* Durham, NC: Duke University Press, 2001.

Williams, Dee Mack. *Beyond Great Walls: Environment, Identity, and Development on the Chinese Grasslands of Inner Mongolia.* Stanford, CA: Stanford University Press, 2002.

Williamson, Fiona. "Weathering the Empire: Meteorological Research in the Early British Straits Settlements." *British Journal for the History of Science* 48 (2015): 475–92.

Willis, Edmund P., and William H. Hooke. "Cleveland Abbe and American Meteorology." *Bulletin of the American Meteorological Society* 87, no. 3 (2006): 315–26.

Wilson, Jon. *India Conquered: Britain's Raj and the Chaos of Empire.* London: Simon and Schuster, 2016.

Winder, Gordon M., and Michael Schmitt. "Geographical Imaginaries in the *New York Times*' Reports of the Assassinations of Mahatma Gandhi (1948) and Indira Gandhi (1984)." *Journal of Historical Geography* 45 (2014): 106–15.

Winslow, C-E. A., and L. P. Herrington. *Temperature and Human Life.* Princeton, NJ: Princeton University Press, 1949.

Wintersteen, Kristin. "Fishing for Food and Fodder: The Transnational Environmental History of Humboldt Current Fisheries in Peru and Chile since 1945." PhD dissertation, Duke University, 2011.

Wood, Alan. *The Groundnut Affair.* London: Bodley Head, 1950.

Wright Mills, C. *The Sociological Imagination.* Oxford: Oxford University Press, 1959.

Yusoff, Kathryn, and Jennifer Gabrys. "Climate Change and the Imagination." *WIREs Climate Change* 2, no. 4 (2011): 516–34.

Zeller, Suzanne. *Inventing Canada: Early Victorian Science and the Idea of a Transcontinental Nation.* Toronto: University of Toronto Press, 1987.

CONTRIBUTORS

George Adamson is a senior lecturer in geography at King's College London, UK. He conducts interdisciplinary research on climate history, climate-society interactions, and the history of climate science and knowledges. He has published widely in geography, climatology, and environmental history and recently coauthored, with Richard Grove, the book *El Niño in World History.* He is currently working on the science and politics of El Niño prediction and on the role that understandings of historical climate-society interactions can play in informing contemporary climate change adaptation.

Katharine Anderson is professor in the Department of Humanities at York University, Canada. She is the author of *Predicting the Weather: Victorians and the Science of Meteorology* (University of Chicago Press, 2005) and coeditor with Helen Rozwadowski of *Soundings and Crossings: Doing Science at Sea, 1800–1970* (Science History Publications, 2015). Recent articles examine Victorian hydrography and voyage narratives, following an annotated edition of the captains' narratives in the surveying voyages of HMS *Adventure and Beagle* (Pickering and Chatto, 2011). She is currently working on a book on scientific expeditions in ocean environments in the 1920s and 1930s.

Daniel A. Barber is an associate professor of architecture and chair of the Graduate Group in Architecture at the University of Pennsylvania Stuart Weitzman School of Design. His research explores the historical relationship between architecture and global environmental culture with a particular focus on climate-focused architectural design. He is the author of *A House in the Sun: Modern Architecture and Solar Energy in the Cold War* (2016), and he is currently working on a second manuscript, "Modern Architecture and Climate: Design before Air-Conditioning" (expected 2020). Daniel has published numerous other articles, including papers in *Technology and Culture*, *Public Culture*, and *Log*.

Georgina Endfield is a professor in the Faculty of Humanities and Social Sciences at the University of Liverpool. A historical geographer and environmental historian, she has published widely on climatic history and historical climatology, on human responses to unusual or extreme weather events, conceptualizations of climate variability in historical perspective, and the links between climate and the healthiness of place. She is the author of *Climate and Society in Colonial Mexico: A Study in Vulnerability* and coeditor of the recent volume *Cultural Histories, Memories and Extreme Weather: An Historical Geography Perspective.*

James R. Fleming is the Charles A. Dana Professor of Science, Technology and Society at Colby College. His research interests include the history of the geophysical sciences, especially meteorology and climate change, and he has written extensively on the history of weather, climate, technology, and the environment, including their social, cultural, and intellectual aspects. His books include *Meteorology in America, 1800–1870* (Johns Hopkins, 1990), *Historical Perspectives on Climate Change* (Oxford , 1998), *The Calendar Effect* (American Meteorological Society, 2007), *Fixing the Sky* (Columbia, 2010), and *Inventing Atmospheric Science* (MIT, 2016).

Matthew Goodman is a research manager for the Kew Foundation, which is part of the Royal Botanic Gardens, Kew, London. His PhD was titled "From 'Magnetic Fever' to 'Magnetical Insanity': Historical Geographies of British Terrestrial Magnetic Research, 1833–1857" (University of Glasgow, 2018). He has published articles on aspects of this research in *Notes and Records* (2016, 2019) and *Scientia Canadensis* (2016–2017).

Mike Hulme is professor of human geography and fellow of Pembroke College, University of Cambridge, UK. His interests include relationships among climate, history, and culture and the historical development of cultural and scientific knowledge of climate and its changes. He is the author of *Contemporary Climate Change Debates: A Student Primer* (Routledge, 2020), *Weathered: Cultures of Climate* (SAGE, 2016), *Can Science Fix Climate Change? A Case against Climate Engineering* (Polity, 2014), and *Why We Disagree about Climate Change* (Cambridge, 2009).

James Kneale is an associate professor in the Department of Geography, University College London. He works on the cultural and historical geographies of consumption, health policy, and insurance as well as on literary geographies and representations of space in nonrealist genres such as horror and science fiction. His work has been published in the *Journal of Historical Geography, Society & Space, History of Meteorology*, and *Cultural Geographies.*

David N. Livingstone is professor of geography and intellectual history at Queen's University Belfast, UK. He works on the histories of geographical knowledge, the spatiality of scientific culture, the social history of environmental determinism, and the historical geographies of science and religion. He is the author of *Dealing with Darwin: Place, Politics, and Rhetoric in Religious Engagements with Evolution* (Johns Hopkins, 2014), *Adam's Ancestors: Race, Religion and the Politics of Human Origins* (Johns Hopkins, 2008), *Putting Science in its Place: Geographies of Scientific Knowledge* (Chicago, 2003), *The Geographical Tradition* (Blackwell, 1992), and *Nathaniel Southgate Shaler and the Culture of American Science* (Alabama, 1987).

Martin Mahony is a lecturer in human geography in the School of Environmental Sciences, University of East Anglia, UK. He has published widely on the intersections of environment, science, and society, with a particular focus on the cultural politics of climate and the history of the atmospheric sciences. He is coeditor of *Cultures of Prediction in Atmospheric and Climate Science* (Routledge, 2017) and has published in journals such as the *Annals of the American Association of Geographers*, *Social Studies of Science*, and the *British Journal for the History of Science.* He is currently working on a manuscript on atmospheric science and technology in the interwar British Empire.

Meredith McKittrick is an associate professor in the Department of History and Edmund A. Walsh School of Foreign Service at Georgetown University. Meredith has a particular research focus on southern Africa and the climates, waters, and environments in historical colonial and postcolonial contexts. She is the author of *To Dwell Secure: Generation, Christianity and Colonialism in Ovamboland, Northern Namibia* (2002) and numerous book chapters and articles in journals such as *Environmental History, Journal of African History*, and *Journal of South African Studies.*

Ruth Morgan is a senior research fellow in history at Monash University, Australia. She is an environmental historian and historian of science with a particular focus on Australia, the British Empire, and the Indian Ocean. Her first monograph, *Running Out? Water in Western Australia* (2015) was awarded the State Library of Western Australia Prize for Western Australian History.

Simon Naylor is a senior lecturer in historical geography in the School of Geographical and Earth Sciences, University of Glasgow. His research interests encompass the histories of science, technology, and exploration with a particular focus on the historical geographies of meteorology. He is the author of *Regionalizing Science* (2010) and coeditor, with James Ryan, of *New Spaces of Exploration* (2010). He recently coedited, with Simon Schaffer, a special issue of *Notes and Records* on the nineteenth-century survey sciences (2019). He is currently writing a book about Victorian weather observatories.

Samuel Randalls is an associate professor in geography at University College London. His research explores the interconnections between science, commerce, and weather/climate, with a particular focus on historical and contemporary weather risk management through insurance and finance. He has coedited *Future Climate Change* (Routledge, 2012), *The Routledge Handbook of the Political Economy of Science* (Routledge, 2017), and *Just Enough* (Palgrave Macmillan, 2019), as well as papers in journals including *Nature Climate Change*, *WIREs Climate Change*, *Osiris*, and *Social Studies of Science*.